INTERNATIONAL ENVIRONMENTAL POLITICS

Also by Lorraine M. Elliott

PROTECTING THE ANTARCTIC ENVIRONMENT
Australia and the Minerals Convention

International Environmental Politics

Protecting the Antarctic

Lorraine M. Elliott
Lecturer in Political Science
Department of Political Science
Australian National University, Canberra

St. Martin's Press

First published in Great Britain 1994 by
THE MACMILLAN PRESS LTD
Houndmills, Basingstoke, Hampshire RG21 2XS
and London
Companies and representatives
throughout the world

A catalogue record for this book is available
from the British Library.

ISBN 0–333–59449–5

Printed in Great Britain by
Antony Rowe Ltd
Chippenham, Wiltshire

First published in the United States of America 1994 by
Scholarly and Reference Division,
ST. MARTIN'S PRESS, INC.,
175 Fifth Avenue,
New York, N.Y. 10010

ISBN 0–312–12136–9

Library of Congress Cataloging-in-Publication Data
Elliott, Lorraine M.
International environmental politics : protecting the Antarctic /
Lorraine M. Elliott.
p. cm.
Includes bibliographical references and index.
ISBN 0–312–12136–9
1. Environmental law—Antarctic regions. 2. Conservation of natural resources—Law and legislation—Antarctic regions.
3. Environmental law, International. I. Title.
KWX705.E43 1994
344.989' 046—dc20
[349.890446] 93–48285
CIP

For my parents

Contents

List of Tables

Tables

Acknowledgements

During research for this book, I have accumulated debts to a number of people in several countries. I am delighted to have the opportunity to thank them.

Friends and colleagues at the Scott Polar Research Institute in Cambridge were always helpful. I should like to thank in particular William Mills, Janice Meadows, Shirley Sawtell and Bob Headland. Other Antarctic scholars were also willing to discuss our mutual interests. I have enjoyed and benefitted from conversations with Gillian Triggs, Deborah Shapley, Jack Child, Bruce Davis, Don Greig, M. J. Peterson and Alan Henrikson. In particular, thanks are due to Stuart Harris and Peter Beck for their time and insight. Representatives of Antarctic NGOs have responded willingly to requests for information and were available to be interviewed about their organisations and activities. I should like to thank Margaret Moore, Kaye Dyson, Alan Hemmings, Cath Wallace, Sue Sabella, Jagdish Patel, Lena Hagelin and Kelly Rigg. I am particularly grateful to Lyn Goldsworthy for her support and time. Several members of the Antarctic diplomatic community have been generous with their time. Thanks are due to Brendon Doran, Jean Page, John Heap, Tucker Scully, Chris Beeby and Vladimir Golitsyn. Particular thanks must go to Bill Bush. Marie Kawaja of Australia's Department of Foreign Affairs and Trade was especially helpful in facilitating access to documents. Many other people agreed to be interviewed during the course of research for this book. Therefore I should like to thank also Robert Hofman, Dale Crane, Robin Tuttle, Michael Tillman, Al Fowler, William Westermeyer, Scott Hajost, Jack Talmadge, Sherburne Abbot, Lee Kimball, Bruce Manheim, Peter Zoller, Danny Elder, Peter Clarkson, Nigel Bonner and Inigo Everson. Clare Andrews at Macmillan provided encouraging editorial support.

My thanks are especially due to friends and colleagues who provided intellectual and emotional support while this project was completed and who provided the opportunity for me to test some of the ideas incorporated in this book – Diane Stone,

Frances Daly, Steve Bates, Greg Fry, Richard Higgott, John Handmer, Paul Keal, Rob Hall, Marcus Haward and last, but by no means least, Nancy Viviani.

LORRAINE M. ELLIOTT

List of Abbreviations

AAT	Australian Antarctic Territory
ACF	Australian Conservation Foundation
AEIMEE	SCAR Group of Specialists on Antarctic Environmental Implications of Possible Mineral Exploration and Exploitation
AMIC	Australian Mining Industry Council
ASOC	Antarctic and Southern Ocean Coalition
ATCM	Antarctic Treaty Consultative Meeting
ATCP	Antarctic Treaty Consultative Party
ATS	Antarctic Treaty system
BAS	British Antarctic Survey
BAT	British Antarctic Territory
BIOMASS	Biological Investigation of Marine Antarctic Systems and Stocks
CAN	Campaigners Antarctic Notes
CCAMLR	Convention on the Conservation of Antarctic Marine Living Resources
CCAS	Convention for the Conservation of Antarctic Seals
CEE	Comprehensive Environmental Evaluation
CEMP	CCAMLR Environment Monitoring Program
CEP	Committee for Environmental Protection
CLASP	Centre for Law and Social Policy
COMNAP	Council of Managers of National Antarctic Programs
CRAMRA	Convention on the Regulation of Antarctic Mineral Resource Activities
CSAGI	Comité Spéciale de l'Année Géophysique Internationale
DASETT	Department of Arts, Sport, Environment, Tourism and Territories
DFAT	Department of Foreign Affairs and Trade
EAMREA	SCAR Working Group of Specialists on the Environmental Impact Assessment of Mineral Resource Exploration and Exploitation in Antarctica

ECOSOC	Economic and Social Council
EEC	European Economic Communities
EEZ	Exclusive Economic Zone
EIA	Environmental Impact Assessment
EPA	Environmental Protection Agency
FAO	Food and Agriculture Organization
FoE	Friends of the Earth
FRG	Federal Republic of Germany
GDR	German Democratic Republic
G77	Group of 77
IAATO	International Association of Antarctic Tourist Operators
ICAO	International Civil Aviation Authority
ICJ	International Court of Justice
ICSU	International Council of Scientific Unions
IEE	Initial Environmental Evaluation
IGY	International Geophysical Year
IHO	International Hydrographic Organisation
IIED	International Institute for Environment and Development
IMO	International Maritime Organisation
IOC	International Oceanographic Organisation
ISBA	International SeaBed Authority
IUCN	International Union for the Conservation of Nature and Natural Resources
IWC	International Whaling Commission
MARPOL 73/78	International Convention for the Prevention of Pollution from Ships (1973) and its Protocol of 1978
MEPC	Marine Environment Protection Committee
MPA	Multiple-use Planning Area
MSY	Maximum Sustainable Yield
NCP	Non-consultative Party
NERC	Natural Environment Research Council
NGO	Non-governmental organisation
NIEO	New International Economic Order
NSF	National Science Foundation
PCB	Poly-chlorinated biphenyls
SCALOP	Standing Committee on Antarctic Logistics and Operations

SCAR	Scientific Committee on Antarctic Research
SCOPE	Scientific Committee on Problems of the Environment
SCOR	Scientific Committee on Oceanic Research
SCM	Special Consultative Meeting
SMA	Specially Managed Area
SMP	Special Meeting of Parties
SPA	Specially Protected Area
SPRI	Scott Polar Research Institute
SRA	Specially Reserved Area
SSSI	Site of Special Scientific Interest
TAC	Total Allowable Catch
UN	United Nations
UNCLOS	United Nations Convention on the Law of the Sea
UNCTAD	United Nations Conference on Trade and Development
UNDP	United Nations Development Program
UNEP	United Nations Environment Program
UNESCO	United Nations Education, Scientific and Cultural Organisation
UNGA	United Nations General Assembly
USMMC	United States Marine Mammal Commission
WMO	World Meteorological Organisation
WRI	World Resources Institute
WTO	World Tourism Organisation
WWF	World-Wide Fund for Nature

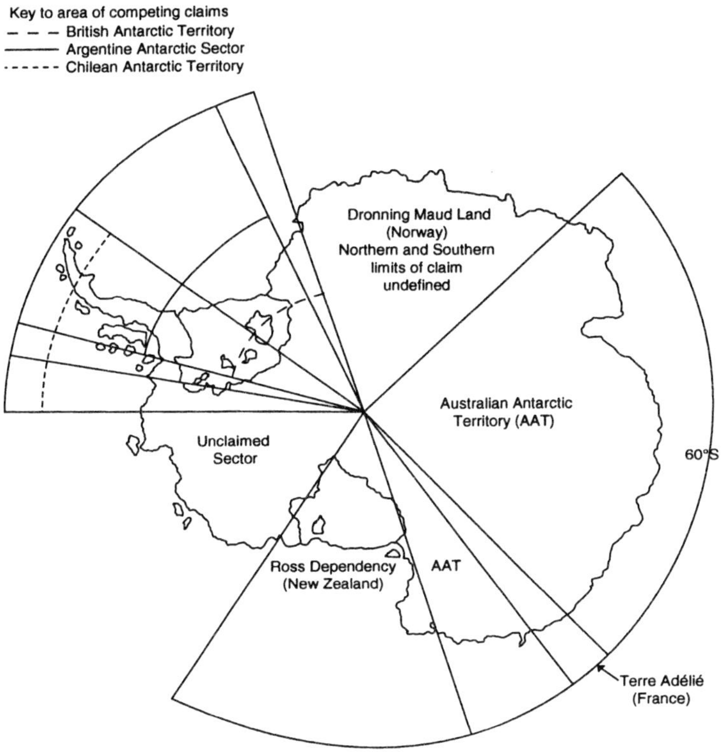

Antarctic Territorial Claims
(60°S – Limit of Antarctic Treaty)

Introduction

On 22 May 1989 the Australian government announced that it would not sign the Antarctic Minerals Convention, which it had helped to negotiate, because it now believed that minerals activity was totally incompatible with protection of the Antarctic environment. Further, the government argued, the approach to environmental protection that the Antarctic Treaty parties had adopted since the Antarctic Treaty came into effect in 1961 was inadequate. A new approach was needed, one based on a comprehensive new agreement under which the Antarctic would be declared a wilderness reserve, off-limits to minerals exploration and any other activity which threatened the fragile environment of the southern continent. This announcement was a key event in the history of the Antarctic regime and one which threatened, or appeared to threaten the stability of inter-state cooperation in the Antarctic. It was also an important turning point in the protection of the Antarctic environment. Despite initial strident opposition to the Australian proposals (in which Australia was soon joined by France), within two and a half years the Treaty parties had adopted a new agreement – the Protocol on Environmental Protection to the Antarctic Treaty – which, while flawed in many ways, represents a change of values and ideas about how best to protect the Antarctic from the impact of human activity.

This book is concerned, first, with an examination of the record of environmental protection under the Antarctic Treaty and an explanation of the Treaty regime's change of direction. While there is now a substantial body of literature on the Antarctic, no text to date has taken as its theme the international politics of environmental protection within the Treaty system, nor examined in depth the more recent events surrounding the Minerals Convention, Australia's (and France's) defection and the subsequent negotiation of the Madrid Protocol as it is usually called. This book has a second concern, one which is less explicit but nevertheless equally important. The Antarctic case study contributes to the now burgeoning international relations literature on the environment. While the Antarctic has sometimes been styled 'a pole apart' (Quigg, 1983), the insights to be gained from how states have managed environmental protection issues there are relevant to our understanding of the difficulties of addressing other international environmental issues. In part, those insights confirm the difficulties of securing environmental protection in a decentralised world of sovereign states.

The norms and values that have shaped decision-making on international environmental issues have derived from interstate politics more relevant to traditional military and territorial security issues and, in spite of trade and capital liberalisation regimes, the management of the international political economy. In spite of a recognition of increasing interdependence among states and peoples, in particular ecological interdependence which transcends state borders, sovereignty norms have continued to dominate international relations. That has engendered a reluctance to relinquish national authority to centralised institutions with decision-making powers or to agree to effective monitoring, compliance or enforcement mechanisms. Regimes are also biased in that the hierarchy of values which underpins them privileges political and economic interests over environmental ones. Together these factors mean, first, that environmental protection rules adopted, both in the Antarctic and elsewhere, are often inadequate to the task identified for them because they are compromised by other interests and, second, that compliance is often minimal in the absence of any strong or enforceable sanctions.

This book also explores questions of how those normative strait-jackets might be undone. The Antarctic case study provides some further examples of what Nye (1988, p. 238) has called one of the most thought provoking questions in international relations – how states learn, that is, how their interests are defined and redefined. There is now a considerable body of analysis which suggests that if we want to overcome international environmental degradation, if we want to achieve environmental security along with the other securities that international politics has always valued – territorial and economic – then we may well have to change the way we practice international politics. The state-centric, competitive, zero-sum game approach to international politics (even if we recognise that our world is an interdependent one) may not only be unable to address global environmental concerns adequately, it may well stand in the way of achieving environmental security. If that is the case, and I believe that it is, then what is required is the much vaunted, but poorly understood paradigm shift. That paradigm shift involves an emphasis on interdependence norms, a recognition of the importance of environmental values that is then translated into action, and the participation of other actors besides states. It involves also the application of new international norms, many of which are still in a somewhat nascent stage, including the precautionary principle, intergenerational equity, and ecological accountability and responsibility. The question is how does such a regime change occur, how can it be mobilised and by whom.

The analysis offered here, both of the Antarctic study and its relevance to international environmental politics generally, draws on international relations theory. It takes, quite consciously, a specific focus which is elaborated more fully in chapter one. That focus is provided by the literature on international cooperation and, in particular, on international regimes and institutions. It is a focus which also rejects an analysis of international relations as simply state-centred. Indeed, a third theme implied in this book is that because much of the discipline of international relations has marginalised environmental concerns, its paradigms have often been less than helpful in addressing some of the key dimensions of international environmental politics. In particular non-state actors, especially the scientific community and non-governmental environmental organisations, have become important and influential actors in international environmental debates, including those on the Antarctic. Yet international relations is only now beginning to explore how we might theorise non-state actors and what that might mean for existing paradigms.

Chapter 1 examines these theoretical questions in more detail to develop a framework within which the Antarctic case study can be understood, and which also enables connections to be made between that study and other environmental issues generally. Chapter 2 introduces the Antarctic regime and, in particular, the history of its development and the entrenched political and strategic interests which states carried with them into the Antarctic Treaty negotiations. It was those interests which shaped the regime and decision-making on environmental protection in the Antarctic. Chapter 3 moves to the more specific focus of this book – an examination of the environmental sub-regime within the Antarctic Treaty system. In particular, it explores how both policy and practice on environmental protection were constrained by the political interests which dominated the regime. Chapter 4 looks at the first two resource conventions negotiated by the Antarctic Treaty parties – the Convention for the Conservation of Antarctic Seals (CCAS) and the Convention on the Conservation of Antarctic Marine Living Resources (CCAMLR). It does so again from the perspective of the politics of environmental protection because both conventions were characterised by the Treaty parties themselves as important contributions to the development of the environmental regime under the Antarctic Treaty. The third resource convention negotiated by the Treaty parties was the controversial Minerals Convention or, in full, the Convention on the Regulation of Antarctic Mineral Resource Activities (CRAMRA). Chapter 5 discusses the early debates about minerals activity. Chapter 6 looks more specifically at the negotiation of the Convention in the 1980s, and the political and other values that shaped it. Again the Convention is assessed particularly for its contribution or otherwise to environmental protection. Chapter 7 traces the more recent events surrounding the unravelling of the Minerals Convention and the negotiations of the new Environmental Protocol. In chapter 8, that Protocol is assessed as an environmental protection agreement. Since the Protocol was adopted in 1991 the Antarctic Treaty parties have, at time of writing, met twice more as part of their normal round of consultative meetings. The conclusion provides a brief postscript on those meetings and the progress made on environmental issues since the Protocol was adopted.

There was not a paradigm shift in the Antarctic regime, but there was a degree of change that was not, indeed could not have been anticipated in the late 1980s. That change provides a basis for further change and for improved environmental protection. This study does not provide definitive answers on how to achieve systemic change in international relations. It does, however, make a contribution to debates in which many international relations scholars, and others, are now engaged – how can we make better environmental regimes and upon what basis do we judge them.

1 International Cooperation and Change

Understanding cooperation in international relations has drawn scholars to the study of regimes and institutions. The purpose of this chapter is not to provide a critical exposition of regime theory.[1] Rather it is simply to identify those dimensions of regime analysis which may provide some insight into the nature of environmental politics in the Antarctic and which may also be useful in exploring international environmental politics on a more general level. In that context, then, this book does not set out to test theories of regimes or insitutions, but simply to adopt some of its explanatory tools. The concepts of international regimes and institutions were developed in the 1970s to explain patterns of international behaviour that did not seem to fit with the more traditional realist interpretation of a world determined by power politics and, for the most part, some degree of competition. In particular, the post-World War II world had come to be marked by an increased level of policy coordination and cooperation among states on a variety of issues beyond those of war and peace. While the concepts of regime and institution are contested ones, as are many concepts in the study of international relations, they have been widely adopted by international relations scholars to examine and explain patterns of international cooperation in an interdependent world.

Technological developments, revolutions in communications and transportation and the expansion and liberalisation of world trade and finance have fostered a changing world in which states do not function, except in theory, as independent and fully autonomous actors. In an interdependent world, not only are states not the only actors in international relations but they may not always be the only powerful or influential actors. In this view, first cogently argued by Keohane and Nye (1977), contemporary international politics is characterised by multiple channels of communication between societies, which challenges the realist conceptualisation of states as unitary and predominant actors. Force lacks utility as an instrument of policy, a factor which casts doubt on the centrality of the 'power' factor in international politics. An interdependent world is also marked by disagreement on the hierarchy of issues which suggests, in contrast with realist interpretations, that security issues, traditionally understood as territorial and military concerns, do not necessarily dominate international politics. Keohane and Nye (1977, p. 3) acknowledged that interdependence was a 'vague phrase' but one that nevertheless expressed 'a poorly understood but widespread feeling that the very nature of world politics [was] changing'.

While not all issues might satisfy the criteria of complex interdependence, international environmental issues can be so categorised. They are characterised by multiple channels of communication, especially through the activities of non-governmental organisations and the scientific community. The use of force is infeasible as a solution to the global threat of ecological insecurity. Further there is little agreement among states, or among states and other actors, on where environmental issues should be located within the hierarchy of international concerns. Concepts which seek further to understand the nature of international politics in an interdependent world are therefore likely to be of use not only in understanding international cooperation but also to understanding the nature of international cooperation on environmental issues. Both are relevant to this book. Indeed, Young is convinced that the tools of regime analysis offer an appropriate vehicle for tackling problems of an environmental kind and for devising institutional arrangements to cope with transboundary environmental problems (1989a, p. viii).

Nevertheless, assertions that states will cooperate under conditions of complex interdependence are as inherently deterministic as the realist assertion that they will not, or at least not for any period of time, under conditions of anarchy. The study of regimes and institutions has brought a more nuanced understanding of cooperation to international relations. It has also enabled a more critical focus on cooperation as both a process and as an outcome. In particular, these ideas provide some insight into a series of questions which are relevant to the two connecting themes here – cooperation in the Antarctic and cooperation on environmental issues. Why do states cooperate and how? In particular, how and for what reasons might cooperation be maintained over a period of time? This analysis of cooperation is important. If we are to meet the environmental challenge then, as Keohane argues (1989, p. 11) understanding 'which patterns lead to more rather than less cooperative behaviour' is imperative.

COOPERATION WITHIN REGIMES

States are likely to cooperate for a variety of reasons and cooperation may take a variety of forms, from short-term bilateral accords to multilateral agreements that may require little further policy action to comprehensive decision-making institutions. Some degree of collaboration can reduce uncertainty and minimise risks. Governments may recognise that they have a shared interest in avoiding a particular outcome, even if they do not agree on what should happen instead. Joint management of an issue may well be more beneficial to governments than unilateral action. A particular problem may not be able to be managed without some degree of collaboration among states. Cooperation does not, however, imply a harmony of interests. Indeed it suggests that processes will be required to manage competing interests and to address collective action problems. Those processes arise within the context of regimes. The most widely used definition describes regimes as 'sets of implicit or explicit principles, norms, rules and decision-making procedures around which actors' expectations converge in a given area of international relations' (Krasner, 1983b, p. 2). Keohane has defined institutions more broadly, but drawing on the same characteristics. They are, he suggests 'persistent and connected sets of rules, formal and informal, that prescribe behavioural roles, constrain activity and shape expectations' (Keohane, 1990, p. 732). Regimes and institutions may be informal, where a particular pattern of behaviour has developed over a period of time, or formal, that is grounded in some kind of international agreement.

Regimes are usually characterised as either negotiated or imposed. The idea of an imposed regime, which is of less interest here, has arisen in connection with hegemonic stability theory.[2] In this view, the establishment and maintenance of a regime requires the presence of a hegemonic power, that is a state with both military and economic resources and the will to undertake a predominant role in providing international order. Hegemonic stability theory suggests that cooperation would be inherently unstable without the presence of a hegemonic state. A hegemonic, or imposed regime is defined by the hegemon either in its own interests (coercive hegemony) or in the interests of providing international order from which all benefit (benign hegemony). In particular, of course, theories of hegemonic cooperation have arisen to explain and sometimes to justify the predominant role of the United States in both military and economic regimes in the post-World War II era.

The analysis here is concerned with negotiated rather than imposed regimes for two reasons.[3] First, the Antarctic regime is a negotiated regime, a point which will be explored further in chapter two. Second, the management of international environmental issues is, in the long run, unlikely to work well under a system of imposition or coercion. Even under conditions of 'benign' hegemony, no single state or group of states is likely to be able to impose or provide a public or common environmental good without the willing participation of other states. Imposed or hegemonic regimes are also more susceptible to free rider problems in which actors benefit from access to the public good that the regime provides without contributing to the provision of that good. The costs of providing the good, in this case environmental security, are therefore disproportionately distributed, but the benefits are not, although with ecological security the good is underprovided if some states do not contribute. The hegemonic state may well be a participant in the negotiation process, but only as one bargaining actor among many, rather than in such a way as to have its preferences shape the regime. Nevertheless, it is sometimes difficult to be certain whether a state, particularly the United States, is engaged in hegemonic behaviour or whether it is constrained to act as a bargaining power. Therefore understanding the processes that underpin the construction of negotiated regimes, and cooperation within those regimes through compliance and convergence of interests is helpful to this study.

INSTITUTIONAL BARGAINING: NEGOTIATING A REGIME

Negotiated regimes, and the agreements upon which they are based, arise through institutional bargaining.[4] This process is characterised by what Young (1989b, p. 361) calls integrative bargaining whereby 'negotiators do not start with a common understanding ... [of] the contract curve'. In other words, there is rarely agreement or understanding on where the intersection of interests will lie. For this reason the bargaining process is more likely to be successful when it proceeds on the basis of unanimity among the negotiating parties. In spite of the emphasis on non-hegemonic behaviour, institutional bargaining is nevertheless more likely to be successful, Young suggests (1990b), when some form of leadership is exercised. In this context, leadership also describes a type of power relationship where power is defined, or may well be defined, as other than the disproportionate possession of traditional military and economic resources. Leadership is also not the sole prerogative of states – it may be exercised by non-state actors or by individuals or, indeed, by a combination of actors. Leadership takes a variety of forms (see Young, 1989b; 1990b; 1991). It may be structural, entrepreneurial or intellectual. It may involve a variety of negotiating skills and behaviour – the use of bargaining power which is more common to structural leadership, imagination in inventing institutional options and skill in brokering the interests of other actors to line up support for those options which characterises entrepreneurial leadership, or the use of the power of ideas through intellectual leadership.

In the absence of a hegemonic power, attention has been paid also to the potential importance of middle power leadership and the role of middle-powers in constructing and shaping negotiated regimes. Middle powers, in this characterisation, are identified attitudinally rather than through some kind of 'objective' measure of their economic or military capability (see, for example, Higgott and Cooper, 1990). In particular, middle powers emphasise the politics of persuasion and negotiation rather than coercion in the conduct of international relations. They seek to employ networks of influence rather than hierarchies of power and to build coalitions of support for their policy initiatives. Middle powers are likely to engage in niche diplomacy, using their resources where they determine they can have some effect on the outcome (Evans and Grant, 1991, p. 322). Nevertheless, institutional bargaining and the emphasis on unanimity rather than majority rule as the basis of decision-making does not suggest that all states will be equal in the negotiations, nor that power relationships of various kinds are irrelevant. Some states may well have more issue-specific resources to bring to bear on influencing other participant states. Neither does it preclude the use of payoffs, incentives or moral suasion. Institutional bargaining also assumes neither equity in the outcome, nor even efficacy. States which are affected to some degree by the rules of a regime might have little or no say in its content.[5]

Because the bargaining process necessarily involves a degree of compromise, the rules of the regime may not adequately address the task for which they were designed. It is this problem of regime inefficiency, and possible reasons for it, that is of particular relevance both to this Antarctic case study and to environmental regimes generally. Even when regime rules are negotiated specifically to address problems of environmental degradation, they often do not do so in any comprehensive way, and this proved to be a problem with the Antarctic environmental regime. More generally, in spite of an increase in the number of international agreements designed to manage environmental and conservation concerns, environmental degradation proceeds apace and, indeed, the problems are increasing rather than decreasing. These difficulties arise in part because of a lack of congruence or coherence between the norms and values which prevail in international politics, and the institutional bargaining process they generate, and the norms and values which are required if environmental issues are to be seriously and successfully addressed. We need to question, then, why this incongruence arises. Some answers, at least, can be found in the characteristics of regimes in a decentralised world.

REGIMES AND THE PROBLEMS OF SOVEREIGNTY NORMS

The basic characteristics of a regime derive from its principles and norms. Principles are those 'beliefs of fact, causation and rectitude' that inform states' attitudes on an issue (Krasner, 1983b, p. 2). Norms, on the other hand, identify rights and obligations and, in that context, specify standards of behaviour expected by participants within a regime. Neither may be explicitly stated but rather implicitly understood through socialisation processes. Regime values may derive generally from the international system, or they may be specific to a particular issue area. The functionally specific component of a regime is identified in its rules and procedures which are, nevertheless, determined by the norms and principles, that is the values that underpin a regime. The better the fit between the two, the greater the internal coherence of a regime and the more likely it is to persist over time. Therefore in examining regime behaviour, such as that found in the Antarctic, it is important to identify the key values, principles and norms which shape a regime and the kinds of rules and procedures those values engender, in order to offer propositions about regime effectiveness, especially with regard to environmental protection.

Norms may be identified, in line with Finlayson and Zacher (1981, p. 564) as sovereignty or interdependence norms. Sovereignty norms characterise 'business as usual' politics or relationships defined primarily by short term interests. Interdependence norms, on the other hand, 'incline states to maximise welfare through collaboration' (Finlayson and Zacher, 1981, p. 564) and are therefore understood to arise in connection with the construction of a regime. This distinction is particularly helpful in examining and possibly explaining the problem of incongruence raised above. When sovereignty norms prevail, governments will still seek to maintain as much control as possible over important decision-making processes rather than relinquish authority to an international bureaucracy or institution. Monitoring and compliance rules or any kind of centralised enforcement agency are likely to be resisted. However it is not clear that these norms and values have been able to engender adequate international environmental protection rules. Gurtov suggests (1988, p. 19), and he is not alone in this, that the more transnational politics becomes, and the more global some of the issues, the less relevant and effective are policies and interpretations based on a narrow concept of national interests. Cooperation and regimes which are based on sovereignty norms will continue to focus on managing competing national interests in the institutional bargaining process. Environmental security, on the other hand, requires a focus on shared and common long-term interests. Effective management of international environmental problems may therefore require some surrendering of sovereignty, particularly to independent monitoring and compliance institutions with some degree of authority to make decisions on rules and to impose sanctions. A reliance on sovereignty norms is also increasingly incongruous as the basis for managing problems which are no respectors of territorial integrity or state borders.

Because norms and principles reflect the values held by states, regimes will also involve some degree of bias in that, as Puchala and Hopkins suggest (1983, p. 66), they 'establish hierarchies of values, emphasising some and discounting others'. That hierarchy may reflect a general bias in the international system which, for example, privileges states' political and economic interests even when environmental issues are under discussion. Value hierarchies may also relate more specifically to an issue area or regime where effective environmental rules are compromised by the need to maintain consensus or protect political compromises as has been the case in the Antarctic regime. Such a hierarchy of values cannot ensure that environmental concerns rather than political or economic concerns will be the primary determinants of environmental regimes.

CONVERGENCE AND COMPLIANCE: MAINTAINING A REGIME

One of the key propositions advanced by regime theory is that regimes shape states' behaviour, that there is some degree of convergence of interests over time. In other words, states not only create regimes, consciously or otherwise. They will then modify their behaviour, and the way in which they perceive their interests, to take account of the regime. In this way, regimes acquire a dynamic of their own. The question of determining whether a regime exists and whether it acts as an intervening variable in this way is a contentious one.[6] As Keohane observes (1990, p. 738) 'causal inference is difficult where experimental or statistical research designs are infeasible'. However, Zacher suggests (1987, p. 174) that a 'stable pattern of collective behaviour, coupled with statements by at least the most powerful actors in a system that they support certain rights and obligations, provides strong evidence for the existence of regime injunctions'. What is of particular interest in negotiated regimes is how that process of convergence, which is essential if a regime is to persist, arises. It arises not through the activities of states per se, but through decision-makers within and across states. Puchala and Hopkins identify (1983, p. 63) 'elites who are the practical actors' within a regime. It is these transgovernmental bureaucratic elites which are crucial to the maintenance of a regime. Bureaucrats and policy-makers in participant states become attuned to the pattern of a regime and its standard operating procedures such as annual consultative processes. Policy decisions are framed within that context and are contingent upon previous decisions. This socialisation process, a process that Haas (1980, p. 368) describes as 'cognitive congruence', creates 'informal coalitions of like-minded officials' which arise within a regime to achieve common purposes (Keohane, 1984, p. 101).

The strength and stability of a regime over time is measured by the degree of adherence to its norms and rules through that process of convergence of interests. The success of a regime is often judged as much by that stability as by its ability to manage the task identified for it. The importance of compliance with both the norms and rules of a regime is an inherent part of the consolidation of a regime. If the most important aspects of an agreement are not adhered to, then there is no regime. The judgement as to what constitutes the most important aspects of a regime is, of course, a subjective one but one which implies that there is some agreement among the key participants on those aspects. Compliance requires also that specific rules and measures be implemented, that mechanisms exist for monitoring compliance and that sanctions may be enacted in the event of non-compliance. This can be a problem in a highly decentralised system which is less likely to rely on formal compliance mechanisms (see Young, 1980, p. 339) and it has been a problem both for the Antarctic and international environmental politics. It is important, then, to consider why compliance and non-compliance might occur.

Imperatives for compliance with regime norms, rules and procedures can be motivated by a number of factors – a sense of obligation, a recognition of the value of particular rules and procedures in managing an issue area, fear of repercussions in the event of non-compliance, or concern that defection or non-compliance may raise doubts about the trustworthiness of that actor in other regimes. Thus compliance rests on diffuse reciprocity.[7] Cooperation will persist, and the regime will be strengthened, when participants are persuaded that the benefits of participation continue to outweigh the costs. Regimes reduce transaction costs by providing some degree of certainty about decision-making on an issue area, and providing actors within information which might otherwise be unavailable to them. Nevertheless, commitment to the norms and rules of a regime may vary over time. As Young notes (1983, p. 94) 'deviance or non-conforming behaviour is a common occurrence in connection with most institutions'. The test of the importance of a violation or non-compliance rests on the extent to which other participants in the regime feel that their own interests are threatened by the violation, or the comparative value they attach to the norm or rule that has been violated. Non-compliance will be judged more seriously if the erring state or states possesses some degree of issue specific power within the regime where its actions may contribute to regime instability or undermine the effectiveness of the regime in managing the issue area.

Putnam (1988) identifies two basic types of non-compliance, or defection as he terms it – voluntary and involuntary. Voluntary defection refers to 'reneging by a rational egoist in the absence of enforceable contracts' (1988, p. 438). Defection therefore, appears more profitable or less costly than compliance where there are no enforcement rules or strong sanctions against the violating actor. Involuntary defection 'reflects the behaviour of an agent who is unable to deliver because of failed ratification' (1988, p. 438), in other words a lack of domestic support. Both are relevant to this case study. The first, voluntary defection, arises in the context of non-compliance with environmental protection rules and measures adopted by the Antarctic regime. The second, involuntary defection, describes the decision by Australia and France not to sign a convention negotiated and agreed to by consensus despite the norms of the

Antarctic regime that consensus, once reached, should not be undermined. This problem of compliance also bears directly on international cooperation on environmental issues and successful management of environmental insecurity. Non-compliance with rules established to manage and overcome environmental degradation can have cumulative and long-term effects, not just on the defecting state but on other states and peoples as well. The environmental imperative for compliance nevertheless raises a difficult problem where sovereignty norms prevail – who has the authority to monitor and who has the authority to impose sanctions?

BEYOND 'POWER' AND 'RATIONAL ACTORS': LEARNING AND KNOWLEDGE

A cognitive approach to regimes posits that convergence of interests and compliance with regime injunctions might well rest on something other than power or rational self-interest. What might also be important, Smith suggests (1987, p. 255) is 'knowledge and the processes by which states learn', that is the way in which states (or actors within states) define and redefine their interests. Haas suggests (1980, p. 390) that 'we know learning has taken place when actors adopt new rules and behaviour that make use of information and knowledge, or adopt new ways to search for knowledge'. Haggard and Simmons also suggest (1987, pp. 509–10) that, under cognitive models, 'cooperation cannot be completely explained without reference to ideology, the values of actors, the beliefs they hold ... and the knowledge available to them'. Knowledge can be defined in a number of ways. It is, understandably, most often used to identify scientific and other kinds of technical information. The principles, norms and rules of a regime can also be identified as a relevant body of knowledge. More broadly, ideas and values about an issue area, in this case about environmental protection, are important to defining knowledge. Knowledge is relevant to the pattern of cooperation and to policy-making when it is accepted as legitimate and credible by those with formal decision-making authority. As Krasner observes (1983b, p. 20) without consensus, knowledge can have little impact on regime development, nor on learning in a world of sovereign states. This has been a particularly difficult problem in the construction and maintenance of international environmental regimes where scientific information is often limited or in dispute. International environmental regimes have been only partly knowledge-driven and actors have often used the problem of scientific uncertainty to defer difficult decisions, especially where those decisions involve economic or political costs. Therefore an analysis of the Antarctic regime and environmental regimes generally needs to identify sites of knowledge and to explore further how that knowledge is used and how it might change power relationships within international relations.

This emphasis on knowledge suggests that 'power' could be understood as epistemic (that is knowledge-based) power, which may well support the proposition that power within a regime or within international relations generally does not rest solely on the relative possession of military or economic resources. It is not clear, for example, that power is fungible when it comes to the successful management of transboundary or global environmental degradation. Traditional power resources, especially economic ones, may shape the regime in the interests of a group of states but the regimes so defined may be ineffective to the task of managing environmental degradation. Further, because of the nature of global environmental problems in particular, such as ozone depletion or the potential for climate change, possession of traditional power resources is no guarantor that any one state can secure itself, in the long term, against environmental degradation if other states, traditionally less powerful, are not prepared to participate in any regime which is developed. This is not to suggest that territorial, military or economic power is no longer relevant, even to the management of environmental degradation. However it does suggest that states can also bring epistemic power, based on ideas, values, information, scientific proposals to bear on the international relations of environmental concerns although just how that process might occur requires considerable further investigation.

NON-STATE ACTORS

The concept of epistemic power is also useful in understanding how non-state actors can be 'powerful' international actors. It is axiomatic that states, and 'official' policy-makers as agents of the state, are not the only participants in regimes or international relations especially in international environmental politics. Any analysis of cooperation or attempt to understand international environmental relations which assumes otherwise is going to be limited, and this insight is crucial to understanding the recent politics of the Antarctic regime. Nevertheless, regime theory has not paid a great deal of attention to non-state actors, although that oversight is now being addressed specifically within the context of exploring environmental issues. Two important and sometimes competing groups of non-state actors can be identified – the scientific community and non-governmental environmental and conservation organisations. What is of interest to our understanding of global environmental politics is how those actors function transnationally and internationally as actors in their own right and in their attempts to influence states.

The impact of the scientific community, which has become influential as a source of knowledge in international and national environmental debates, has often been overlooked because of misplaced perceptions, often supported and fostered by scientists themselves, that they are not political actors and are therefore peripheral to the study and practice of international politics. Their importance to environmental issues in particular stems from the need, not always taken up in practice, for decisions to be based on sound scientific data and interpretation although that interpretation is rarely value-free. Questions arise, then, about how the scientific community might function as an international political actor and how it might affect the nature of cooperation on environmental issues. The most comprehensive analysis of these questions has characterised scientific communities as epistemic communities which Peter Haas has identified as 'transnational networks of knowledge based communities that are both politically empowered through their claims to exercise authoritative knowledge and motivated by shared causal and principled beliefs' (1990, p. 349). While this is a broad definition, the emphasis so far has focused on scientific and technical knowledge and the term has often come to be used simply as a substitute for the scientific community.[8] Nevertheless, at least two key characteristics must define an epistemic community – the consensual knowledge base of such networks and a degree of conscious policy advocacy. Epistemic networks seek to influence decision-makers to adopt policies which reflect and give effect to their knowledge and their beliefs about what that knowledge means. As Young notes (1990a, p. 341) while the concept 'requires further elaboration and may not stand up in its strongest form, there are indications that groups of this kind have played significant roles in the formation of some environmental regimes'. This influence is more likely to be pronounced when that scientific or technical expertise is regarded by decision-makers as important to making sound policy, when technical information is at a premium, when decision-makers defer to experts and when the knowledge is not contested. Epistemic communities may not arise simply as a response to regime demands for information. They may also be important in mobilising states to seek agreements, in influencing the content of those agreements, and in encouraging compliance with regime norms. Haas suggests (1989, p. 399) that the activities of epistemic communities may contradict three common explanations for the development of congruent state policies – foreign pressure, public opinion and the rational anticipation of future benefits by a unitary government. It is probably simplistic to treat these as mutually exclusive explanations, or to suggest that the empowerment and activity of the scientific community may not also interact with any or all of those other explanations. What it does suggest, though, is that analysis of environmental politics and the adequacy of environmental regimes, both in this Antarctic study and elsewhere, needs to explore how, and the extent to which the relevant scientific community has participated in decision-making or in mobilising states on environmental issues.

Any serious investigation of environmental issues in international politics needs also to inquire into the role of non-governmental organisations. They are, as Sands observes (1989, p. 401) 'repositories and developers of ideas. And they are actual international actors'. As domestic actors, NGOs can play an important role in defining national interests and policy on both national and international environmental issues. NGOs also form transnational coalitions to coordinate policies and activities, to facilitate the exchange of information and to strengthen their potential influence and it is that transnational activity that defines them as international actors. NGOs are knowledge-based actors. They contribute to scientific and technical information, often challenging accepted views. They bring to international debates new values and ideas about what is environmentally acceptable. NGOs are often leaders in identifying environmental issues which require multilateral responses. They may initiate proposals and point the way for government action. They are agenda-setters. Caldwell suggests (1988, p. 16) that the 'emergence of an organised international environmental movement and a collateral growth of knowledge regarding the consequences of environmental problems ... have generated an evolutionary progression in international institutional development'. As part of the environmental movement, NGOs are a vanguard for the development of new international norms on the integration of environmental ethics into world views and politics. They play a consciousness raising and value promotion role. In this way they differ from scientists as a potential epistemic community and as knowledge-based actors in that NGOs provide a conscious link between public concerns and voices and policy-makers. NGOs seek to improve the policy and practice of states on environment protection issues. They do so through a variety of tactics, some of which are more akin to pressure group activity and some of which are similar to those of formal diplomatic actors. As NGOs are brought inside the diplomatic process, through participation in national delegations or as observers to intergovernmental meetings, the boundaries between formal and informal actors become increasingly blurred.

NGOs are also important actors in encouraging and, if necessary enforcing accountability and compliance. They perform a 'watchdog' role in monitoring state practice and seeking to reinforce state accountability and compliance. They have been active in identifying and publicising breaches of existing rules and making intergovernmental environmental politics more transparent and accountable. Rowlands suggests (1991, p. 113) that NGOs may also 'create channels of communication that can later be used by states' representatives in order to conduct intergovernmental negotiations'. In this context, NGOs can play an important role in the construction of international environmental regimes and the processes of bargaining, leadership and coalition building that underpin them. Ness and Breckin (1988, p. 271) argue that NGOs 'typically face two tasks in reaching their general goals: they must develop some consensus in the world community for the goal, and they must deliver technical assistance'. Their success in those tasks will depend on their degree of access to decision-makers and their ability to mobilise public support. Their voices are more likely to be listened to when there is a lack of unity among decision-makers on what is the best policy to pursue. At the international level, where NGOs seek to inform and encourage institutional learning, access depends on the nature of the regime. The legitimacy which NGOs acquire in the eyes of state actors is also important in determining access and influence and is strengthened by their development of expertise and knowledge over time. NGOs have been important actors in the Antarctic environmental regime and their relationship with both the bureaucratic elite and the scientific community is an important factor in understanding the nature of environmental politics under the Antarctic Treaty.

CHANGE IN REGIMES AND COOPERATION

The discussion so far, and indeed much of the emphasis in the study of regimes has been on their establishment and consolidation and rather less on the process of change, either within or of a regime. Yet understanding change in international relations is important to the analysis of events within the Antarctic Treaty system and is also a crucial dimension of the international debate on overcoming environmental insecurity. Nevertheless, regimes may be inherently resistant to change even when they produce 'outcomes that are widely understood to be undesirable or suboptimal' (Young, 1983, p. 96). Sunk costs, that is the investment of past efforts, means that regimes are not 'lightly to be changed or destroyed' (Stein, 1983, p. 138), a problem that may well characterise international environmental politics. The costs of establishing a new and unknown agreement are likely to be considered correspondingly high. Transgovernmental bureaucratic elites may well be resistant to anything more than an incremental change. Caution is normally the keyword. This arises in part because of the very basis of regimes or institutions in structuring cooperation on an issue area – they provide a known and understood framework. Thus the processes that strengthen regimes may also undermine, in the long term, their ability to manage the tasks established for them (such as environmental protection) if substantive change in their norms, rules and procedures is required for those tasks to be undertaken successfully.

Change may be functional – states may modify or adopt new rules and procedures. This is a form of incremental or evolutionary change. Puchala and Hopkins describe it as 'cognitive learning and the recasting of goals among dominant elites' (1983, p. 66). Peter Haas refers to it as single-loop learning – the 'pursuit of new instrumental policies while the ends remain constant or unquestioned' (1989, p. 397). Regime participants may also accept the need to moderate sovereignty norms and accept a greater influence of interdependence norms to achieve an outcome that is collectively thought to be desirable. Rules and procedures are then likely to be adjusted to give expression to the altered normative hierarchy. Norms and principles may alter. Haas identifies this as double-loop learning – 'the recognition of new ends and the adoption of new means to accomplish them' (1989, p. 397. This is also more akin to Puchala and Hopkins' 'revolutionary change' (1983, p. 66). Mann (1991, p. 304) characterises it as 'paradigmatic learning', a 'radical transformation of a community's norms, values, beliefs, practices [and] perceptions'. Thus cognitive change, in response to new knowledge, can give way to normative change. Indeed, many environmental NGOs would argue that this normative change is required if environmental security is to be achieved. It is important, then, to explore why and how regime change might come about, to understand the processes of change and to assess the likelihood of normative change.

Most explanations of regime change, especially of change to norms and principles, have focused on power-related factors within a regime.[9] In other words, it reflects a new 'distribution of capabilities (overall or within issue areas) among major actors of world politics' (Keohane and Nye, 1977, p. 54). These explanations have the advantage of parsimony and may well be appropriate in some circumstances but they do not adequately account for change in negotiated regimes such as environmental regimes, where traditional power factors are less relevant.

Other explanations are therefore required. The adoption of new rules and procedures may be a response to new information and technology. Policy objectives may change with a change of government or when one group of decision-makers gives way to another. Policy preferences on international environmental issues, for example, are increasingly the outcome of domestic political contests in response to international debates. Traditional foreign ministry elites may be challenged in their policy recommendations by bureaucrats from other ministries, especially the newer environment ministries.[10] Environmental organisations, the scientific community and commercial interests can also determine this domestic process. Regime change may also occur as a strategic response to pressures from outside the regime. The appearance of new actors, including non-state actors, within a regime may generate new demands and different expectations of behaviour (Ernst Haas, 1983, p. 217). Keohane and Nye (1977, p. 55) suggest also that 'power over outcomes [can] be conferred by *organisationally dependent capabilities* such as voting power, ability to form coalitions, and control of elite networks' (emphasis in original).

The imperative for normative change or paradigmatic learning as a basis for the pursuit of environmental security arises because of the cumulative impact of environmentally irresponsible behaviour and the irreversibility of environmental change. Astrophysicist Carl Sagan, in testimony to a 1985 US Senate Committee hearing on legislative responses to the threat of excessive global warming, warned that

> nations ... have to make a change from their traditional concerns about themselves and not about the planet and the species, a change from the traditional short-term objectives to longer-term objectives ... [W]hat is essential is a global consciousness (cited in Seyom Brown, 1988, p. 184).

Conventional international relations wisdom would argue, however, that change in a state's national interests or in the way states cooperate is unlikely to be based on normative change, or a paradigm shift. So called altruistic values (such as support for human rights, or for the protection of nature) have generally been thought subordinate to political and economic values in determining the way any one state conducts its foreign policy or defines its national interests. Yet the Antarctic example shows that some degree of altruism, or enlightened self-interest at least, can be a factor in changing the nature of cooperation on environmental issues and in undermining the dominance of sovereignty norms. The catalyst for such a change is most often to be found within the scientific community or the non-governmental environmental community. Nevertheless, the point at which cognitive learning, that is a reassessment of policy goals in response to new knowledge, becomes a normative change based on a new set of values and principles, is unpredictable and the process is not easily amenable to analytical explication. Indeed, it may be, as many have argued, that self-interest will, in the final analysis, be the motivating factor. Gurtov argues that

> realist decision-makers may be driven toward globally meaningful change by self-interest ... [A]s state leaders come to appreciate the non-military threats to national security they may invest ... more in dealing with them and eventually more in cooperative responses to problems that are transboundary in scope (1988, pp. 170–71).

Seyom Brown (1988, p. 306) also argues that there is 'no requirement for altruism on the part of the relevant actors, only that they act consistently with their enlightened self-interest'. However the critical voices have not yet developed, as Falk observes (1983, p. 15), a 'convincing base for a conception of transition from the world that is to the world that might be, and needs to be, if current challenges are to be met'. The Antarctic case study may shed some light on the constraints upon or possibilities for normative change.

CONCLUSION

The analysis in this chapter has been consciously directed to an identification of propositions which will be useful in understanding and offering explanations for the nature of environmental politics and cooperation in the Antarctic Treaty system, and making judgements about the effectiveness of environmental rules. What needs to be explored in the following chapters, on the basis of this discussion, are the imperatives for cooperation on Antarctic issues, the institutional bargaining process which shaped the Antarctic regime and the environmental regime nested within it, the exercise (or otherwise) of leadership in building those regimes and the nature of the regime and the rules and procedures which arose within that framework. In particular, the imperatives for and the record of compliance, both with regime norms and environmental rules is one which needs to be examined more fully. The framework sketched here also suggests that an understanding of the relationship between actors within the regime may provide a more nuanced analysis of cooperation, the management of environmental protection in the Antarctic and the process of change within the regime. That examination needs to consider not only the role of bureaucratic actors but also of the scientific community and environmental non-governmental organisations, particular from the perspective of whether the relationship between those actors can be understood in the context of epistemic power and sites of knowledge. The discussion of change in this chapter has been directed particularly towards the development of possible explanations for the rejection of the Minerals Convention and the negotiation of the Madrid Protocol. Understanding change requires an understanding, first, of just what was changed. The next chapter, then, begins this exploration by investigating the establishment and consolidation of the Antarctic regime.

2 Regime Building: the Antarctic Treaty System

International cooperation in the Antarctic is structured by a regime based on the Antarctic Treaty of 1959. The Treaty's purpose was, nevertheless, a limited one – to find some kind of solution to conflicts over sovereignty and to prevent conflict erupting in the future. It was intended to manage problems of the past rather than to provide any blueprint for the future. Antarctic history prior to the Treaty was characterised by imperial, commercial, and scientific interests which shaped political and strategic tensions. Those interests and tensions did not disappear with the negotiation of the Treaty. They were crucial to it. For that reason, they were decisive factors in decision-making on environmental issues. This chapter examines those political and strategic interests, and the way in which they shaped the Treaty and the Antarctic regime.

THE HISTORY OF ANTARCTIC POLITICS

Antarctic history prior to the mid-1950s can be conveniently divided into two slightly overlapping periods – the so-called Heroic Age of exploration (1770 to 1915 or thereabouts) when the Antarctic was sought, discovered and explored, and the Imperialist Era (1908 to the mid-1950s) when territorial claims, and the potential for conflict in the region, were established. The Heroic Age began with Captain James Cook's 1772–75 circumnavigation of the Antarctic although, as far as we know, he never saw it. Cook's legacy was commercial over-exploitation of the Southern Ocean. Attracted by his reports of abundant wildlife in sub-Antarctic latitudes, sealers proceeded to decimate seal populations to such an extent that by 1830 the industry was at an end. The first sighting of the Antarctic sometime in 1820 or 1821, and the right to claim discovery, is disputed. The British, Russians (and later the Soviets) and the Americans have all fielded candidates.[1]

Scientific interests spurred further Antarctic exploration in the late 1830s. Discoveries made by scientific expeditions had implications for later claims to territorial sovereignty.[2] Interest in Antarctic exploration waned after 1830, but was revived again in the 1890s.[3] In the early years of the twentieth century, Antarctic expeditions began to explore inland in the race for the South Pole.[4] From the turn of the century, whaling became the dominant resource industry in the region, repeating the over-exploitation of the sealing industry. Norway and Britain were the major whaling nations, although Japan, the Netherlands, Russia and Argentina were also active. By the end of the Heroic Age, while very little was known about the Antarctic and very little of it had been explored, thirteen countries had established some degree of interest there, based on discovery, exploration, commercial activity and scientific investigation.[5]

Antarctic Imperialism

The territorial expansion of the first half of the twentieth century, driven by imperialist motives, placed the Antarctic firmly and permanently on the international political agenda and was to shape attempts to negotiate a regime for the Antarctic. The Antarctic was judged to be *terra nullius*, land owned by no-one, and therefore properly open to territorial claims. By the mid-1940s seven states – Argentina, Australia, Britain, Chile, France, New Zealand and Norway – had laid claim to parts of the Antarctic. An unclaimed sector, Marie Byrd Land, was tacitly 'reserved' for an American claim.

The British, arch-imperialists, were the first to lay formal claim to Antarctic territory. In 1908, the Governor of the Falkland Islands was charged with the administration of undefined lands lying to the south. There were no official protests from other states. In 1917, the claim was further delimited to describe a sector with its apex at the South Pole, thereby establishing a precedent for other claims which were to follow.[6] In pursuit of its aim to acquire the whole of the Antarctic for the British empire, the British government claimed the Ross Dependency for New Zealand in 1923. In 1933, 42 per cent of the continent, much of it unexplored, was placed under Australian control.[7] France laid formal claim to Terre Adélie on the grounds of discovery and exploration in 1924, although no

French national had ever set foot on the coastline sighted by Dumont d'Urville in 1840.[8] The French claim was further defined in 1938, but France did not attempt to revisit its claim until 1949. In 1939 Norway, fearing an imminent German claim,[9] asserted sovereignty over Dronning Maud Land, although the southern and northern boundaries of this claim are undefined.[10] The Soviet Union, hitherto silent on the Antarctic, refused to recognise the Norwegian claim and reserved judgement on the ownership of any Antarctic land discovered by Russian explorers (Wolk, 1958, p. 44). Argentina and Chile responded with claims of their own, based on geographical contiguity and propinquity and on rights of legal inheritance derived from Spain, although neither had been particularly active in the Antarctic to date and did not send expeditions until after their claims were made.[11] The Chilean claim, which substantially overlapped that of Britain, was promulgated in 1940 by a Presidential Decree which sought to define what the President argued were Chile's historical rights in the Antarctic. Japan responded to the Chilean claim with a diplomatic note reserving all Japanese rights in the Antarctic. Argentina's assertion of sovereignty over the area already claimed by Chile and Britain is usually dated to 1943 although 'as a matter of principle Argentina has not issued any formal claims document. Title ... is viewed as original and continuous from the beginning of the Republic' (Hayton, 1960b, p. 379).[12] United States policy in this period was characterised by ambivalence and vacillation.[13] In 1924, Secretary of State Hughes refused to recognise existing territorial claims on the grounds that discovery had to be followed by settlement to perfect claims to sovereignty over newly found territory. The revival of US interests began with Byrd's 1928 expedition which was followed by others in which Americans saw and mapped (and secretly claimed for the US) more of the Antarctic than nationals of any other state and established the basis for later US assertions of interest in the whole continent. While the US government was reluctant to declare a formal claim, it did support efforts by private expeditions to map and claim parts of the Antarctic for the US. Secret instructions encouraged expeditions to do whatever they might to assist a possible US claim.[14]

The claims were and are contentious. Non-claimant states argued that the claims, and claimant exercise of sovereignty, were invalid because there had been no effective occupation of the claimed territory.[15] The use of the sector principle for Antarctic claims has been contested by international lawyers. In the Arctic, where it was developed to support Canadian claims, it relies on geographic continuity and contiguity. Hayton (1960b, p. 398) has described the application of this principle to large areas of undiscovered land (and much of the land claimed in the Antarctic had not been discovered when the claims were made) as a 'flagrant flaunting of international law standards [which] should be exposed as such'. There were also tensions among the claimants. Only five of the seven claims (Britain, Australian, New Zealand, France and Norway) were mutually recognised. Argentina, Britain and Chile claimed substantially the same area. While Argentina and Chile had been unable to reach a boundary agreement on their competing claims they did agree that there was a South American Antarctic and only they had exclusive sovereign rights over it.[16] The British were not inclined to be excluded from 'their' Antarctic territory in this way.[17] Britain sought to strengthen its presence in the Antarctic through its naval operation *Tabarin* of 1944–45. In the 1946–47 and 1947–48 seasons, Argentina and Chile sent expeditions into the area also claimed by Britain. The British government formally protested at what it saw as provocative acts of trespass. To complicate the situation even further, both Japan and the Soviet Union had reserved their rights to Antarctic territory and the US position was still not clear. The potential for conflict was further heightened as the bipolarity of the post-war international order, dominated by the US and the Soviet Union, threatened to extend to the Southern polar region. In 1946–47 the US despatched 13 ships and over 4000 service personnel to the Antarctic on *Operation Highjump* to train personnel and test equipment under polar conditions.[18] Soviet presence was confined, at this stage, to whaling and some scientific activity but was unlikely to remain at that level in the face of US activity.

This potential for conflict led the US government to initiate informal discussions with the claimants on ways to manage their interests in the Antarctic and to avoid confrontation among them. Various proposals were canvassed with little success. In August 1948, the US government floated the idea of a multiple condominium, stressing the need to solve the problem of the conflicting sovereignty claims and the importance of scientific investigation and research in the Antarctic.[19] Responses from the claimants indicated no common ground among them except for their opposition to any internationalisation of the Antarctic. Argentina rejected the US proposal out of hand on the grounds of 'uncontestable sovereignty' (Hanessian, 1960, p. 444). New Zealand and the UK thoughtthe US proposal a useful basis for discussion. France and Australia saw no need to discuss the sovereignty issue at all, although they supported the principle of scientific investigation. Norway was somewhat equivocal. Chile proposed a five-year freeze of all claims and rights, during which time further scientific cooperation would be encouraged (see Hanessian 1960, p. 444). A second proposal that the Antarctic be established as a United Nations Trust Territory was also not well received.

Discussions stalled in the early 1950s. This faltering of interest cannot be described as a failure to establish a regime. Discussions in the late 1940s were little more than an informal exchange of views. There was no imperative for cooperation. The scope and nature of an Antarctic problem was not clearly defined. The claimants perceived no urgent need for an agreement, in terms either of benefits to be gained or losses to be minimised. US presence in the Antarctic was generally accepted by them and, in the absence of *active* political interest from other states, any territorial or legal challenge to their interests did not seem imminent. While rivalry between the overlapping claimants was potentially a matter for concern, it had been contained to some extent by the 1949 Tripartite Naval Agreement (renewed in subsequent years) in which the three agreed not to send warships south of 60° south latitude. The only point on which there was agreement was the exclusion of the Soviet Union. The Soviet response was to make it clear that it would not countenance domination of the Antarctic by the US and its allies. In 1949 the semi-official Geographical Society asserted that the Soviet Union had an 'indisputable right ... to participate in the solution of questions concerning the Antarctic' (Wolk, 1958, p. 45). In June 1950 (just before the outbreak of the Korean War) the Soviet government addressed a note to the US and six of the seven claimants[20] declaring that any discussion on the future of the Antarctic should not take place without Soviet involvement.

Science and Politics: Towards a Regime

It was science which changed Antarctic politics and provided the ground-work for the Antarctic Treaty and the Antarctic regime. Between 1945 and 1950, at least ten nations sent scientific expeditions into sub-Antarctic and Antarctic waters.[21] These expeditions were now almost always sponsored by governments, rather than privately sponsored as most had been before World War II. This ad hoc scientific activity became more organised during the International Geophysical Year (IGY) (1957-58). Twelve countries (the seven claimants, the US, Belgium, Japan, South Africa and the Soviet Union) undertook scientific programs in over 60 research stations in the Antarctic.[22] Planning for IGY scientific activity in the Antarctic, which was coordinated by the Comité Spéciale de l'Année Géophysique Internationale (CSAGI), a special committee of the International Council of Scientific Unions (ICSU), was designed to marginalise politics. Military activity was banned although logistical support by the military for scientific expeditions was permitted. Mapping activities were considered inappropriate in the Antarctic because of the political implications for claims. Cooperation among scientific expeditions and exchange of scientific personnel was encouraged.[23] Meteorological and rescue services were pooled. Scientists lobbied their governments to secure free access to all parts of the Antarctic. The claimants agreed to refrain from their usual diplomatic and sometimes active attempts to control access to their claimed sovereign territories. While this agreement was formulated in a resolution of CSAGI, the tacit support of the claimant governments was required to put it into effect. Scientific cooperation during the IGY was not totally unaffected by political concerns. For example, the decentralised mother–daughter system for communication of essential weather data was constructed so that no 'daughter' station was reliant on a politically unsympathetic 'mother' (Shapley, 1985, pp. 87–8). Scientific activity also had direct political implications, particularly for the claimants and the superpowers. The siting of the seven US scientific stations, for example, was influenced by the Department of Defence and a 1954 National Security Council memorandum and was intended to insure that the US was second to none in the Antarctic (Quigg, 1983, p. 49).[24]

The IGY provided governments with an opportunity to break the political impasse of the late 1940s and early 1950s on the sovereignty question. Territorial concessions were made and political arrangements tested without formal or long-term commitment. However, this managing of the claims and conflict over them and the emphasis on cooperation through science was scheduled to last only for the duration of the IGY. The potential for a resurrection of prior conflict over territory, as well as the spectre of future tensions between the superpowers which followed the Soviet announcement that it would remain in the Antarctic after the IGY, encouraged the Antarctic states to think once more about a treaty for the continent.[25] Discussions on a political settlement resumed during the IGY, with a greater sense of urgency than in 1948.[26] The political landscape had changed since then. The claimants and the US could no longer negotiate an exclusive settlement. The Soviet Union could not be excluded from any agreement and the other Antarctic IGY states (Belgium, Japan and South Africa) were unlikely to accept any agreement made without their participation. This increase in the number of states with well-defined interests in the Antarctic contributed to the political imperative for an agreement and, potentially, a regime.

In May 1958, well before the IGY finished, US President Eisenhower formally invited the eleven other Antarctic IGY states to participate in a conference to conclude an agreement for the management of Antarctica. The US suggested that any agreement to continue scientific cooperation 'could have the additional advantage of preventing unnecessary and undesirable political rivalries ... the uneconomic expenditure of funds to defend individual national interests and the recurrent possibility of international misunderstanding' (cited in Hayton, 1960b, p. 371). This 'additional advantage' really went to the heart of the matter: avoiding conflict was the fundamental reason for negotiating a treaty. As British Antarctic diplomat, John Heap puts it (1983, p. 105) 'few, if any, of the governments invited were attracted by the positive aspects of the Treaty. The crucial stimulus was ... fear. Each government had its own scenario of the chaos it foresaw if the Treaty was not successfully concluded'. The US also argued that a treaty could be concluded 'without requiring any participating nation to renounce whatever basic historic rights it may have asserted' (in Hayton, 1960b, p. 371). It is unlikely that a treaty could have been concluded otherwise. This provision was the *quid pro quo* for claimant participation in the negotiations and was to be a fundamental principle of the Antarctic regime.

The parties were motivated by self-interest in seeking to negotiate a treaty. In the late-1940s, the claimant governments and the US had not been convinced that any agreement could protect their interests. Indeed the claimants were concerned then that their interests might be compromised by a treaty. By the late 1950s, however, the parties, in particular the claimants, took the view that their interests could best be protected, indeed could *only* be protected by a treaty. As well as a shared interest in avoiding conflict, the participating governments had other interests which could now best be met by some degree of formalised cooperation among them. All seven claimants were faced with the potential erosion of their influence in the Antarctic. The likely costs of conflict and the difficulties of militarily protecting territorial interests were high, with little chance of success. The chances of legal settlement of title in favour of any of the claimants were low. The claimants also sought, to varying degrees, to contain the influence of the superpowers and the three non-claimant IGY states. In spite of their ideological rivalry, the US and the Soviet Union had similar interests in the Antarctic. Neither had made a formal claim but each had reserved its right to do so in the future. Each wanted to prevent the other gaining a dominant foothold. Both were also cognisant of, and wished to avoid the logistic and financial costs of conflict between them in the Antarctic.

Treaty Negotiations: 1958–59

Negotiations for the treaty were conducted in two stages. Over 60 informal, secret meetings, which produced a series of draft provisions, were held between 13 June 1958 and 13 October 1959.[27] In November 1958, the US representative submitted draft articles as a basis for further discussions. By mid-March 1959 there was agreement that the first three articles would provide that Antarctica should be used for peaceful purposes only, for freedom of scientific research in the area, and for international cooperation in the conduct of this research. By mid-May there was some measure of agreement on draft provisions on the territorial claims, inspection, jurisdiction, the relationships between signatory and non-signatory countries, accession, the zone of application, periodic meetings of signatories, settlement of disputes, and the revision and ratification of the Treaty (Department of State, 1964, p. 9). Formal negotiations (the Washington Conference) began on 15 October 1959. Again the more than 50 formal and informal meetings were held behind closed doors, except for the opening and closing sessions. The precedent was thus set for what Davis (1987, p. 27) has called the 'closed circle mentality' which has characterised the Antarctic regime. The main areas of contention were the area of application, jurisdiction, dispute settlement and membership (Hanessian, 1960, p. 467). On 23 October delegates announced that they had reached agreement on the principle of international cooperation in scientific research and exchange of information and on the principle of non-militarisation. This is not surprising, given that these issues had been effectively settled during earlier discussions and were based on IGY precedent. By mid-November 1959, the USSR and the US had accepted an inspection system. The Antarctic Treaty was concluded and signed on 1 December 1959, less than seven weeks after the formal conference began.

Power-based explanations for the successful negotiation of the Treaty, which would suggest a link between overall and issue-specific power, are unsatisfactory. While the US took a leading role in mobilising the Treaty negotiations it was not fulfilling the role of a hegemon. Peterson's suggestion (1988, p. 85) that most of the provisions of the Antarctic Treaty fully reflected the shared preferences of the superpowers is misleading in that she later observes, correctly, that on further investigation most of those interests were also shared by the other parties. Krasner (1985, p. 252) suggests that the 'basic American objective was to stabilise strategically the Antarctic area' but he neglects to mention that this goal was shared by the other states, especially the claimants. Neither the US nor the USSR were able to impose their preferences on the negotiations.[28] Further, five of the claimants (Australia, New Zealand, Norway, Argentina and Chile), while important Antarctic powers with considerable influence over the Antarctic negotiations, had little overall power in the international system.

The Treaty is a negotiated rather than an imposed agreement – the outcome of institutional bargaining. This process rests on integrative bargaining, rather than the pursuit of relative gains, in a situation in which there is no 'fully specified zone of agreement' (Young, 1989b, p. 366). While there was some common agreement on the general purpose of a treaty, there was little agreement prior to the negotiations on what issues should be addressed and what specific provisions should be included, or omitted. The sovereignty question was always going to be a difficult one. The claimants, especially Chile and Argentina, would not agree to any provision which derogated from their rights as territorial sovereigns. Thus questions of administrative arrangements, compulsory dispute settlement and jurisdiction were contentious. There were significant differences of opinion on the issues of demilitarisation, the related question of inspection, the membership scope of the Treaty and the exploitation of resources.

The Treaty is the product of an internal accommodation among the parties over their differing and entrenched views on territorial sovereignty and the exercise of jurisdiction. Indeed, Young (1980, p. 354) has typified the Antarctic Treaty negotiation process as a 'small group of actors with strong or generally compatible interests banding together to work out an acceptable set of arrangements'. Along these lines Kremenyuk (1991, pp. 30–1) suggests that the Antarctic Treaty is the result of a joint-problem solving approach, by which the negotiators sought to find an acceptable solution to a problem the nature of which was agreed upon, a process he too distinguishes from a relative gains model of negotiation.[29] The original signatories were able to establish a strong founding coalition *vis-a-vis* other states because of their issue-specific power rather than their overall power, which was enhanced because there was little interest in Antarctic issues from other states in the late 1950s. Young (1990b, p. 32) argues that we can expect to encounter leadership in action whenever we observe success in the formation of international institutions. Yet it is not clear what leadership role was exercised, and by which state or states, in the Treaty negotiations. There is not a strong case for arguing that the US, or any other state, acted as a structural leader by employing 'bargaining power to bring pressure to bear on others to accede to the terms of the proposed constitutional contract' (Young, 1990b, p. 10). It is more likely that the negotiations were characterised by entrepreneurial leadership. Entrepreneurial leaders (Young, 1990b, p. 17) are most often individuals who 'employ negotiating skills to put together deals that would otherwise elude participants'. It may be that Paul Daniels, chief US negotiator at the informal meetings in Washington at which considerable agreement was reached on draft treaty provisions, acted in this way.[30] Young suggests also (1989b, p. 373) that CSAGI and SCAR had a leadership role in forming the regime for Antarctica in 1959. While it is the case that cooperation under the IGY provided an example for the diplomats and international lawyers to adopt as the basis for the Antarctic Treaty, the scientific community was not directly instrumental in its negotiation.

THE ANTARCTIC TREATY[31]

The Treaty establishes a regime of mutual security by proscribing potentially conflictive activities. It is characterised by a contrived ambiguity, and paucity of definitions, which leaves much open to interpretation. While the Treaty was hailed as an innovative agreement for managing international cooperation, very little in the Treaty was new. For the most part it codified existing arrangements. It does not substitute for national interests. It complements them and could not have done otherwise. It satisfies individual and collective national goals which were based on the need for differing positions on territorial claims to be recognised and for influence over decision-making on Antarctic issues to be guaranteed. Freedom of access to all parts of the Antarctic satisfied those who did not accept the exercise of sovereign rights by the claimants. The claimants, on the other hand, were able to accede to freedom of access, to inspection, to territorial overflight, and to the fact that the Treaty did not formally recognise their claims, as a concession to scientific cooperation and the pursuit of peace, protected by provisions which meant that those concessions could not later be used as legal argument against their claims.

The Treaty reflects three fundamental principles about the basis for cooperation in the Antarctic. The first acknowledges the value of the compromise on sovereignty, the second recognises the importance of avoiding conflict in the Antarctic and the third emphasises scientific research and cooperation as the explicit face of political cooperation on the continent. All three have their roots in the politics of the early twentieth century and the experience of the IGY.[32] The more specific rules and procedures embodied in the Treaty are designed to support these three principles. Even those articles which deal with the practical aspects of activity in the Antarctic were shaped by the sovereignty question, as much in what was omitted as what was included. The hierarchy of values implicit in the Treaty places political and security concerns before economic or environmental ones where those interests are seen to be in conflict. Although Finlayson and Zacher suggest (1981, p. 564) that some degree of commitment to interdependence norms is required for the construction of a regime, in that parties forego a degree of independent decision-making, the Antarctic Treaty is characterised by a substantive commitment to sovereignty norms.

The Compromise on Sovereignty

The unstated objective of the Treaty was to find, and maintain, a successful compromise on the exercise of sovereign rights in order to protect the interests of all parties, claimant and non-claimant. Only if this political bargain was maintained could other interests be met. While the claimants recognised that their interests could best be protected by a treaty, they would not agree to anything in that treaty which prejudiced their exercise of territorial sovereignty. Non-claimants, on the other hand, would not agree to *any* provision that gave legal status to the claims. Without some accommodation on this issue, there would have been no agreement. The essence of the internal accommodation between the original signatories is found in article IV. It does not solve the sovereignty problem. It puts it to one side by confirming the existence of the territorial claims, but making no statement as to their legal validity. It is non-discriminatory in that it applies equally to the claimants, those with a basis of claim (the US and the Soviet Union) and the non-claimant parties. Article IV states that the position of *any* party on the recognition or otherwise of the claims to territorial sovereignty will not be prejudiced by the Treaty. Nothing in the Treaty can be taken as a renunciation of any claim or basis of claim. Further, no act or activity by any Treaty state can be used to support, deny or expand an existing claim or to venture a new claim.[33]

The sovereignty dispute was also connected with the question of applicable jurisdiction. Argentina, France and Chile opposed any suggestion that national jurisdiction should apply on the grounds that, in spite of the provisions of article IV, it would undermine their claims to sovereignty. Territorial jurisdiction, on the other hand, would imply that claimants had the right to exercise jurisdiction over nationals of non-claimant states. In the absence of any middle ground the issue was simply omitted. Article VIII of the Treaty rules only that observers, scientific exchange personnel, and their staffs will be subject to the jurisdiction of the Contracting Party of which they are nationals. It says nothing about jurisdiction over other personnel (military, scientific and support) in the Antarctic, whether part of a national scientific program or not.[34]

The Peaceful Purposes Principle

The imperative for avoiding conflict over sovereignty gave rise to a more general peaceful purposes principle which embodies a broad commitment to maintaining the Antarctic as a zone of peace. The preamble to the Treaty recommends that the Antarctic should continue for ever to be used exclusively for peaceful purposes and should not become the scene or object of international discord. Peaceful purposes are not defined although scientific investigation clearly meets the test. Thus the Treaty parties can determine what constitutes a peaceful purpose at any time while the Treaty is in force. The Treaty contains fairly extensive rules and procedures to support this principle. 'Measures of a military nature' are prohibited in article I.[35] These measures are not defined exclusively, but do include 'the establishment of military bases and fortifications, the carrying out of military manoeuvres, as well as the testing of any type of weapon'. As with the IGY, military personnel or equipment may be used in support of scientific research or any other peaceful purpose. Nuclear explosions and the disposal of radioactive waste in the Antarctic are prohibited under article V.[36] Although there are no sanctions to give effect to these provisions, the

Treaty does demilitarise an entire continent and its surrounding oceans, a unique achievement in international relations. However, in spite of the acknowledged value of avoiding conflict, there are only limited disputes resolution procedures and these have never been invoked. Argentina, Chile and the Soviet Union opposed any form of compulsory arbitration (Hayton, 1960a, p. 363). The Treaty provides only that parties may choose a method of resolution after consultation among them and that a dispute may be referred to the International Court of Justice only if *all* parties to the dispute agree. The importance of managing conflict engendered an unspoken understanding that disagreement on Antarctic matters would be managed with the confines of the Antarctic Treaty, and differences on other issues would not affect Antarctic cooperation.

Scientific Cooperation

Scientific cooperation provides an acceptable rationale for continuing political control in the Antarctic. Freedom of scientific investigation is emphasised. In support of this, governments are encouraged to exchange, among themselves at least, information on their scientific programs in Antarctica (article III) and to provide information on expeditions, stations and military personnel and equipment in the Antarctic (article VII). Exchange of scientific personnel is also encouraged. Parties are also exhorted to encourage 'cooperative working relationships' with United Nations specialised agencies and other international organisations 'having a scientific or technical interest in Antarctica' (article III(2)). Despite this normative emphasis on openness, those cooperative relationships have not been widely encouraged by the parties. The exception to this is the close connection between the Treaty and the Scientific Committee on Antarctic Research, a relationship which is explored below.

Membership and Decision-Making

Membership of the Treaty proved contentious. In theory, the Treaty establishes an open membership agreement. Any member of the United Nations may accede to the Treaty under article XIII. Non-UN members may accede on the invitation of the decision-making parties.[37] This two-tier system of membership was a compromise between parties (such as the Soviet Union) which advocated open membership and those (such as the US) which wanted a limited membership. In practice, participation in decision-making is restricted. While any state may accede, only the original signatories and those who undertake substantial scientific research in Antarctic, collectively referred to as the consultative parties (ATCPs), have decision-making authority.[38] There is no formal decision-making role for those acceding states which do not conduct research – the non-consultative parties (NCPs). Thus science provides the basis for political privilege in the Antarctic regime and for membership of the Antarctic 'club'. The differences over sovereignty and jurisdiction ensured that the administrative machinery established under the Treaty (article IX) was a generally weak arrangement. Most of the parties would not agree to the establishment of a secretariat or international organisation.

The consultative parties may meet 'at suitable intervals' (article IX) to consult and make recommendations on issues relevant to the principles and objectives of the Treaty but decision-making remains, in effect, in the hands of national governments. While the Treaty established no voting rules, article IX requires that any measure adopted at a consultative meeting will become effective only when it has been approved by all participating governments. This effective veto ensures that no state can be outvoted. The requirement for consensus is implicit. The rule that recommendations must also be *adopted* by consensus was made explicit in the Rules of Procedure drawn up prior to the first consultative meeting. The importance of consensus was further incorporated in the Treaty's amendment procedures (article XII) which ensure that no changes can be made to the Treaty without the agreement of all consultative parties. Amendments or modifications may be made at any time by unanimous agreement and ratification by the consultative parties. After thirty years, any consultative party may request a conference 'to review the operation of the Treaty'.[39] While review conference amendments require the support only of a majority of Treaty signatories, including a majority of the consultative parties, they enter into force only when *all* consultative parties have ratified them. Consensus is therefore still likely to be sought in spite of the majority rule.[40]

Area of Application

The area of application of the Treaty and, in particular, the designation of high seas within that area, was another source of disagreement. Antarctica is not defined in the Treaty, even though the word is used forty times. The Treaty applies to the area south of 60° south, including all ice shelves (article VI) but without prejudice to the high seas rights of any state under international law.[41] However, this provision was drafted 'so as to leave indefinite the question of what was the high seas' (Phleger cited in Bush, 1982, p. 144). In particular, it provides no assistance as to whether the continental shelf or territorial waters are high seas or not. For the claimants, both were part of their sovereign territory and therefore under national jurisdiction: for the non-claimants they were not.[42] Article VI is also ambiguous on whether *any* high seas rights can be restricted under the Treaty. Military activity and the disposal of radioactive waste (other international agreements notwithstanding) are high seas freedoms, in which case the Treaty rules which prohibit those practices would not apply when they are conducted on the high seas within the Treaty Area. In practice only those high seas freedoms which are not explicitly proscribed by any other article of the Treaty may be exercised freely (on the high seas) as long as they meet the test of a peaceful purpose.

Compliance Rules

Adherence to the principles, norms, rules and procedures of an agreement is as noted in chapter 1, crucial to the development of a regime. The Antarctic Treaty contains weak compliance rules. Inspections procedures designed to 'ensure the observance of the provisions of the present Treaty' (article VIII) are limited. Only consultative parties may conduct inspections. No agreement could be reached on provisions for multilateral or collective inspection or the establishment of an independent inspectorate because of Soviet fears of setting a precedent for inspection in arms reduction agreements. While all areas of Antarctica, including all stations, installations and equipment and all ships and aircraft at points of discharging and embarking cargoes or personnel, are to be open at all times to inspection, the Treaty provides no guidance on sanctions or procedures to be followed if an inspection reveals violations of the Treaty. The main significance of the inspection provisions were that they represented the first mutual inspection procedures agreed by the Soviet Union and the US although admittedly not on their territories.[43]

Omissions

Two key omissions were to become main points of contention in the development of the Antarctic regime and later were focal points for change. Any mention of resource exploitation (especially mineral resources) was deliberately avoided because of its implications for the contentious sovereignty question – that is, who owned the resources.[44] As resource exploitation was not considered to be an imminent activity, and not fundamental to securing an agreement, it was left to one side. The Treaty also contains almost no reference to environmental protection. This omission relates not to the sovereignty issue but to the timing of the Treaty (environmental issues were not of high salience in the late 1950s) and to the focus of the Treaty as a security agreement. References to the conservation and preservation of living resources in article IX were added as an afterthought at the insistence of Chile because of concern over the economic impact of unregulated exploitation than the intrinsic protection of wildlife. Peterson suggests (1988, p. 45) that the prohibition on nuclear explosions and the dumping of radioactive waste reflects an implicit concern with the Antarctic environment. However, the southern hemisphere governments who introduced this topic were more concerned (at this stage) about the effects of radioactive fallout on their own environments than they were about protecting the Antarctic environment.

CONSOLIDATING A REGIME

The Treaty came into effect on 23 June 1961.[45] It was hailed as a model for international cooperation and, in the context of the Cold War, a significant step in the 'construction of a durable peace' (Hanessian, 1960, p. 461). However, it was not clear just what would eventuate from the operation of the treaty. International lawyers were concerned with its legal weaknesses, including the ambiguities of article IV and its failure to solve the sovereignty question, and the absence of sanctions, formal compliance mechanisms and any organisational structure. The Treaty was a political agreement among the parties to solve problems of the past, rather than a blueprint for future action. It provided only a basic framework for the development of a regime. Indeed, some of the parties were insistent that the Treaty should not 'institute regimes or create structures ... It is not its mission to change or alter anything' (ATS, 1959, p. 31).

In the absence of formal compliance rules, the founding coalition, together with states which have acceded to the Treaty, have nevertheless maintained a high degree of internal cohesion. As chapter 1 suggested, regime strength relies on states' commitment to regime norms and rules, and we need to examine the imperative for that commitment in the Antarctic. The 'habit of cooperation [that] has been maintained and developed' (Beeby, 1972, p. 4) is generated by self-interest. This process rests on more than a calculation of short-term interests. Rather it reflects the realisation of long-term individual *and* joint gains. Once agreed to and in force the Treaty became the guarantor of the parties' interests in the Antarctic.[46] As long as the Treaty remained in place, the spectre of conflict could be subdued and other national interests in the Antarctic could be met, or at least not undermined. Further, sunk costs (that is, the investment of past effort) made any alternative to the Treaty system, especially one which might erode the influence of existing consultative parties and the claimants in particular, inherently unpalatable. The principles and norms inherent in the Antarctic Treaty dictated the nature of the decision-making process and informed standards of behaviour which became both valued and expected by the key actors as the Treaty system developed. Similarly, as that decision-making process was seen to work over time in protecting interests and preventing conflict, its value was reinforced. Thus a learning process underpinned the development and strengthening of the regime. Reciprocity, that is a shared expectation that all parties would adhere to the norms of the Treaty, served to build and reinforce confidence in the Treaty system as a guarantor of states' interests.

There has also been a remarkable continuity and cohesion of Antarctic decision-makers who became well socialised into the norms of the Treaty system. Decision-making within the regime has been dominated by bureaucrats, predominantly foreign ministry officers, scientists and representatives of national Antarctic operators, that is, the organisations responsible for the conduct of Antarctic programs.[47] Through consultative meetings a close network of individuals with experience of the Treaty system was established with an 'inner circle' of perhaps a dozen who, in the absence of a secretariat or formal minutes, carried with them the institutional memory of the regime. This transgovernmental elite, socialised into the norms of the Treaty, was important in maintaining and defending the Antarctic regime. They were also key actors in defining national Antarctic interests and policies. For the most part, during the first two decades of the regime, they were able to do so without challenge from other national or international actors. The low political salience of Antarctic issues, nationally and internationally, meant that there were few domestic constituencies on Antarctic issues (with the exception of scientists) to challenge this decision-making authority.

However, as chapter 1 argued, state actors may well not be the only actors involved in regime consolidation. Scientists have been important actors in the Antarctic regime, confirming the intrinsic but ambivalent relationship between science and politics in the Antarctic. Science was confirmed in the Treaty as an important component of the agenda of Antarctic politics although the Treaty was not, as Zumberge (1987, p. 4) and Gould (1971, p. 54) erroneously suggest, negotiated to protect a scientific programme. Rather, science was 'a means to an end by providing an excuse to remove a potentially controversial region from international conflicts' (Laws, 1987, p. 250). Political influence and prestige within the Treaty system is linked directly to the conduct of scientific research. Therefore, as Walton observes (1987b, p. 61) 'it must be expected that political considerations will inevitably intrude, in some countries, in the selection of which scientific activities to support'.

Scientific cooperation occurs through practice 'on the ice' and through the coordination of scientific research by the Scientific Committee on Antarctic Research (SCAR). While SCAR is not mentioned in the Treaty, even though it predates it, the consultative parties have come to rely on it for advice on all manner of scientific and logistic issues, including protection of the environment. It has effectively become the Treaty system's scientific secretariat and an important source of knowledge in the regime. Yet there is no *direct* relationship between SCAR and the consultative process.[48] The consultative parties have also sought to ensure that SCAR's participation is advisory only and make no formal commitment to take its advice. It is extraordinary that it was only in 1987, after 24 or more years of providing scientific advice to the consultative process and essentially being responsible for the substance of a number of scientific and environmentally-related aspects of the system, that SCAR was invited to send an observer to the consultative process to provide assistance on specified issues under consideration. What is more extraordinary is that the item SCAR was asked to assist with was that on air safety in Antarctica, at a meeting where its report on the Protected Area system was also under consideration.[49]

Senior scientists have been included in consultative delegations, even though lawyers and diplomats predominate at consultative meetings. In fact, scientists have complained that they 'have second-class status, although almost invariably they represent the only practical experience there' (Laws, 1987, p. 251). Generally, however, the Antarctic scientific community and SCAR has sought to distance itself from the politics of the Treaty system. SCAR has neither sought nor undertaken an advocacy role but has concentrated on coordinating scientific research and responding to requests for information and advice from the ATCPs. For this reason, it is not clear that the scientific community can be identified as an epistemic community within the Antarctic regime, especially when we recall that the definition assumes a degree of conscious policy advocacy. In spite of this apparent indifference to politics, the scientific community is an interested actor in the regime. Scientists have clearly benefited from the successful maintenance of the Treaty system and therefore have some vested interest in its successful implementation. As Laws notes (1987, p. 250) 'Antarctic scientific research has profited from its role in perpetuating this political truce by continuing to receive financial support from governments'.

The consolidation of the Treaty system in the 1960s and 1970s was achieved through the incremental adoption of new rules and procedures in response to policy proposals, but always guided by the need for unanimity and consensus. The range of topics addressed by the consultative parties has expanded to include telecommunications, meteorology, postal services, emergency assistance, air safety and, of course, a number of issues related to environmental protection. The consultative process has given rise to a complex network of recommendations, supplementary procedures and separate agreements collectively known as the Antarctic Treaty system (ATS). The decision-making rules and procedures in the Treaty were sufficiently general as to be susceptible of interpretation, a compromise between those who, with an eye to the future, wanted some form of administrative machinery, and those who wanted none. The view that prevailed about the procedural implementation of the Treaty is best represented by the eloquent counsel at the first consultative meeting of the head of the Chilean delegation (who had also been present at the Washington Conference). He argued that governments should 'proceed cautiously ... to avoid ... misunderstanding and harmful friction'. They should avoid 'all matters on which we are divided'. Finally, he counselled, they should 'avoid the formation of bodies which could create the impression that we are establishing a supra-national administration for Antarctica' (ATCM, 1961a, p. 25).

This emphasis on caution, on avoiding conflict and seeking consensus shaped the Antarctic regime. The procedural framework which was based on those values resisted institutionalisation and reinforced the consensus rule and the authority of the consultative parties (and the bureaucratic elite) over Antarctic decision-making. In the hierarchy of values which Puchala and Hopkins (1983, p. 66) suggest exemplifies all regimes, these values took precedence over others such as scientific cooperation and environmental protection. However, this bias engendered a functional, ad hoc, piecemeal and ultimately unsatisfactory approach to decision-making and management. As Kimball notes (1988c, p. 15), 'every time [the consultative parties] seek agreement on a new measure, they must once again find a balance that preserves the positions of ... countries claiming territory in Antarctica and those that do not recognise any claims'. Thus decision-making is often slow. Inherently conflictive issues – such as the issue of jurisdiction –

or issues upon which consensus could not be reached because of their implications for the political and legal compromise of the Treaty were quickly removed from the consultative agenda.[50] There has been a strong tendency to avoid enforcement mechanisms and sanctions. Recommendations adopted at Consultative Meetings are hortatory rather than mandatory – they urge, suggest, and recommend rather than require or enforce action.

New institutions and decision-making procedures, under the continuing authority of the consultative parties, were adopted '*only as and when necessary* ... in response to [a] well-defined need and well-defined problems' (Scully, 1986, p. 405; emphasis added). Special consultative meetings (SCMs) have been convened to deal with single issues including applications for consultative party status and the difficult resource management problems.[51] Meetings of experts have been convened from time to time to discuss technical problems relating to Antarctic activities, such as telecommunications. In spite of the emphasis on expertise, only experts from consultative parties may attend unless unanimous agreement is reached to invite other experts who have no voting rights.[52] Where relationships have been established with UN agencies or other international organisations, the consultative parties have made it clear that they will not accede to the authority of any other organisation on Antarctic issues.[53] They have sought also to keep Antarctic matters off the agenda of other international institutions.

There was continued resistance to any formal institutionalisation of the regime. The British had hoped that 'some form of permanent administrative machinery' would be established 'to facilitate inter-governmental cooperation in the intervals between Meetings of this group' (ATCM, 1961a). This was not to be the case.[54] Administrative responsibility for consultative meetings was devolved onto the host government, chosen by alphabetical order from within the ranks of the consultative parties (with the exception of South Africa). Communication between meetings has been conducted through bilateral diplomatic channels in accordance with recommendation I–XIV.[55] Proposals for a secretariat were discussed at the second consultative meeting and then dropped from the agenda of the third consultative meeting.[56] The issue was not formally raised again until the 1980s and has still not been resolved at time of writing.

Limited Change in the Antarctic Regime

In the 1980s, in response to pressures for change from within the Treaty system and from actors outside the regime, there was a cautious reassessment of the procedural dimensions of the Treaty system. Even then, the response to those demands was defined and limited by the normative framework of the Treaty system and there was no change to the principles and norms, nor to the values which had shaped the regime in the 1960s and 1970s. It was, therefore, a limited cognitive change. The first pressure came from the increase in membership of the Antarctic 'club'. Between 1961 and 1980 there were only nine new signatories to the Treaty. The decision-making club did not change until 1977 when Poland became a consultative party. Since 1980, twenty-one more states have acceded to the Treaty. States have acceded for a variety of reasons including the desire to participate in scientific investigation (and the prestige that accompanies it), to have a voice in Antarctic decision-making and discussions on the future of the Treaty system, and the possibility of access to resources.[57] The 'granting' of consultative party status was designed in part to strengthen the Treaty system against outside challenges. Existing consultative parties sought to co-opt developing states to counter criticisms of the Treaty system as a club for western, industrialised states. The number of consultative parties has increased by fourteen to include countries such as India, China, Brazil, Uruguay and the Republic of Korea. New consultative parties benefit from access to decision-making on an equal basis and from the scientific expertise of the more experienced Antarctic countries.[58] These accessions, and applications for consultative party status provide a measure of the legitimacy and strength of the Treaty system.[59]

While the participation of non-traditional Antarctic states such as India and China has altered the balance of interests within the system, it has not altered the perceived need to maintain the political compromise on sovereignty. Indeed the newer states have become rapidly socialised into the traditional norms of the Treaty system and have become staunch supporters of the Treaty system. Nevertheless, as Antarctic diplomat Christopher Beeby has observed (1990, p. 16) 'it is [now] much less easy to generate a dynamic that will lead to a consensus acceptable to all'. If the numbers continue to increase, he suggests, 'there might need to be a search for new methods of arriving at a consensus'. One consequence of the increase in parties, many of whom could not afford to undertake scientific research, was that the non-consultative parties (NCPs) became increasingly dissatisfied at being excluded from the decision-making process and lobbied for greater involvement. As a result, in a 'major departure from prior practice' (Kimball, 1983, p. 5) the NCPs were invited to attend consultative meeting, as observers, only from 1983 although the invitation was initially extended on a one-off basis only.[60]

While this internal expansion of the Treaty system could be easily managed, external criticisms provided a different kind of challenge to the regime. From the late 1970s, non-governmental environmental organisations began to focus on the consultative parties' record in protecting the Antarctic environment and found that record unsatisfactory. They also challenged the closed and secretive nature of the Antarctic regime and called for a greater degree of transparency and accountability. Third-world countries led the challenge in the UN General Assembly where, beginning in 1983, they accused the consultative parties of constituting an exclusive, secretive club managing the Antarctic for their own ends. They argued that the management of the Antarctic should become the responsibility of the whole international community under the auspices of the United Nations. For the most part, the consultative parties did not pay a great deal of attention to the Malaysian-led criticisms. They noted that any state could join the Treaty, that the members of the Treaty represented a large proportion of the world's population and included states from the first, second and third worlds.

In response to criticisms about the lack of transparency in particular, the parties took steps to improve the 'openness' of the regime although no quick decisions were made. Documents tabled at consultative meetings were progressively made publicly, although not easily available. The final reports of consultative meetings began to include more information and often contained the substance of agreements and programs of further action which were not incorporated into recommendations. From 1987, experts from relevant international organisations were invited to consultative meetings to assist the consultative parties with their deliberations although under strict and restrictive rules of participation.[61] The vexed question of administrative arrangements and a secretariat was again raised although it still has not been settled in the face of opposition from Argentina with the support of some of the newer consultative parties such as India and (prior to 1992) East Germany.

By the end of the 1980s, the Antarctic Treaty system had changed to the extent of participation, with some adjustments to procedure. Those changes were incremental and limited, and did not undermine the cohesion of the Treaty system, nor its principles or norms. It was within this framework that the Treaty parties addressed environmental issues – a process explored in the next chapter. There was no change in the hierarchy of values or in the dominance of sovereignty norms. That change was to come with the overturning of the Minerals Convention and the negotiation of the Madrid Protocol.

3 Environmental Protection: Politics and Outcomes

THE ANTARCTIC ENVIRONMENT

The Antarctic terrestrial ecosystem is marked by limited biological diversity. By contrast the marine ecosystem of the Southern Ocean is teeming with life, including seals, whales and over forty species of birds and more than 200 species of fish. There are four species of Antarctic seal – the crabeater, Weddell, leopard and Ross – and two more species, the Southern fur seal and the Southern elephant seal, which breed in the sub-Antarctic islands and range across the whole Southern Ocean (see Laws, 1983, p. 609; Roberts, 1977, p. 99). The southern oceans are also host to six species of baleen (or filter-feeding) whale – the blue, fin, sei, humpback, minke and southern right whale – and two species of toothed whale – the sperm and killer whales (see Tierney, 1978, p. 8). Forty species of breeding birds, of which penguins are by far the most numerous, inhabit the Antarctic region although only sixteen of them actually breed on the continent itself (Tierney, 1978, p. 8). In spite of this abundance of life, the marine food-chain is a remarkably short and simple one, depending primarily on a small shrimp-like crustacean, the Antarctic krill, which is the main food of whales, seals, penguins and sea-birds (see Lucas and Dingwall, 1987, p. 226). Both ecosystems are vulnerable to, and slow to recover from, disturbance. The simple terrestrial ecosystem lacks the regenerative powers of more elaborate ecosystems (Rudolph, 1970, p. 53; Holdgate and Tinker, 1979, p. 19).[1] The short food-chain makes the marine ecosystem vulnerable to perturbations to any constituent part.

In its undisturbed state, the Antarctic environment is a valuable scientific laboratory. Disturbances to the Antarctic environment as a result of human activity are likely to erode this scientific value. Scientific knowledge is also a necessary component of sound environmental and conservation decision-making. As well as increasing our knowledge about the Antarctic, research there can reveal a great deal about past climatic changes and about the impact of past global pollution. The Antarctic also provides an important baseline for measuring the effects of contemporary industrial pollution, in particular the impact of ozone-depleting and greenhouse gases.[2] The Antarctic plays an important role in global weather patterns and acts as a heat sink for the planet by reflecting incoming solar radiation (IUCN, 1984b, p. 5; Drewry, 1988, pp. 7–8). Severe changes to this albedo (reflective) factor 'are likely to have global impacts' (Laws, 1990, p. 10).

Human impact on Antarctic ecosystems can take a variety of forms (see Benninghoff and Bonner, 1985). Scientific activity can have, and has had, localised impact on the Antarctic environment, particularly on terrestrial and fresh-water ecosystems on the continent and in-shore marine habitats. The most suitable locations for scientific stations are the predominantly ice-free coastal regions required by bird and seal populations for breeding grounds.[3] Environmental impact arising from human activity includes disruption to breeding colonies, the potential introduction of alien diseases and viruses, localised marine and terrestrial pollution as a result of sewage and waste disposal (where there is little natural biodegradation) and the burning of fuel for heat and energy, and the impacts of the construction of stations and support facilities. The increase in the number of consultative parties in the 1980s brought with it an increase in the number of scientists and support personnel in the Antarctic and a consequent increase in the potential for adverse environmental impact.[4] As most countries rely on ships to transport personnel to the continent, this increase in activity also brought with it an increase in the potential for accidents and marine pollution. Similar impact problems arise from tourist activity, especially in those areas where repeated visits can have a cumulative impact (see Hart, 1988). In the past, economic activity has resulted in the near decimation of particular species (first seals and then whales) and in severe disturbances to the balance of the ecosystem. The impact of the exploitation of krill and fish stocks remains a central debate of the Antarctic Treaty system.

In the 1980s, debates about environmental protection in the Antarctic focused also on the continent's intrinsic wilderness and aesthetic values and the importance of maintaining the integrity of the environment as a whole as well as preventing localised impacts. In this view, environmental protection is a valuable goal its own right, not just as a contributing factor to resource sustainability and scientific investigation. Benninghoff and Bonner (1985, p. 21) identify two important components of environmental protection in the Antarctic – the maintenance of the high productivity and ecological relationships in the southern ocean and the maintenance of the unspoiled environment and fragile ecosystems of the land, inland waters and ice-covered areas of the continent. The consultative parties have addressed both aspects in the development of the environmental protection rules.

However, the Antarctic Treaty is not an environmental treaty. Protection of the environment is barely mentioned. References to the preservation and conservation of living resources were as much to do with economic sustainability as with the intrinsic protection of Antarctic species. Nevertheless, through the adoption of consultative recommendations and other instruments, the Treaty parties developed an environmental regime nested within the Antarctic Treaty system. That regime was not, on the whole, an effective one.

THE ANTARTIC ENVIRONMENTAL REGIME

Understanding the environmental regime in the Antarctic requires, as Chapter 1 indicates, exploration of a number of issues – the purpose of the regime, the actors involved, the values that shaped its rules and procedures, and the strength of the regime, in particular the degree of compliance. The environmental regime was developed so that Antarctic fauna and flora, and the environment generally, would be minimally disturbed. Consideration of environmental issues initially focused on the conservation and preservation of living resources. The effects of tourism was inscribed on the agenda in 1966 and the preservation of historic monuments in 1968. 'Man's [sic] impact on the Antarctic environment' (now 'human impact') has been a standing agenda item since 1970, and environmental topics now include marine pollution, conduct of scientific expeditions, waste disposal, the development of a protected area system, the siting of scientific stations, environmental impact assessment and the effects of the exploitation of mineral resources. As well as the increasingly complex network of rules and procedures adopted in consultative recommendations, which are discussed in this chapter, the Treaty parties have negotiated three separate conventions. Those conventions, which attempt to balance environmental and use values are the Convention for the Conservation of Antarctic Seals, the Convention on the Conservation of Antarctic Marine Living Resources and the Convention on the Regulation of Antarctic Mineral Resource Activities. They are examined in the following chapters.

The consultative parties have characterised their approach to regulating environmental impact in the Antarctic as a 'prevention rather than cure' approach which sought to regulate activities that 'a reasonable person would look upon as being at least regrettable' (Antarctic Treaty System, 1989, p. 1101). However, they also balanced the need to minimise environmental impact against scientific research, tourism and resource exploitation. Environmental needs were not only set against other uses of the Antarctic. They were also subordinated to the political concerns of the regime. The hierarchy of values implicit in the Antarctic regime did not favour environmental protection. As Auburn suggests (1982, p. 274) 'sovereignty, politics, logistic convenience and the facilitation of scientific research prevail over environmental considerations'.

The process of decision-making on the environmental subregime rested, as it did in the Treaty system generally, on negotiation rather than imposition by a hegemonic state. Consultative parties were engaged in a joint problem-solving exercise focused on common, but limited, environmental objectives. The lack of minutes from any of the consultative meetings means that it is difficult to assess whether any state, or states, took a leadership role on environmental issues. Nevertheless the documents tabled at these meetings make it possible to suggest that a number of countries were at the forefront of putting proposals to meetings. Those countries were the UK, the US, Australia, Chile and, to a lesser extent, New Zealand. Of those, Britain seems to have seen itself as the foremost advocate of 'quiet diplomacy' as the basis for negotiation and consensus. Its approach to decision-making in the Treaty system generally was to take a long-term view and seek to build consensus incrementally. The British tactic was to define problems clearly in the belief that agreement on the nature of the problem would advance consensus on the nature of the solution. At least one observer has argued that the British were the only party with concern for the environment which was able to effectively initiate solutions to environmental problems.[5] Nevertheless, because of the perceived need to balance political interests with environmental ones those solutions were increasingly out of step with protection imperatives. This approach also precluded long-term planning and the elaboration of clear goals for environmental management in the Antarctic.

The development of the regime effectively resulted from incremental change within the Treaty system. By the late 1970s, rules and procedures were in place for the protection and conservation of fauna and flora and the designation of Specially Protected Areas (SPAs) and Sites of Special Scientific Interest (SSSIs). The consultative parties had begun to consider how to regulate tourist activity (but made little progress) and put in place a general code of conduct for scientific expeditions. While these recommendations imposed obligations, they incorporated no enforcement rules, reflecting the dominance of sovereignty norms in the Antartic regime and identifying a problem common to international environmental politics. Implementation and, to a great extent, interpretation were the responsibility of individual governments and national operators. Most of those issues continued to be of importance in the 1980s. The agenda of the 1980s also widened to include a more detailed consideration of waste disposal procedures, environmental impact assessment and the adequacy of the Antarctic protected area system. New issues were added to the agenda in response to the increase in human activity as well as to new knowledge about the Antarctic environment. The consultative parties sought to regulate human activity not only because of the environmental impact of such activity, but also because of fears that unregulated activity might have the potential to upset the Treaty system and its political compromises.

The main features of this environmental regime are the lack of institutional framework and linkage between recommendations, implementation based on self-restraint rather than compliance rules and slow decision-making as a result of the political need to maintain a consensus and avoid conflict. The process of negotiation and compromise which underpinned the incremental approach to decision-making under the Treaty proved to be a problem when activities on the ground moved far more quickly than the consultative process. Years sometimes passed before substantive rules and procedures could be agreed upon. Draft recommendations were often watered down so that agreement could be reached. In spite of the often acknowledged importance of protecting the Antarctic environment, contentious issues were often deferred for political reasons. SCAR's advice was not always acted upon. In other words, decision-making lagged behind knowledge. The decentralised nature of the Treaty system also worked against consistency of implementation practice across the Treaty parties. Decision-making on environmental issues in this context was issue-specific and thus piecemeal. The environmental measures established by the Treaty system were disparate and fragmented. The regime was incomplete and ad hoc. By the late 1970s it was clear that 'the legal and political arrangements for governing ... the Antarctic lag disturbingly behind the human ability to affect [this] region' (Schatz, 1978, p. 477).

Compliance mechanisms were avoided in accordance with the political demands of the system which resisted anything that seemed to erode national autonomy and sovereignty over the conduct of activity in the Antarctic. The consultative parties argued that Treaty's inspection procedure was an adequate mechanism for ensuring compliance with environmental rules. However, those rules were included in the Treaty primarily to oversee the demilitarisation provisions. Not only did the potential for inspection fail to ensure compliance, no steps were taken by the parties when violations of the rules were revealed by inspections. Because of the reliance on sovereignty norms there were no enforcement mechanisms or procedures to deal with violations. Parties were also unwilling to bring them to the attention of the consultative process in any formal way. As Kimball notes (1988c, p. 18) 'parties to the treaty have traditionally been reluctant to "rock the boat" by asking too many questions about each others activities in Antarctica'. Violations of the environmental rules were not considered destabilising to the regime whereas disputes over them might well have been. Many of the consultative parties were, at some time, in violation of the environmental protection rules. Thus there was, in effect, an 'in-club' conspiracy of silence.

The consultative parties argued that the rules and procedures adopted to protect and manage the Antarctic environment were adequate to the task and provided evidence of the strength and flexibility of the Treaty system in that it could respond to changed demands and needs. However, this judgement of success rests on political criteria in that environmental regime has been developed in such a way as to not undermine the compromise on sovereignty. The environmental rules adopted did not compromise the fundamental national interests of the Treaty parties outlined in chapter 2 and thus they contributed to the internal coherence of the regime. From the perspective of their ability to protect the Antarctic environment from adverse effects of human activity, a different judgement is made: the environmental regime was, in the final analysis, flawed.

In the 1960s and 1970s the smallness and cohesiveness of the Antarctic club and the general lack of international attention to the Antarctic, as well as the low level of international consciousness about environmental protection issues generally, meant that potential difficulties over the inadequacies of environmental rules and implementation could be masked. The principles and norms of this environmental regime were generally not questioned by either the bureaucratic elite or the scientists who were the major actors within the regime. In the 1980s, the environmental regime was shaped not only by changes in the Antarctic but also by factors outside the Treaty regime. The number of people in the Antarctic each year increased as the number of consultative parties grew. Scientific research contributed to the growing (although still limited) knowledge about the Antarctic environment and the impact of human activity upon it. Protection of the environment became an international agenda item and non-governmental organisations became active and increasingly experienced informal participants in the Antarctic regime. The late 1970s and the 1980s were marked by a growth in environmental consciousness on the part of the public and, to some extent, policy-makers. In the 1980s there were demands, by NGOs in particular, for an end to the piecemeal and disjointed approach to environmental decision-making and the permissive approach to compliance which characterised the environmental sub-regime. This push for a more comprehensive approach took place almost entirely outside the formal bounds of the Treaty system.

Formal and Informal Actors

As chapter 1 noted, understanding the dynamics of a regime is helped by identifying participating actors and their roles as agents for stability and agents for change. International environmental politics has been typified by the involvement of the scientific community and non-governmental environmental organisations as knowledge-based actors, as well as the central role of bureaucratic actors. The bureaucratic actors who were the core of Antarctic decision-makers applied the broader norms of the Treaty system, into which they were socialised, to environmental issues. It was these actors who effectively 'owned' the issue. The scientific community was an important participant as a source of knowledge and advice but it did not adopt an advocacy position. Further it was an interested actor in that most of the rules were designed to regulate the conduct of scientists and scientific research. In this respect the role of the scientific community, as a potential epistemic community, in encouraging compliance with environmental rules must be questioned.

Environmental rules were adopted through the closed consultative process dominated by the same people who were responsible for Antarctic decision-making generally. Few delegations included representatives from environment ministries or individuals with environmental expertise. By contrast, representatives from energy and industry ministries (and private enterprise) were included once minerals issues came onto the agenda. The same is true of the fisheries issue. United Nations or other intergovernmental organisations with expertise or interests relevant to protection of the Antarctic ecosystem were rarely called upon for advice although this began to change in the late 1980s. The scientific community, particularly through the agency of the Scientific Committee on Antarctic Research, had a major role in providing the knowledge upon which the rules and procedures of the environmental sub-regime were based. Environmental issues were most often dealt with by the conservation sub-committee of the SCAR Working Group on Biology and later by the Group of Specialists on Antarctic Environmental Affairs and Conservation.[6] However, in line both with the consultative parties' determination to maintain final authority over Antarctic decision-making and SCAR's determination to separate science from politics, this role was essentially advisory. Although a formal role for SCAR was written into some of the environmental measures adopted by the Treaty parties SCAR chose not to play an advocacy role. Nevertheless SCAR saw itself as the natural, and only legitimate source of environmental advice for the Treaty parties. Scientists were both the users of the Antarctic and the enforcers of the Treaty system's environmental rules. While operationally convenient, there was a potential clash between these roles.

Non-Governmental Organisations

Bureaucrats and scientists had been the only participants in the Antarctic regime in the 1960s and 1970s. In the 1980s, their control of decision-making and the provision of information was challenged by non-governmental environment and conservation organisations. NGO interest in the Antarctic was spurred by concern about the potential exploitation of the marine ecosystem and then by the minerals negotiations. NGOs argued that the environmental rules adopted through the consultative process were inadequate, that they were poorly implemented and that there were, therefore, real threats to the security of the Antarctic environment. In the closed atmosphere of the Antarctic regime, NGOs sought to impose transparency and accountability on the Treaty parties. Environmental groups became 'important sources of lateral and vertical communication within the regime' (Shapley, 1988, p. 329).

NGO activity was initially organised on a national basis, in the US, Australia, New Zealand and the UK.[7] A transnational focus was provided in 1977, with the founding of the Antarctic and Southern Ocean Coalition (ASOC) to provide coordination between national NGOs so that their developing expertise and knowledge could best be used collectively.[8] This transnational network is important also in that information is exchanged more quickly than it is among the consultative parties through traditional bilateral diplomatic routes. ASOC is primarily a clearing-house for information: it does not direct members activities.[9] Greenpeace International, the largest member organisation of ASOC, on the other hand provides policy direction and coordinates action among its national offices in support of its Antarctic campaign which began in 1983. While ASOC has a large organisational membership the number of experienced Antarctic lobbyists is still relatively small. In this respect NGOs mirror the formal Treaty system, where a small group of people with considerable 'institutional memory' have dominated the decision-making process.

The International Union for the Conservation of Nature and Natural Resources (IUCN), which was established in 1948, is another environmental organisation with a well-developed interest in Antarctic issues. Like ASOC, it is an umbrella organisation. However, IUCN includes both government and non-government organisations, including Greenpeace, in its membership. The IUCN is not a lobbying or pressure group in the style of ASOC or Greenpeace. While IUCN does share with the more advocacy-oriented NGOs a concern about the adequacy of environmental rules, it sees its role rather as a one of providing advice and expertise to promote conservation objectives.

NGO activity has been directed towards influencing environmental policy and practice in the Antarctic. They have maintained a lobbying presence at Treaty meetings since the late 1970s during which time they have sought a greater degree of formal participation, as observers, in the consultative process. Coordinators of the NGO campaign usually liaise with the temporary secretariat in advance of a Treaty meeting. 'Accredited' NGO representatives – who are nevertheless 'outsider' participants – are usually guaranteed entry to the meeting venue although not, of course, to the meeting itself. This nevertheless enables them to lobby delegates at coffee breaks. Lobbying activities are supported by press conferences and publication of the newspaper *ECO* which is circulated to delegates and the media. In reporting the meetings and educating the public NGOs 'create and inform a wider constituency for Antarctic matters' (Kimball, 1990, p. 32). NGOs have also hosted dinners and other social functions at consultative meetings, strategies which are more familiar in formal diplomatic circles and they are now regularly invited to attend functions held by Treaty delegations.[10] Many of the NGO representatives now have long-standing connections with Antarctic diplomats which further blurs the boundaries between formal and informal actors. This international and, for the most part, behind the scenes activity has been complemented by campaigns designed to mobilise public opinion and maintain public pressure on the consultative governments.

Antarctic bureaucrats have displayed a somewhat ambivalent attitude towards NGO representatives. Quigg notes (1983, p. 179) that

> [w]hile showing proper respect for conservationists in public, officials involved in Antarctic affairs in the US and Britain particularly are in a state of more or less perpetual irritation with the environmentalists for what is often seen as their demand for absolutes, their unwillingness or inability to comprehend the reality of international politics, and the processes of negotiation among sovereign states.

Few consultative parties have included NGO representatives in their delegations. By the early 1980s only three – the US, Australia and New Zealand – had adopted this practice.[11] This practice is still limited and NGOs have had a lower profile in Latin America and the newer consultative parties.[12] However, 'inside' NGOs have become an important source of information for 'outside' NGO lobbyists at Treaty meetings. Until recently, NGOs had little success in gaining *independent* access to the formal mechanisms of the Treaty system. ASOC and Greenpeace International first applied for observer status to meetings of the Commission and Scientific Committee of CCAMLR in 1983.[13] ASOC was successful in 1988.[14] ASOC was also granted observer status to the special consultative meeting on comprehensive measures in 1990, discussed in chapter 7, and was invited as an expert organisation to the 16th and 17th consultative meetings in 1991 and 1992. Antarctic decision-makers were more responsive to the IUCN to which some of their governments belonged. The IUCN was invited as an observer to the diplomatic conference at which CCAMLR was signed and was given observer status to institutional meetings under that convention.[15] Since 1987, an IUCN expert has been invited to assist in consultative meeting discussions on human impact on the Antarctic environment.

NGOs are knowledge-based actors within the Antarctic regime. Their often critical analyses of environmental measures and their alternative policy proposals are not constrained by the political dictates of the Treaty system. They frequently provide the only alternative source of information for delegates, especially those new to the consultative process or, from 1983, for non-consultative parties.[16] Their participation has been characterised by increasing diplomatic and scientific expertise and they have become an important knowledge base within the Treaty system, sometimes in opposition to the scientific community. NGOs have drawn on 'in-house' scientific expertise to put cogently argued policy alternatives to the Treaty parties.[17] This has been useful in establishing credibility with the scientific community, many of whom were equally concerned about environmental protection but were often less than complimentary about the environmentalists. A former director of the British Antarctic Survey observed that while 'scientists tend to avoid publicity and are constrained to be truthful about their work [i]t is self-evident ... that environmentalist ... propaganda tends to be blatant and less scrupulous' (Laws, 1985, p. 55).

The IUCN has been somewhat more successful in establishing a working relationship with the scientific community although, as Kimball notes (1988d, p. 40) it was a tentative and somewhat wary collaboration in its early stages. While the IUCN saw SCAR as the best channel through which to approach the Treaty parties to promote their conservation objectives 'SCAR's relatively closed network of old Antarctic hands was somewhat sceptical of IUCN's lack of Antarctic credentials' (Kimball, 1988d, p. 40). However, in late 1985 a joint IUCN/SCAR working group on long-term conservation in the Antarctic was established, although some form of collaboration had been suggested as early as 1982.[18]

The Greenpeace Antarctic Campaign

The public face of NGO activity was the Greenpeace International Antarctic campaign and, in particular, its construction of a scientific station at Cape Evans (in territory claimed by New Zealand) in the Antarctic in February 1987. The Greenpeace Antarctic expeditions and the World Park Base were intended to serve both a political and a scientific purpose.[19] Greenpeace sought to demonstrate that expeditions and scientific research could be conducted in an environmentally sound manner in accordance with the various environmental protection measures in place.[20] They also sought to raise public awareness of Antarctic issues and to acquire expertise to support their application for observer status to the Treaty. They were, they argued 'looking over the shoulders of bases, literally ... [and] morally' (Wilkinson, 1988). In many respects, the activities of Greenpeace were no different from those undertaken by Antarctic programmes run by many of the consultative parties. By establishing a permanent base and conducting scientific research, Greenpeace International had effectively met the criteria applied to the granting of consultative status to states. During its 1987–88 resupply voyage, Greenpeace 'inspected' several bases in the vicinity of Cape Evans and the Antarctic Peninsula to identify and publicise violations by consultative parties of environmental protection measures. Its inspection report was distributed to all Treaty states. In the 1988–89 austral summer, the Greenpeace expedition visited 25 scientific bases and again drew attention to violations of the codes of conduct. By 1990 Greenpeace had conducted environmental inspections at 39 Antarctic stations (de Poorter and Schmidt, 1990, p. 39), probably more than any consultative party. Greenpeace has recently dismantled its base to concentrate on monitoring environmental practice in the Antarctic.

The Greenpeace presence in the Antarctic put increasing pressure on the consultative parties to respond to their charges of environmental mismanagement, in a world where public concern about environmental issues was increasing. In other words, a learning process took place – the Treaty parties changed their behaviour (admittedly slowly and in many respects in a limited way) in response to knowledge, to information that was generated in part through the activities of NGOs. The consultative parties were keen to assure the world of their responsible stewardship of the Antarctic at a time when they were under some pressure in the United Nations for being secretive and exclusive and where electoral pressure could be mobilised. Non-governmental activity focused public attention both on the decisions the consultative parties made and the way those rules were implemented. The parties have accepted that they must be seen to respond positively to the increased public attention generated by NGOs. The New Zealand head of delegation at ATCM-X in 1979 noted that 'the international community at large, international organisations and private groups interested in the environment are looking to the consultative parties to live up to the high ideals' upon which the environmental regime was supposedly based (ATCM, 1979a, p. 78).

Principles, Rules and Procedures: Environmental Protection in Practice[21]

The rules of the environmental regime which are discussed in more detail here were characterised by a number of features which were shaped by the broader Antarctic regime. In that respect, then, there was a degree of internal coherence. SCAR had a key role as a provider of information, and sometimes as the initiator of proposals but, for all that, its advice was often watered down when recommendations were finally adopted. Decision-making was, in line with the consensus imperative, often slow. Concerns over sovereignty engendered a continued tendency towards decentralisation, and compliance was characterised by voluntarism and was, therefore, often unsatisfactory. The precautionary principle was not well adopted in practice. The first articulation of environmental principles came in a recommendation (VI-4) adopted at the sixth consultative meeting in 1970 (see ATS, 1989, pp. 1101–2). This recommendation acknowledged that the Antarctic ecosystem was particularly vulnerable to human interference, that the Antarctic's scientific importance derived from its uncontaminated and undisturbed condition and that there was, therefore, an urgent need to protect the environment from human interference. It added to the consultative parties' obligations under the Treaty by suggesting that it was they who should 'assume responsibility for the protection of the environment and the wise use of the Treaty Area' (ATS, 1989, p. 1101). The consultative parties took the view that, because of their Antarctic experience and knowledge, they were best placed to undertake this task.[22] It was, nevertheless, a vague statement of intent rather than of clear and rigorous guidelines. There is nothing to say how that obligation should be met. At the 1975 meeting, the consultative parties agreed that

> no act or activity having an inherent tendency to modify the Antarctic environment over wide areas within the Antarctic Treaty Area shall be undertaken unless appropriate steps have been taken to foresee the probable modifications and to exercise appropriate controls with respect to the harmful environmental effects such ... uses may have (recommendation VIII-13, ATS, 1989, p. 1102).[23]

These general environmental principles and injunctions were not translated into a comprehensive set of environmental standards against which human activity in the Antarctic could be measured.

Protection of Fauna and Flora, and Development of the Protected Area System

The first environmental issue addressed by the consultative parties was that specifically mandated by article IX(f) of the Treaty, the conservation and preservation of living resources.[24] The consultative parties adopted measures for the protection of fauna and flora along with a protected area system based on specially protected areas and sites of special scientific interest. While the advice of the Scientific Committee on Antarctic Research (SCAR) was sought on these issues, all final decisions rested with the consultative process and those decisions were guided by political considerations. The provisions of the rules on protecting fauna and flora and the adequacy of the protected area system came under increasing scrutiny in the 1980s from non-governmental organisations and from environmental management experts.

Debate at the first two consultative meetings on protecting fauna and flora culminated in the adoption, at the third consultative meeting in 1964, of the Agreed Measures on the Conservation of Antarctic Fauna and Flora. While the Treaty parties were genuinely concerned to establish conservation measures, the substance of those measures was dictated by sensitivities over the exercise of sovereignty and jurisdiction which engendered an 'unwillingness ... to have any but the most rudimentary institutional structure to oversee implementation of the Agreed Measures' (Bush, 1990, p. 126). The Agreed Measures had their genesis in guidelines for the protection of living resources developed by SCAR at its 1960 meeting. While there was some support for SCAR's guidelines, the consultative parties disagreed over the extent of protection. While they acknowledged the urgent need for measures to conserve the living resources of the Treaty area and protect them from uncontrolled destruction or interference by humans (recommendation I-VIII) governments were urged only to consult on the form in which it would be most suitable to establish 'in due course' internationally agreed measures for this purpose. In the interim, general rules of conduct extracted from SCAR's guidelines were to be observed 'to the extent possible under national legislation' (ATS, 1989, p. 2302). This enabled the rules to be given some status without requiring any formal commitment on the part of the Treaty parties, which fits with the voluntary nature of recommendations that was to become a hallmark of the environmental regime. Little progress was made at the second consultative meeting in 1962. The difficulty in reaching a consensus arose because of political differences over the form of any agreement – that is, whether it should be a code, a convention or a recommendation – and the area to be covered by any such measure. Nevertheless, governments were asked to formulate a draft text as part of the preparation for ATCM-III, a task which was undertaken at a meeting in 1964.

The Agreed Measures for the Conservation of Antarctic Fauna and flora were adopted as an annex to recommendation III-VIII in 1964 (see ATS, 1989, pp. 2303–6). Governments were called upon to approve them as soon as possible and implement them without delay. In fact, this did not happen. The US, for example, did not adopt legislation to give effect to the Agreed Measures until 1978. By 1983 Japan still had not ratified them although the legislation enabling Japan to do so had been passed in 1982 (Bush, 1982, p. 169). Japan's ratification was deposited before the 1985 consultative meeting. In the meantime, the Agreed Measures were to be treated as interim guidelines 'as far as feasible' (recommendation III-IX, see ATS, 1989, p. 2307).[25] The Agreed Measures designated the Antarctic Treaty Area as a Special Conservation Area, but did not define this concept.[26] The measures established rules to protect Antarctic fauna and flora for both scientific study and rational use. Harmful interference with normal living conditions was to be minimised although native mammals or birds could be taken under a permit system. Permits could only be issued for specific purposes, including the provision of food or scientific specimens, and were to be limited to ensure that the natural ecological balance was maintained. The permit system was also designed to limit the introduction of non-indigenous species, parasites and diseases. To assist in the protection of fauna and flora, the Agreed Measures provided for areas of outstanding scientific interest to be designated Specially Protected Areas (SPAs) and given special protection to preserve their unique ecological systems.[27] Strict guidelines were established on the types of activity which would be permitted in such areas. The Measures also provided for the designation of Specially Protected Species (SPSs). Although the draft recommendation contained suggestions for both, no SPAs or SPSs were designated at the 1964 meeting.

While the Agreed Measures established quite rigorous guidelines, they were flawed. In particular, the permit system allowed each state to operate unilaterally. There was no central authority for issuing permits and therefore no way of monitoring the cumulative impact of taking fauna and flora. Argentina and Chile opposed anything in the Measures that would 'hint at a lessening of national sovereignty' (Myhre, 1986, p. 53). While the format of the Agreed Measures allowed governments to take appropriate legislative steps to give effect to the conservation principles, there was no way to enforce the Measures in practice. No compliance mechanisms were adopted. Bush suggests (1990, p. 127) that the Agreed Measures imposed a 'relatively low level of obligation'. Participating governments are simply urged to 'take appropriate measures' to carry out these Agreed Measures (ATS, 1989, p. 2304).[28] Amendments in 1970 and 1975 instituted a permit system for entry into SPAs and only for a compelling scientific purpose. No guidelines were given on how such a purpose might be defined.

Fifteen SPAs were adopted in 1966. Although these areas were based on proposals put forward by SCAR at its September

1966 meeting, former Antarctic diplomat Hugh Wyndham suggests (1980, p. 187) that a 'careful analysis of the specially protected areas adopted ... will show signs of continuing sensitivity' on the claims issues. Australia proposed three SPAs, the United Kingdom eight, and the NZ proposed four (jointly with the US). The claimants (NZ, UK and Australia) proposed only areas within their claims. Under recommendations VI-16 and 17 adopted at this meeting, Fur Seals and Ross Seals were designated Specially Protected Species. After this initial flurry of activity, only two more SPAs and no further SPSs were designated between 1966 and 1983 even though this category of protected area was central to the protection of the Antarctic environment. Indeed, when the protection of an SPA conflicted with human activity, the latter use seemed to prevail. Three SPAs were terminated in the same period. In 1968, only two years after its designation, an SPA on the Fildes Peninsula of King George Island in the South Shetlands was modified to make it much smaller (recommendation V-5) because of the construction of two scientific stations in the area.[29] Construction had begun *after* the adoption of the Agreed Measures (although they were not yet formally effective) and thus violated at least the spirit of the Measures. In 1975 this designation, along with that of two other SPAs, was terminated under recommendation VIII-2 and downgraded to a lesser protective category of Site of Special Scientific Interest.[30] Thus, within ten years of the adoption of the Agreed Measures, areas deemed in need of protection because of their unique ecological nature were being denied that protection. A SCAR report on the purposes and designation of SPAs which was presented to the seventh consultative meeting in 1972 expressed concern that existing SPAs were not fully representative of major Antarctic land and freshwater ecological systems (see ATCM, 1972c; 1972d). The consultative parties agreed that SPAs should be reviewed ('in due course' as they put it) to include areas with unique complexes of species, areas which were the only known habitat of a particular species, or contained interesting breeding colonies, or areas which should be kept 'inviolate' for future comparison with disturbed areas. SCAR was invited to conduct this review.[31] However, after this meeting the preservation and conservation of fauna and flora effectively dropped off the agenda until 1981.

A second type of protected area was established in 1972 with the adoption of the category of Site of Special Scientific Interest (SSSI), designed to protect from wilful interference areas which were important to scientific investigation. SPAs could not be used for this purpose because they were designed specifically to protect unique fauna and flora habitats rather than broader scientific values. For this reason SSSIs were not incorporated into the Agreed Measures which focused specifically on fauna and flora. Each proposal for an SSSI was to be accompanied by a management plan based on recommendations from SCAR. Seven sites proposed by SCAR were established in 1975.[32] An eighth site was designated in 1979. Thirteen sites approved by SCAR were adopted in 1985 under recommendation XIII-8. A further four terrestrial sites, proposed by the United Kingdom, were adopted at ATCM-XIV in 1987. At the first consultative meeting in 1961, the consultative parties also established the category of historic sites and monuments (recommendation I-IX, see ATS, 1989, p. 2405) which were consolidated into a single listing in 1972. Many of these sites have been designated for political purposes, to make a statement about national activity in the Antarctic, as well as for conservation purposes.

The coverage of SSSIs was restricted by political difficulties. In particular, marine sites proved controversial because of concerns held by some consultative parties about infringements of high seas rights under article VI of the Treaty. Because of their differences on coastal state jurisdiction, the parties could not agree whether immediate off-shore areas were high seas or territorial seas. The question of marine SSSIs therefore sat on the agenda for twelve years before agreement could finally be reached. Interest in proposing marine SSSIs was first raised at the 13th SCAR meeting in 1974. It was inscribed on, and then deleted from, the agenda for the 8th consultative meeting in 1975. At the following meeting in 1977 Chile proposed two marine sites although Boczek suggests (1986, p. 96) that Chile's motivation was not so much environmental as the political consideration of asserting its jurisdiction over offshore waters. Advice requested from SCAR was discussed at the tenth consultative meeting in 1979, along with Chile's proposals, but disagreement over the political and legal status of off-shore waters precluded any consensus, notwithstanding the importance of protecting marine ecosystems.

In 1983, Chile, Argentina and the UK jointly proposed a site that was partly terrestrial and partly marine, indicating some desire to manage their competing claims in the interests of protecting scientific sites. The final report of the consultative meeting includes a weak reference to the importance of careful examination of such proposals in the light of article VI of the Treaty on high seas freedoms (see ATCM, 1983a, p. 9). Delegates decided to defer the issue, *again*, until SCAR had approved SSSIs which were partly or wholly in the marine environment although SCAR had already advocated such sites at its 17th meeting in 1982. In spite of SCAR's concern that this issue had been on the table for several years, three proposals tabled in 1985 were again rejected although several delegations did stress the importance of the designation of these sites.[33] At least one delegation continued to argue that marine SSSIs contradicted freedom of the high seas. Three marine SSSIs were finally adopted at ATCM-XIV in 1987, twelve years after the issue first appeared on the agenda.[34]

These rules were not adequate to the task of protecting fauna and flora. In spite of the Agreed Measures, 'there can be no doubt that breeding populations of Antarctic birds are adversely affected by direct exposure to man's [sic] activities in Antarctica' (Muller-Schwarze and Belanger, 1978, p. 380). In a single incident during the 1970–71 season a helicopter taking off over the Cape Crozier rookery caused several thousand incubating Adelie penguins to leave their nests in panic. Their eggs rolled out of the nests and were lost. This incident was either in direct contravention of the Agreed Measures, or it points to the inadequacy of the Agreed Measures in protecting fauna and flora. Navy helicopter flight patterns are argued to have caused a 50 per cent reduction in breeding at one penguin rookery close to McMurdo (Manheim, 1989). The construction of Hallett station in 1968 resulted in the displacement of a penguin rookery covering two acres (Quigg, 1983, p. 50–1). The protected area system was unable to 'adequately protect the Antarctic environment over the full range of human activities and their consequent environmental pressures' (Kriwoken and Keage, 1989, p. 36).[35] The designation of SPAs and SSSIs was limited. In 1985 the protected area system covered only 0.15 per cent of the *ice-free* part of the continent (Lucas and Dingwall, 1987, p. 234), the area where most human activity and therefore most environmental impact was likely to take place. SPAs and SSSIs were susceptible to human interference. The construction of two bases on the Fildes Peninsula (King George Island) resulted in a decrease in the size of the SPA designated there and, finally, in the termination of that designation. The biologically important melt-lake which was included in that SPA was used by the Chilean base as a dump site (Anon, 1988g, p. 19). There were also problems with compliance. The most publicised example of violation of the Agreed Measures has been the construction of an airstrip by the French at the Dumont D'Urville station at Pointe Géologie.[36] The initial environmental impact report was inadequate, a point made by French scientists themselves. The construction involved blasting to create a runway strip between five small islands, disturbing and killing birds, something of an irony as the station had been constructed there specifically because of the scientific value of the penguin rookeries nearby. During construction of the Chinese 'Great Wall' station on King George Island, penguins were chased with sticks and glass and metal objects were placed in nests, in contravention of the Agreed Measures. Considerable disruption was also caused to scientific experiments being conducted by German hydrologists.[37]

By the time any new Specially Protected Areas were established – three were designated at the 13th consultative meeting in 1985 – the consultative parties were caught up in a more general debate about the adequacy of conservation measures in the Antarctic and the exposure by NGOs of violations of the Agreed Measures. At the 1985 consultative meeting, Australia proposed a new category of protected area on the grounds that the existing arrangements (that is, SPAs and SSSIs) were not sufficiently flexible to provide protection for areas with aesthetic or wilderness values but with no immediate scientific interest.[38] Australia acknowledged that its proposal was a response to concern in the international community for increased levels of protection for Antarctic environments (in effect, a form of cognitive learning at least). While some delegations were supportive, others thought such measures unnecessary (ATCM, 1985a, p. 20). Several (unnamed) delegations argued that the establishment of a new category of protected area could bring political problems, although it is not clear what those problems might be (ATCM, 1985a, p. 21).[39]

SCAR was asked to review the various mechanisms for environmental protection and to provide scientific advice on the system of protected areas in the Antarctic. In particular, SCAR was asked for its opinion on whether an additional category of protective area was necessary. SCAR's report, which was tabled at the 14th consultative meeting in 1987, was the first comprehensive review of the protected area system in the twenty-year life of the Agreed Measures. SCAR made five recommendations (see SCAR, 1987). Protected areas should be visited periodically to assess whether their protection objectives were being met. Information from those visits, SCAR suggested, should be circulated among consultative parties *and* made publicly available. SCAR proposed also that management plans should be mandatory for SPAs and that additional protective areas should be designated to provide for representative examples of Antarctic terrestrial, inland water and marine ecosystems. Finally, SCAR recommended that a new category of protected area should be established which would 'to the greatest extent possible, provide for buffer zones, multiple use, management plans and values other than those of a purely scientific nature' (SCAR, 1987, Appendix 6).[40] If implemented, these recommendations would have substantially improved the protected area system. In spite of SCAR's report, no recommendation was adopted. Several delegations were of the view that the concept of multiple use was not sufficiently developed, conceptually or operationally although the meeting 'accepted in principle the *need* to agree on additional protective measures' (Kimball, 1987c, p. 10, emphasis added). The final report (ATCM, 1987a, pp. 41–2) suggested that provisional management plans for seven specified areas might be prepared for the preparatory meeting for the next consultative meeting but noted, however, that this suggestion implied no commitment to the eventual acceptance of a new category.

When the issue of the protected area system came onto the agenda of ATCM-XV in 1989 it was in the context of broader debates about environmental protection following the Australian/French decision not to sign the Minerals Convention and the stated intention of those countries to pursue the negotiation of a comprehensive environmental protection regime.

The parties moved with uncharacteristic speed at this meeting to give effect to SCAR's recommendations (discussed further in chapter 7), although this might well have been to defuse the Australian and French proposals and to foreshadow anything that might be on the agenda of the special consultative meeting.

Managing Science: Code of Conduct for Antarctic Expeditions and Station Activities

Scientific activity in the Antarctic has given rise to a number of environmental problems, particularly related to station construction, logistic support and waste disposal. The rules adopted by the consultative parties were inadequate and were, in any case, poorly implemented. In 1970, the parties asked SCAR to identify the types and extent of human interference with the Antarctic environment and to propose measures to minimise this. In the meantime, governments were urged to reduce known causes of harmful interference. In 1972, SCAR's recommendations were proposed as voluntary guidelines for the conduct of expeditions and stations, 'as far as feasible and practicable' (recommendation VII-1, ATS, 1989, p. 1102). At the 1975 meeting, delegates adopted a short *Code of Conduct for Antarctic expeditions and station activities* based on SCAR's report (see ATS, 1989, p. 2101–2). The Code lacked detail and went no further than to make suggestions rather than require action. There was no formal compliance obligation. The parties were to observe the Code only 'to the greatest extent feasible'. There were no procedures to monitor whether the Code was being implemented or not. The sections on the introduction of alien species and minimising disturbance to breeding colonies and concentrations of birds and mammals simply drew attention to the Agreed Measures. The brief section on waste disposal, which set out recommended but not required procedures, permitted damaging practices such as the incineration of plastic and rubber products and the disposal of non-combustible products at sea.[41] The Code also suggested that national Antarctic operators should conduct an evaluation of the environmental impact of major operations, but provided only two brief paragraphs on the content of an evaluation. The Code did not require, or even suggest, that operations should be cancelled or modified if a negative evaluation was made.[42] There was no requirement for evaluations to be made public or available for comment, although they *may* be circulated through SCAR channels to other states engaged in Antarctic activities.

A number of proposals made by SCAR, which would have strengthened the Code, were not adopted. In particular, SCAR's suggestion for a 'comprehensive statement of anticipated short-term and long-term effects on the environment and its ... associated macro- and micro-biota, together with their primary, secondary and tertiary consequences; a delineation of all probable unavoidable adverse effects, with suggestions for means of minimising them' was omitted (Auburn, 1982, p. 287). Some (unnamed) delegations did express reservations about the Code's weaknesses (ATCM, 1975a, p. 8).[43] Nevertheless, this Code provided the only guidelines for the conduct of expeditions for almost a decade. The rules on waste disposal and environmental impact assessment were not discussed again by the consultative parties until 1983.

The problem of disposing of waste in the Antarctic increased with the growth in human activity in the 1980s. Most of the material brought into the Antarctic stayed there. Waste included not only sewage and food scraps, but fuel, batteries, chemicals, packaging, old machinery and equipment.[44] Over the years the consultative parties have 'spilled oil into the seas, dumped untreated sewage off the coasts [and] burned garbage in open pits' (Lemonick, 1990, p. 30). The cumulative effects of waste disposal impacted on local habitats as well as being aesthetically unacceptable. The enormously accelerated rates at which materials were introduced into the Antarctic were irreconcilable with ecological reality (Campbell and Claridge, 1987, p. 332). For those reasons the Code's rules became increasingly inadequate for managing waste disposal in the Antarctic. Yet the consultative parties did not respond to changing perceptions of what constituted pollution, to the increasing environmental impact of human activity in the Antarctic, or to new analytical techniques for monitoring environmental damage. In other words, new knowledge did not give rise to a learning process or to policy change.

Not only were the rules limited, but, as with the protection of fauna and flora, there were compliance problems. As two senior scientists with the British Antarctic Survey have noted, little effort was made at many research stations to observe the Code of Conduct (Bonner and Walton, 1985, p. 357). Even the consultative parties acknowledged that their record was 'longer on intention than action' (ATS, 1989, p. 2103). Poor waste disposal practices and minimal compliance with waste disposal rules were exposed by NGOs, especially by Greenpeace during its inspection visits undertaken as part of its Antarctic programme. For example, Greenpeace found 47 drums of discarded fuel (many of them leaking) near the airstrip of an abandoned UK base on Deception Island (Anon, 1988k:18). At the US Palmer station, raw sewage, plastics, rusting metal and laboratory chemicals were disposed of into the sea. The waters near McMurdo (the main US base and the largest in the Antarctic) were found to contain higher levels of polychlorinated biphenyls (PCBs) and heavy metals than would be permitted in the US.[45]

At the 1983 consultative meeting, in spite of draft amendments to the Code tabled by Australia, governments were recommended only to seek the advice of their national Antarctic operating agencies on any problems which might have arisen in implementing the Code of Conduct and whether revisions, especially in relation to the rules on waste disposal, were desirable. SCAR also undertook a preliminary review of this matter and it was on the agenda for the 13th consultative meeting in 1985. SCAR was asked to undertake a comprehensive review of the waste disposal aspects of the Code of Conduct. SCAR's efforts in this task were hindered by 'variable and insufficient information provided by national program cperators' (Kimball, 1987c, p. 10) and no recommendation could be adopted in 1987. The final report of the 14th consultative meeting (ATCM, 1987a, p. 31) made a point of noting that delegates were concerned at the limited response to date. In an unusual step, the final report also included a detailed statement of action on waste disposal indicating that there was a serious commitment to improving both the guidelines and their implementation (ATCM, 1987a, pp. 32–4). A number of goals were identified: cleaning up existing sites, minimising the amount of waste, recycling wherever possible and removing waste which could not be disposed of in an environmentally sound manner.[46] In 1989, the consultative parties went even further than the SCAR recommendations in their revision of the Code of Conduct in their attempts to improve waste disposal practices in the Antarctic. These changes are discussed further in chapter 7.

Debate about the Code of Conduct also required an examination of the rules on environmental impact assessment. In spite of a general injunction in the Code, activities in Antarctica were not, as the UK and Australia noted in papers tabled at the 1983 meeting, generally subject to prior scrutiny of their potential environmental effects.[47] The British delegation told the 1985 consultative meeting that the purpose of EIAs was 'to ensure that forethought is exercised ... *not* ... that any activity which is likely to have a significant adverse impact should not go ahead for that reason alone' (ATCM, 1985b). Debate on this issue continued to be marked by resistance to centrally imposed rules which might be seen to compromise national autonomy over the conduct of Antarctic activities. The debate was also hampered by the fact that many of the consultative parties had little or no domestic experience of EIA procedures – that is, knowledge was scarce – even though they had become a more widely-applied practice in some countries during the 1980s.

Recommendation XII-3 adopted in 1983 required parties to undertake detailed assessments for any research and logistic activities with the potential for significant impact, but only in accordance with *national* procedures (see ATS, 1989, p. 2105). There was no requirement for assessments to be circulated to other parties or made available for public comment, no way to ensure consistency of assessment across parties or to monitor the cumulative impacts of Antarctic activities. While the parties noted that a measure of comparability between national procedures might become desirable *in the future*, there was clearly no intention at this stage of imposing any comprehensive procedures. Indeed, some delegates urged caution in the search for any element of comparability (ATCM, 1983a, p. 6). Some resistance to EIAs also arose from within the scientific community because, they argued, such procedures could interfere with freedom of scientific research.

SCAR was invited to offer advice to the consultative parties on categories of scientific and logistic activity which might be expected to have a significant impact on the environment, and on assessment procedures which might be applied 'on an experimental basis'.[48] SCAR's report was discussed at length at the 1985 meeting but no recommendation was adopted.[49] SCAR's categories of activity that were likely to have a significant environmental impact caused some concern among delegates but for different reasons. Some were opposed to any suggestion of imposed requirements whereas others were concerned that any activities which were not included in the list could escape impact assessment procedures. SCAR's list, the British suggested, was much less forbidding if it was taken as indicative only of the sorts of activity for which an assessment should be prepared.[50] On the understanding that any recommendation would not be mandatory, the British argued that any conflict between national legislation and Antarctic procedures could be minimised (ATCM, 1985b). Draft British guidelines were annexed to the final report and a number of delegations indicated their intention to adopt them. However, the final report noted that EIA procedures 'do not presuppose any mechanism for coordination' (ATCM, 1985a, p. 20).

The consultative parties adopted a recommendation (XIII-6) on the siting of scientific stations at this meeting. Where stations were established in the same vicinity, Antarctic operating agencies were urged to consult with each other to avoid operational and logistic difficulties as well as adverse environmental effects arising from cumulative impacts. Close siting of stations had arisen in part because some of the newer parties, in attempting to fulfil the scientific activity criterion for consultative party status, were building stations in the more accessible parts of the Antarctic and duplicating scientific work being done by stations in the same area rather than conducting new work.[51]

The parties finally agreed to new and quite detailed rules on EIA under recommendation XIV-2 in 1987 (see ATS, 1989, p. 2106). These rules required a comprehensive environmental evaluation (CEE) to be undertaken if an initial environmental evaluation (IEE) determined that more than a minor environmental impact was likely. A CEE was to include a description of the activity and possible alternatives and the estimated impact including second order and cumulative effects. Other parties were to be given an opportunity to comment on the evaluation.[52] Considerable attention was paid to what constituted a significant environmental impact. In spite of these improvements, and although some delegates were keen, in the words of the recommendation, to 'introduce a measure of comparability' between assessment procedures the development of appropriate criteria for judging significant environmental impact and assessment procedures and the final decision is left, again, to each national authority. There is to be no centralised oversight of the procedures, no monitoring of individual or cumulative impacts, nor any means to enforce compliance. The British suggestion that SCAR be involved in this process was not adopted. When these procedures were debated again at the 1989 consultative meeting (discussed further in chapter 7), it was in the context of how best to *implement* the procedures. That is, the rules existed in principle but were not yet being put into practice.

Tourism and Non-Governmental Expeditions[53]

The first commercially organised ship-borne tourist expedition visited the Antarctic in 1966 (ATS, 1989, p. 2201).[54] Generally speaking, neither the consultative parties nor the scientific community looked favourably upon tourist and other non-government activity. This was in part because, politically, it reinforced the idea of Antarctica as subject to the law of open spaces and challenged consultative party control over the Antarctic. Nothing in the Treaty or international law could require tourism operators or non-governmental expeditions to ask permission of any state or person to visit the Antarctic. Scientists were irritated by constant interruptions to their work by tour parties wanting to visit stations, and there was the very real problem of ill-prepared expeditions requiring costly and time-consuming emergency assistance. The consultative parties sought to regulate tourism and non-governmental activity on the grounds of protecting the environment but it raised the difficult question of jurisdiction over 'unofficial' persons.[55] This has always been a contentious issue and, in spite of the need to protect the environment against the impacts of increased tourist and non-governmental activity, political and related legal concerns have meant that few substantive rules on tourism have been adopted even though the issue has been on the consultative agenda for over twenty years. Debate about a statement of accepted practices has been protracted and the idea of special tourist areas has not yet been satisfactorily resolved because the parties have no individual or collective jurisdiction to constrain access of non-governmental persons to any part of the Antarctic.

In 1966, the consultative parties adopted administrative rules on the exchange among them of information on tourist expeditions. They recommended that tourist operators should consult governments on the conditions upon which visits to their bases would be permitted. Although this was as much for pragmatic reasons as for jurisdictional ones, there is an implicit recognition here that governments do possess some jurisdictional rights over their stations in the Antarctic. In 1970, suggested restrictions on tourist and non-governmental expeditions to ensure that visitors comply with environmental protection measures already applicable to official expeditions were set out in recommendation VI-7. Visitor entry to SPAs was to be prohibited and Historic Monuments were to be respected. Nevertheless, there was no way to enforce these rules in practice. Proposals for a statement of accepted practice for tourism, in conjunction with the designation of Special Tourist Areas, to which visitors could be encouraged to go so that environmental impact could be minimised and localised, were first discussed at the 1972 meeting. At the 8th consultative meeting in 1975 the parties acknowledged that tourism required regulation (recommendation VIII-9) but they were clearly unsure how to do this. This recommendation refers to an annexed Statement of Accepted Practices and an annexed list of areas to which tourists will be encouraged to go but no annexes were adopted. A draft statement was discussed at ATCM-IX in 1977 but was *again* deferred to the next meeting although it was annexed to the final report of the ninth meeting. Similarly, no action was taken to list or define areas of Special Tourist Interest.

The Statement was finally adopted in 1979 as an annex to recommendation X-8 (see ATS, 1989, p. 2206). However, all it did, after seven years of debate, was to list existing recommendations on environmental protection and restate the existing administrative requirements relating to tourist activity.[56] At subsequent consultative meetings, the question was deferred each time either because a consensus could not be reached, because of insufficient time, or because of concerns about whether it was prudent to proceed further on this question. A draft recommendation suggesting that responsibility for ensuring compliance with the Treaty and its recommendations should rest with 'those States whose physical or juridical persons organise such expeditions' was withdrawn at the 12th meeting in 1983 when no consensus could be reached (see ATS, 1989, p. 2207). At the 1987 meeting, the UK delegation unsuccessfully proposed a code of practice for tourist and non-governmental expeditions (see ATCM, 1987b) which sought to consolidate various existing provisions into a single statement. By the 1989 meeting, North American operators of ship tours, tired of waiting for the consultative parties to act, had released their *own* environmental guidelines for expeditions.[57] The consultative meeting agreed that a comprehensive review of the issue was required, possibly leading to further measures to regulate tourist and non-governmental activities and that such a review should be undertaken by the special consultative meeting to be convened to consider comprehensive measures for protection of the Antarctic environment. The consultative parties had done little *but* review the issue.

Inspections and Compliance

The problems of violations of the various environmental protection rules adopted by the consultative parties raise questions about the adequacy of inspections as the primary compliance mechanism within the Treaty system. It was clear that, while inspections might reveal violations, they provided no sanctions. A couple of examples will suffice. A 1971 US inspection report indicated that at both Casey (an Australian base) and Mirny (a Soviet base) dogs were running free in contradiction of the Agreed Measures. Yet neither the US nor any of the other parties to whom the inspection report was circulated raised this issue in a consultative meeting. In 1985 a US inspection team found hydrocarbon seepage at every station visited, and reported that those stations also burned plastic or used plastic garbage bags for trash disposal at sea in contravention of the Code of Conduct. Again, this violation was not raised at a consultative meeting.

It was the NGOs outside the formal boundaries of the regime who brought questions of compliance to public attention. It is fair to note that Greenpeace has also been at pains to draw attention to instances of compliance with the rules. The consultative parties have responded to this external pressure in what is, for the most part, a state of imposed accountability. For example, after an NGO report detailed inadequacies in US waste disposal practices in the Antarctic (Manheim 1988) – acknowledged by the National Science Foundation to be accurate *(International Herald Tribune*, 17 August 1988) – the NSF took steps to improve practices at US bases and produced its own report on US environmental practice (NSF, 1989b).[58] The Foundation requested increased funding from the US government ($30 million over five years) 'to clean up the debris of past operations and to bring present operations into agreement with current regulations, prevailing attitudes and current technology' (see CAN No. 68, 1989).[59] Nevertheless in doing so they noted that clean-up and compliance with environmental rules should not interfere with or delay scientific work (Anon, 1990c, p. 2).[60] Former Deputy-Director of the British Antarctic Survey, Nigel Bonner has suggested that the improvement in environmental practice found during a joint New Zealand/UK inspection was due in large part to Greenpeace visits the year before (cited in CAN No. 68, 1989). Former President of SCAR, Claude Lorius has argued (1990, p. 8) that conservation organisations 'have sensitised scientists and logistics people to inadequate controls and misbehaviour in the past'. In other words, as noted earlier in the chapter, NGOs were agents for change. The consultative parties changed their practices in response to that knowledge, that is, a learning process took place. Nevertheless the widening of the environmental agenda and improvements in how environmental protection rules were implemented in practice did not overcome the problems of complexity and fragmentation which were at the core of the environmental regime. That required a comprehensive reassessment of the measures. This issue is taken up again in chapter 7.

CONCLUSION

The development of the Antarctic environmental regime, and the problems that arose from it were not unique to the Antarctic. As public environmental consciousness grew, as more information about environmental impact became available, and as ideas about what was environmentally acceptable changed, international rules on protecting the environment in Antarctica and elsewhere were challenged as inadequate. States were challenged to improve the policies adopted and to ensure that rules were implemented properly. Those challenges came, for the most part, from outside the formal boundaries of international diplomacy. In the Antarctic and elsewhere, non-governmental environmental and conservation organisations, organised on a transnational basis, publicised violations of existing rules and drew attention to perceived inadequacies in those rules. The political dominance of sovereignty norms engendered environmental rules that incorporated a permissive approach to compliance and a resistance to any effective form of centralised or independent monitoring agencies. Environmental regimes were not, for the most part, knowledge-driven, even though scientific knowledge was often available to decision-makers. In the Antarctic, SCAR's advice was often sought but not fully incorporated into rules and measures. SCAR itself responded to requests for information, but did not take an advocacy role, nor did it seek to lobby for its recommendations to be given greater influence than they were. Changes to and improvement in environmental rules (that is, the process of learning) were slower than changes to the environment or changes in knowledge. In that way, problems of environmental impact were often increased rather than decreased as the decision-making framework came to be increasingly out of step with changing perceptions of the requirements for good environmental management.

4 Living 'Resources' and the Limits to Conservation

INTRODUCTION

The consultative parties negotiated three conventions on resource use – the Convention for the Conservation of Antarctic Seals (CCAS), the Convention on the Conservation of Antarctic Marine Living Resources (CCAMLR) and the Convention on the Regulation of Antarctic Mineral Resource Activities (CRAMRA). Conventions, rather than consultative recommendations were adopted because these were activities of interest to non-Treaty parties. Rules established under consultative recommendations would have regulated and constrained consultative party activity but not that of non-Treaty states. Rules established under a convention would be legally binding on all signatories and, in theory at least, stronger than recommendations. In the case of seals and marine living resources, the activities to be regulated were also primarily high seas activities which were outside the direct control of Treaty recommendations given the provisions of article VI which protected high seas freedoms.

Both CCAS and CCAMLR, which are discussed in this chapter, were negotiated under the authority of article IX(f) of the Antarctic Treaty which specifically allowed the consultative parties to establish measures for the protection and conservation of living resources. The negotiations were restricted to the consultative parties and were held, as usual, behind closed doors. Both conventions were negotiated in advance of extensive interest in the exploitation of marine resources with the intention of limiting adverse impacts of over-exploitation of the marine ecosystem as well as possible conflict over resource use although in neither case were the consultative parties prepared to *prohibit* the exploitation of the 'resource' in question. The consultative parties were also concerned that agreement among them would be more difficult in the face of entrenched economic interests once commercial activity had begun.

Both conventions extended the legal framework of the Treaty system and contributed to the development of the Antartic regime, through a process of incremental change. They widened the scope of cooperation among the consultative parties, strengthened their hold on Antarctic decision-making and ensured that conflict among them could be avoided. To this extent they met the test of success within the Treaty system. However, both are flawed as conservation instruments. The regime's sovereignty norms shaped the institutional, decision-making and compliance rules adopted. Regulatory and decision-making authority remained in the hands of the Antarctic Treaty consultative parties. Compliance and enforcement rules were limited. The hierarchy of values in the Antarctic Treaty system was reinforced. In both conventions, conservation values were important in principle but the rules and procedures adopted and, in the case of CCAMLR, the record of implementation, did not match this commitment. While knowledge was an important factor in both conventions in principle, in practice neither convention was primarily knowledge driven. Again the role of the scientific community was constrained, in spite of SCAR's formal role in both conventions. While styled as conservation agreements by the consultative parties, both conventions were compromised by political and economic concerns, a factor which has given rise to doubts about their effectiveness as conservation instruments.

CONVENTION FOR THE CONSERVATION OF ANTARCTIC SEALS

The consultative parties negotiated CCAS in the late 1960s and early 1970s. There was, at that time, little international interest in the Antarctic and the NGO presence and the challenges from non-Treaty states that characterised negotiations for later conventions were absent. There was, then, little need for an accommodation with external interests. The internal accommodation between the consultative parties was also less troublesome than it was to be in later negotiations. No consultative state had entrenched commercial interests in sealing, although Norway's experimental catch of crabeater seals in 1964 provided some incentive to advance the negotiations. Sealing activity on the high seas did not impinge upon territorial rights or the exercise of maritime jurisdiction. The bargaining process was therefore reasonably uncontentious. Nevertheless conservation values were set against political and (potential) economic interests during the negotiations. Differences among the consultative parties focused on whether to prohibit or regulate sealing activity and, if it was to be regulated, the extent of compliance and enforcement rules that were acceptable.

The question of protecting seals first arose during the negotiation of the Agreed Measures. While the consultative parties were sympathetic, not all were supportive of a complete ban and debates were characterised by attempts to balance conservation and exploitation values. The UK, for example, thought that 'it would be better to anticipate the establishment of a rationally controlled industry ... rather than to attempt any direct prohibition of sealing' (ATCM, 1964b). At the 1964 meeting, the parties agreed to voluntary regulation but *not* prohibition of pelagic sealing and the taking of fauna on pack ice but only on a national basis.[1] Any suggestion of collective or centralised rules was avoided. In 1966, they adopted Interim Guidelines for the Voluntary Regulation of Antarctic Pelagic Sealing, based on SCAR proposals.[2] These guidelines identified seal stocks as a 'resource of potential value' which should not be depleted by overexploitation (see ATS, 1989, p. 3101). The guidelines also provided for an annex on maximum sustainable yield (MSY) levels for seal catches and the establishment of sealing zones but no MSY levels were included in the annex to the recommendation and no zones were designated at this meeting.[3] SCAR's recommendations on seal catches and sealing zones, which included a suggested permit system for the taking of seals within designated Seal Reserves, were tabled at the fifth consultative meeting in 1968 along with a draft convention in the hope that a binding agreement on the protection of seals could be negotiated.[4] There was no consensus among the parties on the form of an agreement but governments were urged to study the draft prior to the sixth meeting in 1970. Delegates to the sixth meeting decided that this issue should be considered 'outside the framework of the Antarctic Treaty, since [it] ... does not fall within the scope of the latter and is of interest to countries which are not Parties to the Antarctic Treaty' (ATS, 1989, p. 3102). Following a series of informal meetings held concurrently with ATCM-VI in 1970, a draft convention was transmitted to governments. The Convention was adopted at a Conference on Antarctic Seals held in London from 3 to 11 February 1972 and opened for signature on 1 June of that year.[5] Although this diplomatic conference was separate from the consultative process only the twelve Antarctic Treaty consultative parties were represented although the United Nations Food and Agriculture Organization (FAO) was invited as an observer because of its interest in Southern Ocean marine resources. The draft convention had, in any case, already been debated and virtually agreed to *within* the consultative process.

The Convention is designed to 'promote and achieve the objectives of protection, scientific study and rational use of Antarctic seals' (see *Convention for the Conservation of Antarctic Seals* in ATS, 1989, pp. 3104–10). The central principles of the Convention are that Antarctic seals (described as an important living resource) should not be depleted by over-exploitation and that a satisfactory balance within the ecological system should be maintained. In his opening address to the February conference, the British Parliamentary Under-Secretary of State for Foreign and Commonwealth Affairs reassured delegates that the purpose of the convention was to conserve seals, not to encourage commercial sealing. However, an official US document suggests, to the contrary, that 'during the ... negotiations it was very clear that the reason for the Convention was to provide for utilisation of the seal herds but prevent their over-exploitation should harvesting begin' (Department of State, 1974, p. 24). The Convention establishes allowable catch levels for Crabeater, Leopard and Weddell seals. Ross Seals, Southern Elephant Seals and all fur seals are to be fully protected. The Convention sets out rules on closed and open sealing seasons, sealing zones and reserves and provides an indicative list of other protection measures which the parties may adopt. Special permits allow seals to be taken for food, for scientific research, or as specimens for museums, educational or cultural institutions. SCAR has a formal role in providing scientific and technical advice and collating information on seals provided by the consultative parties. A commission and scientific advisory committee will be established only if commercial sealing begins.

The dominance of sovereignty norms in the Treaty regime contributed to weaknesses in the Convention which limit its potential effectiveness as a conservation instrument. There is no collective enforcement mechanism. Each signatory party is to adopt measures for its nationals and flags under its jurisdiction. A system of inspection *may* be established but there is no requirement that it must be in place before sealing commences (article 3(1)(k)). No details are provided on how this might work in practice. Nor is any sealing nation specifically required to advise the other contracting parties of its intention to begin sealing. There are no provisions for national catch allocations and no requirement for permits to be circulated in advance of sealing so that the cumulative impact of exploitation can be assessed. The US argued that opposition to its proposals for stronger protection measures, including observation of operations and enforcement of regulations, was based on juridical rather than commercial interests and went so far as to ensure that a statement to this effect was incorporated into the final Act of the Conference (see Bush, 1982, pp. 261–2). The Convention is closely tied to the Antarctic Treaty. Decision-making remains in the hands of the original consultative parties who negotiated the Convention. Other states may accede to the Convention only on their invitation. Contracting Parties to CCAS must affirm the provisions of article IV (the sovereignty article) of the Antarctic Treaty as they apply to the seas south of 60° south. As Bush argues (1982, p. 249), this represents an attempt to widen adherence to an important aspect of the Treaty regime. Argentina and Chile, the most active of the claimants, also made declarations on signing the Convention to the effect that it did not impair their rights of territorial jurisdiction including jurisdiction over maritime zones (Bush, 1990, p. 134).

The Convention did not enter into force until March 1978 and has never really been put to the test. Its institutional framework has never been activated.[6] There is at present no commercial harvesting of seals in the Southern Ocean, although over the past 21 years about 15 000 seals have been taken (ASOC, 1988i, p. 1), including an 'exploratory' harvest of 4802 seals by the USSR in 1986–87.[7] Several parties have not fulfilled the reporting requirements on their scientific sealing activity (USMMC, 1987, p. 85). The only CCAS review conference to be held to date was convened in London in 1988 in response to the Soviet harvest which, while within permissible catch levels, was thought to be above that necessary for scientific purposes.[8] Two minor alterations to the Annex, proposed by Australia, took effect on 27 March 1990. They encouraged exchange of information with parties to other ATS instruments (such as CCAMLR) and extended protection of Weddell seals to include pups. A suggestion that applications for special permits be circulated in advance was resisted as an attempt to introduce centralised assessment of national science programs under the rules of the Convention.[9]

CCAS figures in the contemporary environment debate in two ways. First, public attitudes towards marine conservation issues have changed to the extent that commercial sealing in the Southern Ocean may now be considered unacceptable. Greenpeace has argued (1988a) that sealing should be totally prohibited. Second, there are inconsistencies between CCAS and other conservation instruments within the environmental regime. For example, the Agreed Measures (now incorporated into the Madrid Protocol) prohibit sealing on land and ice-shelves whereas CCAS permits sealing at sea. CCAS has also not been linked to the ecosystem model of living resource management in the Southern Ocean adopted under CCAMLR.

CONVENTION ON THE CONSERVATION OF ANTARCTIC MARINE LIVING RESOURCES

Once the seals convention was finalised, the consultative parties turned their attention to the possible impact on the marine ecosystem of the unregulated exploitation of other marine resources. They were spurred by a growing commercial interest in krill and by the increasing exploitation of southern ocean fish stocks in the 1970s as traditional coastal fishing grounds in other parts of the world were closed to distant-water fishing fleets by the declaration of exclusive economic zones (EEZs). In negotiating a convention on marine living resources, the consultative parties were keen to demonstrate their continued authority over all Antarctic and Southern Ocean issues. This issue was also a test of their ability to negotiate rules and procedures on a resource issue which had sovereignty implications.

The internal accommodation among the consultative parties and the external accommodation with environmental and non-Treaty state interests were more contentious than in the seals negotiations. The contract curve, or area of potential agreement among the parties, which is crucial to be process of institutional bargaining was much less obvious. The negotiations disclosed serious political and economic differences among the consultative parties even though, as Barnes argues (1982b, p. 272), 'the interests of the [Antarctic] club dominated' those of other states. The claimant/non-claimant split was evident in debates over the exercise of coastal state jurisdiction over maritime areas. Japan and the USSR, the two major fishing states, argued at every opportunity for the utilisation of resources to be given more emphasis and conservation values less (Barnes, 1982b, p. 282n). Non-fishing states, on the other hand, took a generally stronger conservation line. Scientific knowledge was, in theory, an important component of decision-making on marine living resources, but the role of the scientific community and the implementation of the precautionary principle was undermined in practice by political and economic concerns.

Non-Treaty states were also interested in the exploitation of marine resources of the Southern Ocean. This interest, which arose in the context of calls for a new international economic order and the negotiations for the UN Convention on the Law of the Sea, was actively resisted by the consultative parties. Sri Lanka raised the question of Antarctic marine resources at the 1975 session of the UN Economic and Social Council (ECOSOC). Following objections from the consultative parties, an ECOSOC Secretariat report on marine resources in Antarctica was withdrawn. In 1976, on the urging of the Group of 77, the FAO proposed a joint $45 million project in conjunction with the UN Development Program (UNDP) to develop Antarctic living resources for the benefit of all countries. This project was later downgraded following objections by the consultative parties.[10] Delegates to the 1977 FAO Fisheries Committee argued that the resources of the Treaty Area should be shared by the whole international community (see Orrego Vicuña, 1988, p. 474).

The marine resources issue was the one which changed the dynamics of the Antartic regime though the participation, albeit not particularly welcome, of environmantal NGOs. They monitored the negotiations and lobbied for conservation of Southern Ocean marine resources. While the CCAMLR negotiation has been described as the issue on which NGOs 'cut their teeth' (Bush, 1990, p. 143) various accounts of the negotiating process make little reference to NGO activity or argument, with the exception of the participation of the International Union for the Conservation of Nature which is not an advocacy organisation. The US delegation included two NGO representatives but it seems to have been the only one to do so. Barnes suggests (1982b, p. 280n) that the London and Washington-based International Institute for Environment and Development was the main NGO trying to keep the public informed of the negotiations but this does not seem to have extended to a direct lobbying function. For the most part, external interest seems to have had little impact on the bargaining process among the Treaty parties.

At the 1975 consultative meeting, the consultative parties recognised the need 'to promote and achieve, within the framework of the Antarctic Treaty, the objectives of protection, scientific study *and* rational use of [Antartic] marine living resources' (recommendation VIII-10, ATS, 1989, p. 3203; emphasis added). SCAR was asked to convene a meeting to discuss and report on current scientific programmes on the Southern Ocean marine ecosystem.[11] At the 9th consultative meeting in 1977, delegates agreed to convene a special consultative meeting to prepare a draft agreement on marine living resources. Two key *political* principles and one conservation principle, with political overtones, were identified. First, the prime responsibility of consultative parties for the protection and conservation of the environment in the Antarctic Treaty had to be recognised. Second, the provisions of article IV of the Treaty, on territorial claims, were not to be affected. Finally, if necessary, the regime was to extend north of 60° south (without prejudice to coastal jurisdiction in that area) to ensure the effective conservation of species of the Antarctic ecosystem. Governments were urged to observe interim rules which nevertheless required very little of the parties: they should cooperate in exchanging statistics on resource catches and show the 'greatest possible concern and care' in harvesting so as not to deplete stocks (recommendation IX-1, ATS, 1989, pp. 3203–4). A working group on marine living resources convened at this meeting agreed that 'conservation' would include rational use, that harvesting would not be prohibited and that catch allocations would not be included in any regime adopted (ATCM, 1977, p. 340). Thus exploitation and conservation were perceived to be compatible.

Only the consultative parties, now numbering thirteen with the inclusion of Poland, were eligible to participate in the special consultative meeting which began in February 1978. Three formal and two informal negotiating sessions were held prior to the 'decisive' conference at which the Convention was adopted.[12] As usual, the negotiations were conducted in secret although the US argued, with little effort Barnes suggests (1982b, p. 281n), for open negotiations and publication of the negotiating texts. Eight draft conventions (produced by Argentina, South Africa, Poland, Japan, Australia, Chile, France and the USSR) were tabled at the first session in Canberra and a single negotiating text was produced by the end of the session. The next session, in Buenos Aires in July, almost broke down because of intransigence on a number of issues including sovereignty and jurisdiction, conservation standards and decision-making rules (see Barnes, 1982b, pp. 251–5). A draft convention which was acceptable to most delegates, except the French, was produced at an informal meeting in Washington in September 1978. French objections related to its concerns over coastal state jurisdiction over undisputed islands lying just outside the Treaty Area but within the area covered by the draft convention.[13] A second informal meeting in Berne, Switzerland in March 1979 also failed to conclude negotiations but informal consultations at the tenth consultative meeting in 1979 produced a statement which satisfied the French although several other nations were not yet willing to agree to it.

Delegates met briefly in Canberra on 5–6 May 1980 prior to the formal conference which began on 7 May. West and East Germany, which were both non-consultative parties at that stage, were invited because of their interest in Southern Ocean fishing. The Netherlands, a non-consultative party, and the Republic of Korea (not at that stage a Treaty party but fishing in the Southern Ocean) asked to attend, but were blocked by Poland and the USSR.[14] Several relevant international organisations were invited as observers but their participation was constrained by Soviet objections to their presence and they were not permitted to speak at any of the sessions.[15] Barnes' account (1982b, p. 259) suggests that the final conference was dominated by ten days of fruitless negotiations and that a compromise package that would attract a consensus was put together in late-night weekend session. The convention was adopted on 20 May and opened for signature by the fifteen participating parties in September 1980. It entered into force on 7 April 1982.

Leadership

Although leadership is, as chapter 1 suggested, often argued to be crucial to successful institutional bargaining, no individual state seems to have taken a decisive leadership role in the negotiations for CCAMLR. Young suggests (1990b, p. 14) that the US acted as a structural leader in articulating the Convention although he provides no analysis of the negotiations. It is not clear that the US had either more or less to lose than the other participants in the CCAMLR negotiations, a characteristic which often encourages structural leadership, nor that it exercised bargaining power of the sort Young suggests. Indeed, it was unsuccessful, as noted earlier, in persuading other states to adopt its conservation proposals. On the other hand, the main fishing states (the USSR and Japan) and the claimants (especially France) *were* able to exercise issue-specific bargaining power to ensure that their interests were protected in the bargaining process. Their support was essential to a successful conclusion to the negotiations. This represents a negative exercise of structural power in institutional bargaining. Their bargaining positions were marked by a refusal to accept certain proposals rather than a positive leadership role in offering institutional options or brokering the interests of other actors to reach an agreement. Particular individuals may well have exercised entrepreneurial leadership in moving the negotiations forward although, unlike the minerals negotiations which followed, there was no continuity of chairpersons. Ambassador Rowland, of Australia, who chaired the first negotiating session compiled the initial draft which established the shape of the final convention. John Negroponte (of the US) was instrumental in finding a way around the impasse on the northern islands issue and the exercise of coastal state jurisdiction caused by the French.

A degree of intellectual leadership was exercised by the scientific community in its advocacy of the ecosystem model of resource management. Barnes suggests (1982b, p. 250) that the US Marine Mammal Commission (USMMC) first articulated this as an appropriate conservation standard for Southern Ocean fisheries. There was certainly support from the Antarctic scientific community generally (Powell, 1990, p. 61). Richard Laws, from the British Antarctic Survey, was especially effective in convincing non-scientific delegates of the soundness both of the Antarctic Convergence as an appropriate boundary and of the ecosystem approach as a conservation standard (Barnes, 1982b, p. 281n).[16] Thus the scientists had begun to function as a nascent epistemic community. Nevertheless, there were few scientists on delegations and while the scientific debates played an important role in the early negotiating sessions they were soon overshadowed by political and economic concerns.

The Convention

The Convention is shaped by the internal political accommodation on a number of contentious issues including the area of application of the convention, the type of conservation and management principles that would be adopted, the exercise of coastal-state jurisdiction, the nature of decision-making rules and participation in the negotiations and the Convention. On these crucial issues, the hierarchy of values inherent in the Treaty system dominated. Political and economic interests overrode environmental ones and ultimately weakened the conservation potential of the agreement. The decision-making process, which required consensus for all decisions, while politically important, was flawed in terms of ensuring that good environmental rules would be adopted. Again, decentralisation and sovereignty norms shaped the outcome of the institutional bargaining process. The reluctance, on political grounds, of some states to agree to centralised or independent monitoring, enforcement or inspection provisions, meant that the fishing states have been free to act as they will, with little oversight or recourse on the part of the conservation-minded states.

The claimants and non-claimants differed over the rights of the claimants, as coastal states, to exercise jurisdiction in economic zones and to receive benefits from resource activity in those zones. The issue was complicated because there were islands within the proposed convention area over which sovereignty was not disputed and over which the relevant claimant *could* exercise coastal state jurisdiction. The claimants argued that this right should apply to all territory to be covered by the convention. The non-claimants refused to make any concessions on these issues.[17] A 'bifocal' approach to the exercise of coastal state jurisdiction, which reinforced the ambiguity of the Antarctic Treaty on the sovereignty issue, was adopted. Under Article IV(2)(b) nothing in the Convention, and no acts or activities while it is in force, shall

> be interpreted as a renunciation or diminution by any Contracting Party of, or as prejudicing, any right of claim or basis of claim to exercise coastal state jurisdiction under international law within the area to which this Convention applies (see *CCAMLR*, 1989, pp. 3209–17).

The claimants can interpret this as referring to the undisputed islands north of 60° south and the disputed territorial claims south of that area, while the non-claimants can interpret it as referring only to the former. The Convention makes no reference to access to EEZs or to the distribution of any economic benefits of fishing activity.

The Convention is tied to the political compromises of the Antarctic Treaty and, as such, reflects the imperatives of the Antartic regime. Acceding states effectively accept the 'de facto control of the Antarctic Treaty parties over any decisions concerning the ... continent and surrounding waters' (Barnes, 1982b, p. 240). Article V requires those parties to CCAMLR which are not parties to the Antarctic Treaty to acknowledge the special obligations and responsibilities of the consultative parties for the protection and conservation of the Antarctic Treaty Area.[18] Contracting Parties must also agree to observe the Agreed Measures and other measures adopted by consultative parties in fulfilment of this environmental responsibility. Parties to the Convention are required to be bound by articles IV (on the claims) and VI (on high seas rights) of the Antarctic Treaty. The two-tier membership criteria of the Antarctic Treaty is replicated here. All states which participated in the final conference (that is, the consultative parties plus East and West Germany) were eligible to become original signatories to the Convention. Subsequent accession to the Convention was limited to states with an interest in research or harvesting activities related to Southern Ocean marine living resources. However to become a member of the Commission established by the Convention a state has to be *actively* engaged in such activities. Britain, France and Belgium supported full EEC participation in the Convention but were opposed by the Soviet Union and Poland (Barnes, 1982b, p. 256). A compromise solution allowed the EEC to accede to the Convention, if there was a consensus among members of the Commission, and to then be eligible for Commission membership.

The debate about the area of application of the Convention which would best support the ecosystem model, which is discussed below, was susceptible to political interests. The Convention applies to marine living resources south of the Antarctic Convergence (a transition zone where cold southern waters meet, and sink beneath, the warmer northern waters). The exact boundaries of the Convergence cannot be easily defined. The Convention implies that the Antarctic Convergence can be identified by the northernmost limits of the Antarctic ecosystem (article I(4)) but it also sets out a series of coordinates to further define this limit.[19] This was a compromise between those who supported a definition of the Antarctic Convergence based on biological and scientific grounds which tied it to the southern ocean ecosystem (the UK, Norway, New Zealand and the US) and those who favoured geographic coordinates (France, South Africa, Chile and Argentina (Barnes 1982b:281n).[20]

The Convention adopts an innovative ecosystem approach to the conservation of marine resources, rather than a single species model, based on the principles that ecological relationships between harvested and dependent and related populations must be maintained and that harvesting should not undermine stable recruitment levels. Any changes in the marine ecosystem which result from harvesting must be reversible in two to three decades. While these are sound principles, which could be argued to represent something more than simply a single-loop learning process, or the adoption of new means for the same end, other provisions in the Convention did not give adequate support to this approach and the Convention has not been able to prevent over-exploitation. The multi-species approach was spearheaded by the US. The fishing states wanted a weaker management standard based on single-species total allowable catches (TAC) and the USSR and Japan argued strongly against an ecosystem conservation standard. They acquiesced only because of the requirement for consensus in decision-making and the inclusion of an objection provision.

The institutional framework is shaped by sovereignty norms and an emphasis on decentralisation which undermines the environmental protection of the Convention. CCAMLR's requirement for decisions on matters of substance to be taken by consensus (article XII(1))[21] is politically important but it constitutes a potential flaw in the Convention giving, as it does, a veto over conservation measures to any fishing state. The US, UK and Japan wanted a majority-based system, but Argentina, Chile, Poland and the Soviet Union opposed it. Any member of the Commission also has 90 days to advise that it is unable to accept any conservation measure and is therefore not bound by it. This 'opt-out' clause entrenches a double veto. Conservation measures come into effect except for any state which 'opts out'. This limits the potential effectiveness of the conservation measures if the objecting states are the fishing states. There is no provision for a centralised or independent inspection system to monitor either the condition and balance of the Southern Ocean ecosystem or the cumulative impacts of harvesting. Although the Convention includes general principles upon which a system of observation and inspection should be based, it did not establish that system. The UK, US, Norway and New Zealand favoured an inspection and observation system being included in the Convention (Barnes, 1982b, p. 251). However, it was not until the 1989–90 season that a limited and decentralised inspection system was implemented. Inspectors report to their own governments who report in turn to the Commission. Any prosecutions or sanctions are the responsibility of the flag state of the offending vessel. Inspections must be brief (one or two hours) and involve one or two inspectors only (Puissochet, 1991, p. 75). As Powell observes (1990, p. 67), a system that permits foreign nationals to board one's vessels has not been easily accepted by the fishing states.[22] This concern has prevented broader observation rules, which would involve the presence of observers on board *during* a voyage, from being adopted.

Although the Commission has responsibility for adopting conservation measures it is, in practice, a political body. Government officials, in particular foreign ministry representatives, rather than scientists have been the main participants in the Commission. Heads of CCAMLR delegations are usually political rather than scientific experts. In most of the Convention states there is little domestic interest in the Antarctic fisheries resource issue. Where those interests do exist, the political process within the Commission, particularly the emphasis on consensus for the adoption of conservation measures, enables national economic interests to override scientific or environmental interests.

Scientific knowledge is an essential component of successful implementation of CCAMLR. In theory at least this ensures a potentially important role for scientists. However, the Scientific Committee established by the Convention has a limited role. Barnes (1982b, p. 266) suggests that during the negotiations 'no delegation was willing to expend political capital to assure that the Scientific Committee had the power, staff, funding, and independence that arguably would be required to furnish advice grounded in fact and risk assessment, instead of economics and politics'. The Committee is advisory only. The scientists who participate in the Committee's meetings do so ultimately as national representatives, not as independent scientific experts. This constrains the extent to which they can function as an effective epistemic community. Much of the Scientific Committee's early years were spent in disputes over its role[23] and between those states who wanted maximum data to be reported and the fishing states, especially Japan and the USSR, who opposed this.[24] The generation of scientific data, upon which conservation decisions should be made, is constrained by political factors. It requires fishing states to report their statistics accurately and in full which they are reluctant to do. Financial constraints also make the collection of scientific information difficult. The research programs required to compile adequate scientific evidence in support of ecosystem models are expensive. Therefore the excuse of scientific uncertainty can be used to limit the adoption of conservation instruments, rather than underpinning a precautionary approach. Relations between the scientists on the Scientific Committee and government officials in the Commission have not always been as smooth as they might be according to a senior French Antarctic diplomat (Puissochet, 1991, p. 73). It was not until the late 1980s that scientists began to express openly their concern that their views and judgements on conservation measures were being ignored by the Commission.[25]

Success as a Conservation Instrument

The Convention is a flawed conservation agreement. The Convention *permits* fishing activity unless it can be shown on the basis of sound scientific data that it should be controlled. The onus is not upon the fishing states to prove that conservation goals would not be affected by their harvesting activities. The provision and interpretation of sound scientific data has, as noted above, been problematic. The decision-making process is inherently slow and subject to national interests and political bargains. The Commission meets only once a year for approximately twelve days and, as the former Executive Secretary of CCAMLR has observed, it constitutes a collection of national representatives who meet to try to merge their governments' *individual* policies into decisions that accord with the goals of the Convention (Powell, 1990, p. 68). They have not been successful in doing so. Exploitation and environmental interests have clashed. Under the consensus rule, fishing states have successfully opposed conservation measures which run counter to their short-term economic interests.

CCAMLR meetings have focused predominantly on the management of fish stocks. In the first five years only seven conservation measures were adopted and they contributed little to the protection of the marine ecosystem.[26] These measures have not prevented the depletion of fin fish stocks. Indeed, usually measures have not been adopted until stocks are already depleted, thus negating the precautionary approach which the consultative parties argue characterises the Convention. They have differed little from measures adopted under fisheries agreements which are not based on an ecosystem model (Puissochet, 1991, p. 74). The practical problems of implementing an ecosystem approach have been compounded by inadequate and incomplete submission of data which has contributed to uncertainty in establishing conservation measures.[27] This scarcity of information and the related difficulty of establishing what is the best scientific evidence available upon which to make decisions has resulted in less rather than more caution. Where knowledge has been contested, the precautionary approach has not prevailed. The fishing nations in the Commission have often opposed restrictions on catch levels on the grounds of insufficient data existing to support such measures. It is the fishing nations, however, who are one of the major sources of data. Conservation measures for the krill fishery were not addressed until 1989 and measures were not adopted until 1990.[28] Even then, they were opposed by Japan and the USSR on the grounds that adequate scientific data was lacking (Nicol, 1990, p. 6). As an Australian official recently observed, 'it is difficult to argue that CCAMLR is being successfully implemented when there is no agreed conservation approach to the largest fishery [krill] in the area' (Heyward, 1990, p. 5).

Australia and New Zealand have been the leading states in identifying inadequacies in the measures adopted and in pushing the Commission towards the ecosystem approach advocated in the Convention.[29] However, the consensus requirement and lack of issue-specific power which these two parties have within CCAMLR has meant that neither state was particularly successful in the early years of CCAMLR.[30] More wide-ranging conservation measures have been adopted in recent years. Fifteen measures were adopted at the 1991 Commission meeting and nineteen in 1992. At time of writing, forty conservation measures are either in effect or due to soon come into force.[31] These measures have included steps to minimise incidental mortality of seabirds, requirements for notification of new fisheries and procedures to protect sites in which studies are being undertaken for the CCAMLR Ecosystem Monitoring Program (CEMP). The CEMP protection measure (Conservation Measure 18/IX) requires the adoption of a management plan which may include a *hortatory* code of conduct.[32] Nevertheless, these new measures still fall short of what is required to give effect to the ecosystem approach and to ensure that the marine fauna of the Southern Ocean is properly protected.

Non-Governmental Organisations

Although the Convention allows for participation by non-governmental organisations where those organisations can contribute to its work the parties have been slow to include environmental NGOs in that category.[33] ASOC and Greenpeace International first applied to the Commission for observer status in 1983 but it was only in 1988 that ASOC (but not Greenpeace) was granted this status. This delay arose because of doubts expressed by some parties about what kind of contribution ASOC could make to the work of the Commission.[34] The US, Australia, New Zealand and the UK now include NGO representatives on their delegations.

NGOs' role within CCAMLR has been both a lobbying and monitoring one reflecting their participation in the Antarctic regime generally. They send lobbying delegations to Scientific Committee and Commission meetings, and prepare detailed reports on conservation measures and practice under the Convention. NGOs have sought to increase the accountability and transparency of state practice under the Convention. The Greenpeace Antarctic expedition has monitored fishing vessel activity in the southern ocean (see Greenpeace 1990). NGO policy has generally been not to oppose harvesting but to advocate that more stringent conservation measures be adopted and that implementation be improved. In seeking to influence the CCAMLR process in this respect, the NGOs must rely on their presentation of scientific argument: the conservation of fish and krill is an issue on which it is difficult to stimulate public or political attention. Thus they are again knowledge-based actors within the regime, mobilising for normative change.

The changes in conservation measures and, to some extent, inspection rules noted above signal at least a degree of willingness by the fishing states to modify their interpretation of what can best meet, or at least not compromise, their national interests. They can be attributed to an informal and doubtless uncoordinated coalition of the conservation-minded states, the scientific community and the NGOs. The conservation-minded states within the Commission and the scientific community have taken a much firmer line on the need for scientific advice to be accepted and for effective conservation measures to be adopted. Those changes can also be located in the context of debate on the way in which the consultative parties have, or have not, exercised their environmental stewardship of the Antarctic, a debate mobilised to a great extent by the NGOs. This debate was especially important during the minerals negotiations, which are addressed in the next chapter.

CONCLUSION

The experience of CCAMLR in particular points to several factors which have wider relevance to the construction of conservation regimes on so-called 'living resources'. Where species are already being exploited, it is difficult to negotiate effective protection rules in the face of entrenched economic interests. It is also difficult, in a unanimity system, for conservation-minded states to be effective. Such agreements are often bedevilled by a lack of internal coherence. They may well, as with CCAMLR, incorporate extensive and quite innovative protection principles, thus representing a degree of normative change, but political and economic interests constrain the extent to which specific rules and conservation measures are able to give effect to those principles. An agreement such as CCAMLR also points to potential problems for the scientific community in elaborating and maintaining conservation regimes. Scientific data and knowledge is often recognised as important in principle, but the rules and procedures serve to restrict the authority of the scientific community. The precautionary principle, whereby rules anticipate likely scientific outcomes, is not well applied. The experience of CCAMLR also shows that conservation-minded states or NGOs acting alone are often unable to exert political pressure. However when their interests coincide, and when those concerns are further supported by the scientific community, it may be possible to move states towards improved conservation measures. This changes the locus of power, or demonstrates that on environmental issues, where traditional measures of power are inappropriate, empowerment through coalition activity can be important. Such coalitions across state and non-state boundaries were particularly relevant to the question of minerals activity in the Antarctic.

5 Antarctic Minerals: Problems and Processes

INTRODUCTION

The third of the resource conventions negotiated by the Antarctic Treaty consultative parties was the Convention on the Regulation of Antarctic Mineral Resource Activities. The minerals issue brought into the open tensions among the consultative parties which had been more easily managed in the living resource negotiations. It was the issue on which competing national interests were most in evidence, and the issue which throughout its negotiation and afterwards challenged the stability of the Treaty system. It was the pivotal issue in bringing Antarctic environmental protection to the forefront of public debate and the one which generated the most critical attention from states which were not members of the Antarctic Treaty.

The Antarctic Treaty deliberately contains no reference to mineral resources or their management although it was discussed at the 1959 Washington Conference. Sollie suggests (1977, p. 8) that a proposal to include a provision on economic activities in the Treaty was tabled and rejected. Mineral resources involved questions of ownership and rights to exploitation which, in the face of differences over sovereignty, would have rendered agreement between the claimants and non-claimants almost impossible in the late 1950s. Given that minerals activity was unlikely in the near (or even not so near) future, this was an issue that the original signatories felt could be sidestepped in the negotiations in 1959. When the consultative parties first addressed this issue again in the early 1970s they were responding to external pressures, especially commercial interests. Their early discussions raised the spectre of conflict over the exercise of sovereignty and a potential weakening of both the Treaty regime and consultative party authority over Antarctic decision-making. Although the problem of environmental impact was discussed, the minerals questions was perceived primarily as a political rather than an environmental challenge especially as the issue was defined by foreign office bureacrats rather than scientists. The consultative parties had to consider not only what principles should govern their cooperation on minerals activity, that is, how the regime would be extended, but also what rules and procedures could best give effect to those principles and, at the same time, satisfy the matrix of competing individual state interests while not undermining the Treaty system which served their collective interests. There is no doubt that it was the hardest challenge they had faced, not excluding the formulation of the Treaty itself. This chapter explores those early political challenges. The negotiation of the minerals regime and the way in which differences among the consultative parties shaped the Minerals Convention are taken up in chapter 6.

MINERAL RESOURCES IN THE ANTARCTIC

The minerals negotiations took place in the context of limited information. Assessments of Antarctic mineral resources are incomplete and highly speculative. The probability that mineral deposits exist in the Antarctic is quite high: the chances of finding them are considered to be small (Wright and Williams, 1974, p. 1). Speculation on mineral potential in the Antarctic relies substantially on analogy with those continents which, along with Antarctica, were once part of the supercontinent Gondwanaland. The existence of mineral bearing structures in those parts of Australia, South America and Africa which were contiguous with the Antarctic margins has suggested to geologists that similar structures are likely to exist in the Antarctic.[1] Traces of a large number of minerals have been reported in the Antarctic, but no such analogous mineral bearing structures or large deposits have yet been found.[2] Iron ore has been located in the Prince Charles Mountains, and in the Dufek Intrusion in the Pensacola Mountains. Coal occurrences are widespread in the exposed areas of the continent, particularly in the TransAntarctic Mountains where low grade coal deposits are speculated to be extensive. The Dufek Intrusion, which compares geologically with the South African Bushveld (Zorn, 1984, pp. 5–6), has also been identified as a possible site for chromium, platinum, nickel ore, cobalt and copper although only traces of these minerals have been found there. The difficulties of locating extensive deposits, if they do exist, arise because so much of the land mass is buried under ice. Much more attention has been paid to the possibility of finding offshore hydrocarbons. The most favourable locations for hydrocarbon deposits are considered to be the basins of the Weddell and Ross Seas, the continental margins of the Bellingshausen and Amundsen Seas, and the Prydz Bay area off the Australian Antarctic Territory. No confirmed hydrocarbons have yet been located in the Antarctic.[3]

No commercially exploitable deposits of minerals are known to exist in the Antarctic.[4] Nevertheless, the Antarctic is often presented as a storehouse of mineral riches. The most frequently-cited figures are US Geological Survey estimates from the 1970s that the continental margins of West Antarctica might contain discoverable deposits (only one-third of discoverable deposits are considered to be recoverable) in the order of 45 billion barrels of oil and 115 trillion cubic feet of natural gas, quantities which would have almost matched proven US reserves.[5] Although this estimate has been dismissed by subsequent scientific reports as 'based on unacceptable processes of extrapolation' (Holdgate and Tinker, 1979, p. 16), it continues to be cited as authoritative. Similarly impressive and equally hypothetical estimates have been made since.[6] However, none of the investigation to date constitutes a systematic search for mineral deposits of the order that would provide adequate information upon which to commence extensive exploration.

Even if minerals were to be found in large quantities, there are physical, operational, economic and environmental barriers to their exploitation. Land-based mining activity would be difficult because of the continental ice-sheet, the harsh climate and the dangerous and unpredictable weather conditions. All but 2 per cent of the continent's 13.5 million square kilometres is covered by a slowly moving ice-sheet, with an average thickness of 2000 metres. Average temperatures rarely rise above freezing, even in the summer. Antarctic winter darkness would limit the working season and pack-ice, which almost doubles the size of the continent in winter, would restrict shipping access. Mineral processing also requires water which, surprisingly enough, is in short supply in the Antarctic – annual precipitation is less even than in the Sahara. Considerable difficulties would also arise in finding and extracting off-shore minerals. The Antarctic continental shelf is much deeper than in most other parts of the world although the extent to which this would be a major barrier to off-shore drilling is disputed (see Polar Research Board, 1986a, p. 265). Icebergs, which are more numerous and often much larger in Antarctic waters than in the Arctic would present a danger to drilling rigs. Bottom-scouring bergs could severely damage well-heads and any submarine pipelines laid to carry oil.[7]

The distance from markets, high transportation costs and the high salaries that would have to be paid for work in such a difficult environment mean that deposits which might be commercially attractive elsewhere in the world would not be so in the Antarctic. Further, there are known deposits in far more accessible regions of the world. For this reason, for example, only a supergiant hydrocarbon field would ever be commercially attractive. Crabbe suggests (1985, p. 366) that any discovery smaller than a field of at least 500 million barrels would not be economically viable and this view has been supported by industry representatives.[8] In the words of a Gulf Oil official 'we would have to find an awful lot of oil to support commercial production' (cited in *International Herald Tribune*, 31 December 1981).[9]

Even if these difficulties could be overcome, the environmental risks of Antarctic minerals activity are high.[10] Land-based support facilities would compete with wildlife and scientific stations for space on the limited ice-free coastal areas. Increased human activity would cause further disturbance to wildlife, and compound the problems of localised pollution and waste disposal connected with scientific activity. The greatest cause for concern, however, is the potential for and environmental impact of oil contamination. A major oil spill could result from a tanker accident, accident to a drill rig or production pipeline, blowouts during drilling or leakage from or damage to storage tanks. Accidents are likely to occur not only because of the physical obstacles and navigational hazards described above but also because the potential for human error increases in the stressful and rigorous Antarctic environment. The environmental impact of oil spills would be increased because biological decomposition of oil is slow in the cold temperatures of the Antarctic.[11] Antarctic wildlife, much of which lives on or near the sea, could be affected directly and indirectly by oil spills. Oil contamination could affect phytoplankton and krill stocks and thus impact on the Antarctic food chain. Scientific research in the Antarctic, which relies on an unpolluted environment for the conduct of baselines studies and for monitoring global ecological change, would also be adversely affected.

Interest in Antarctic Minerals

In the 1970s, two quite different expressions of interest in Antarctic minerals presented the Treaty parties with a potential threat to the political success and stability of their regime. In this context, change in the Antarctic regime was a learning process – that is the adoption of new rules and procedures – in response to factors outside the regime. The mining industry began to take a much more active interest in the economic potential of the Antarctic. Interest also arose from non-Treaty parties, not over the immediate commercial profit aspect of minerals activity but about the equitable sharing, as they saw it, of the resources of the global commons. In this way, the regime was linked to other international resource and environmental issues. Non-Treaty parties, led by the Group of 77, challenged the rights of the Antarctic club to make decisions on mineral resources in an area which, they argued, was a global commons and therefore, as with the deep seabed, subject to the common heritage principle. These challenges arose in the context of negotiations for the third UN Law of the Sea Convention, which began in 1973 and demands from less-developed countries for a New International Economic Order. While the demands for a new international economic order were directed particularly at the liberal financial and trading institutions of the Bretton Woods era,[12] those parts of the world argued to be global commons and therefore not open to exclusive appropriation by any state or group of states were also the subject of growing interest.[13] Although the Antarctic was not a specific concern during the UNCLOS negotiations, those negotiations had implications for the Antarctic Treaty system. First, it was possible that the International SeaBed Authority established by UNCLOS could have jurisdiction over the Southern Ocean seabed within the Antarctic Treaty area.[14]

Second, the G77 argued that the resources of the Antarctic should be subject to the common heritage principle which they applied to seabed resources. Resources beyond national jurisdiction, they argued, were the property of the whole international community and should be managed accordingly and particularly for the benefit of developing countries.[15] Third, UNCLOS enshrined the right of coastal states to establish an exclusive economic zone beyond their territorial seas, up to 200 miles from the coast (see Oxman, 1983, p. 153). This had implications for the the exercise of coastal state jurisdiction and the territorial claims in the Antarctic.

At the signing of UNCLOS in 1982 (the year the consultative parties began formal negotiations for a minerals convention) the Malaysian delegate argued that it was time to focus attention on 'that other area of common interest ... Antarctica' (cited in Beck, 1984, pp. 169–70). The Seventh Summit Meeting of the Non-Aligned Movement in 1983 called for a comprehensive study of the Antarctic issue and a resolution to this effect was introduced into, and adopted by, the 39th UN General Assembly session that year. The 'Question of Antarctica' has been on the General Assembly agenda each year since, although the resolutions became repetitive and were, for the most part, paid little attention by the Treaty parties.[16] Those resolutions demanded that the whole international community should participate in decisions about the exploitation of Antarctic resources.[17] In response, the Treaty parties adopted the formal position of not participating in UN votes or debates on Antarctic matters.[18]

The Treaty parties do not accept the application of the common heritage principle to the Antarctic. While they differ, obviously, over whether it is or is not subject to national jurisdiction they do accept that it is susceptible to national appropriation (unlike the seabed or outer space) and is therefore not a global commons. They also argue that the Antarctic is already subject to a legal regime which maintains peace and stability in the region and which functions, they suggest, better than any universal regime under the United Nations could. They also suggested that attempts to apply the common heritage principle to the Antarctic would be inherently exploitative of Antarctic resources rather than protective of the Antarctic environment. There is a certain irony as well as possibly some substance in this argument. Environmental concerns, while not ignored by the consultative parties in their negotiation of a minerals regime, were subordinated to political and economic concerns.

Commercial interest in Antarctic mineral resources coincided with the 1973 rise in oil prices and the Arab boycott of oil exports to western industrialised states. Several Treaty parties were approached by companies wanting to undertake prospecting in the Antarctic area and seeking information on the conditions under which they could secure exclusive rights to minerals in the Treaty area although this interest, the British suggested, was 'intrinsically speculative, reflecting the need of prospecting companies to keep occupied rather than a world need for minerals' (see ATCM, 1972e). This commercial interest presented the consultative parties with a dilemma which took some time to resolve. The Antarctic Treaty neither expressly permitted nor prohibited minerals activity in the Antarctic. In other words, the regime provided no guidance on this issue. If commercial interests chose to explore – which was nevertheless unlikely in the absence of any licencing authority or security of ownership over any resources discovered – there were no rules or procedures in place to regulate their activities.

Negotiating a regulatory instrument would mean addressing those contentious issues which the consultative parties had deliberately kept off the consultative agenda. The minerals issue was perceived as one which 'in the future could undermine the foundations of [the] Antarctic brotherhood [sic] which has cost us so much and of which we are proud' (ATCM, 1972a, p. 37) because of its connection with the sovereignty question.[19] However, the consultative parties reasoned, if they did nothing to rectify the absence of rules, unregulated minerals activity could pose both political and environmental dangers. The possibility of unilateral action by any party was unlikely but could not be totally discounted. Any such action by a non-claimant in a claimed area, by a claimant in its own area, or by one of the claimants to the area of overlapping claims could, the parties thought, lead to confrontation between them. Unregulated commercial activity, especially if it involved non-Treaty states not bound by the other provisions of the Treaty, might result in confrontation between competing operators or even between parties and non-parties. If exploitable minerals were discovered, the consultative parties were concerned that it would be difficult for them to reassert their authority over Antarctic matters or to negotiate some kind of regulatory mechanism in the face of entrenched economic interests. Severe environmental damage was also more likely if there were no rules on how minerals activity should be conducted.

The consultative parties' collective judgement was that the spectre of conflict and potential damage to both the Antarctic and the regime raised by doing nothing was worse than that raised if they moved to meet this challenge. It was less likely that the Antarctic 'would become the scene or object of international discord' (in the words of the Antarctic Treaty) and less likely that the Treaty system, which protected their individual and collective interests, would be undermined. The logic of this was summarised by the British (FCO, 1988, pp. 5–6). There was no guarantee that, even in the absence of rules and security of title to resources, risks would not be taken by operators. If such risks were taken, the Antarctic Treaty system would not survive. Consequently, the risks to the consultative parties attendant upon doing nothing were unacceptable. Therefore a minerals regime had to be negotiated. The parties were responding not to knowledge in scientific or environmental terms, but to political pressures and fears. Thus regime maintenance was an important factor in determining the parties' decision. It was a reactive response, rather in the way that the Treaty had been, to deal with a problem without any clear idea of what a regulatory regime would look like. In this way, the institutional bargaining process would proceed with little clear understanding of the contract curve or areas of common interest beyond maintaining the regime itself.

NGOs took a different view. They argued that the exploitation of mineral resources was incompatible with the Treaty parties' obligation to protect the Antarctic environment. That is, they identified what was to them an internal inconsistency in the regime, tensions which in their view could not be reconciled. The consultative parties were not unaware of the potential for adverse environmental impact as a result of mineral exploitation and the need to exercise their responsibility for the protection of the environment. However, their concerns at this early stage were with the political and legal aspects of the issue. The potential for environmental damage did not translate into a comprehensive philosophical debate about whether the Antarctic should be declared 'off limits' to mineral resource exploitation. The inscription of the minerals issue on the consultative agenda, coupled with statements by some Treaty parties and industry sources on the extent of minerals in the Antarctic, created expectations of economic benefits and did nothing to dispel the perception held by non-Treaty parties that the consultative parties were seeking to appropriate the (alleged) wealth of the Antarctic for themselves. Statements to the contrary by the consultative parties had little impact. The Treaty habit of conducting discussions and negotiations behind closed doors, and making little effort to respond to outside criticism, contributed to the impression of exclusive appropriation.

The potential for participating in Antarctic mineral resource exploitation was also one reason for the increase in accessions to the Antarctic Treaty in the 1980s. China's accession to the Treaty in 1983 and its subsequent application for consultative party status was in part, the Peoples' Daily noted, because 'according to preliminary exploration, the Antarctic is rich in petroleum, gas, rare metals' (cited in *Financial Times*, 12 May 1983). South Korea, which acceded to the Treaty in 1986 also observed (incorrectly) in an official report, that 'the Antarctic, man's [sic] last remaining treasure trove of rich natural resources, abounds in minerals deposits' (House, 1988, p. 112).

THE 1970s: PRINCIPLES AND OBJECTIVES

While the consultative parties responded to external pressures in first addressing the minerals issue, subsequent debate was shaped primarily by factors internal to the Treaty system. The consultative parties were concerned to protect the Treaty system from instability and to maintain their authority over Antarctic issues. The early stages of the institutional bargaining process, in effect a pre-negotiation phase, sought only to define broadly the problem and the principles that would guide negotiations for specific rules. In the 1970s, the consultative parties edged only slowly towards a minerals agreement. They sought first to clarify the legal position under the Antarctic Treaty with respect to the exploitation of mineral resources and the regulation of that activity. They then moved to reach a consensus on whether, in principle at least, minerals activity would be banned or permitted. It was only then that the more specific principles and objectives of a minerals regime were addressed in detail. Their debates were characterised by the usual degree of secrecy, by slowness in the pursuit of consensus, and by the general lack of involvement of other organisations. The scientific community provided information on the scientific, technical and environmental aspects of minerals activity and contributed to SCAR reports and consultative working groups, but made no collective judgement on whether or not such activity should occur. Non-governmental organisations were only beginning to focus on the Antarctic as an environmental issue and had little impact on these early debates. The consultative parties were guided by their understanding of the political imperatives of the Treaty system which encouraged them to avoid conflict and maintain consensus at all times. They moved to settle the broad shape of the agreement in the normal consultative process before feeling confident enough to convene a special consultative meeting committed to formal and directed negotiations in which they would bargain over their national political and economic interests.

The minerals issue was on the agenda of every consultative meeting from 1972 to 1981 (see Table 5.1) and a recommendation on this issue was adopted at each of those meetings. The principles and objectives adopted through those recommendations reflected the normative concerns of the Antarctic regime about political stability, the need for consensus and maintenance of the compromise on sovereignty. The process was hesitant and incremental. Several groups of experts were convened at consultative and preparatory meetings. Political and legal issues were always dealt with independently of scientific, technical and environmental issues. This points to the problem of developing environmental rules where knowledge is compartmentalised. In as much as they applied to the technical aspects of minerals activity, decisions were made on the basis of a degree of scientific uncertainty and the parties did take steps to improve their scientific knowledge by commissioning reports from scientific experts. Those reports did not, for the most part, provide that information. Rather they all referred to the continuing lack of knowledge available.

Table 5.1 Debates on minerals activity, 1970–81

Year	*Meeting/action*
1970	ATCM-VI. Informal discussion
1972	ATCM-VII. Recommendation VII-6 adopted
1973	Nansen Foundation Meeting. Working Group on Legal and Political Questions; Working Group on Scientific and Technical Questions
1975	ATCM-VIII. Recommendation VIII-14 adopted.
1976	SCAR Secretariat paper 'Antarctic Resources – effects of mineral exploitation' and SCAR Working Group on Geology paper 'Mineral occurrences and mineral exploration in Antarctic; prepared in response to recommendation VIII-14.
1976	Special Preparatory Meeting, Paris. Legal and Political Working Group; Scientific and Technical Working Group.
1977	Report of the SCAR Working Group on the Environmental Impact Assessment of Mineral Resource Exploration and Exploitation in Antarctica (EAMREA).
1977	ATCM-IX. Working Group on Legal and Political Aspects; Group of Experts on Exploration and Exploitation; recommendation IX-1 adopted.
1979	Preparatory Meetings for ATCM-X. Group of Ecological, Technological and Other Related Experts on Minerals Exploration and Exploitation (held in accordance with recommendation IX-1); preparatory meeting on legal and political aspects of mineral exploitation.
1979	ATCM-X. Working Group on Legal and Political Aspects; Working Group on Scientific and Environmental Aspects; recommendation X-1 adopted.
1980	Preparatory Meeting, Washington DC, December
1981	Report of the SCAR Group of Specialists on Antarctic Environmental Implications of Possible Mineral Exploration and Exploitation (AEIMEE)
1981	ATCM-XI. Claimant caucus meeting; no information on working groups; recommendation XI-1 adopted.

Indeed, the paucity of knowledge on Antarctic minerals and the impact of minerals activity on the environment was a major theme in the debates of the 1970s.

An off-the-record discussion on minerals activity was held at the sixth consultative meeting in 1970 after New Zealand raised it at the preparatory meetings.[20] A more substantive discussion at the seventh consultative meeting focused on the question of urgency which, as Antarctic diplomat Chris Beeby (who was to chair the formal minerals negotiations) has argued (1983b, p. 192) always had more to do with political than economic factors. Chile cautioned against any unseemly pressure to authorise commercial exploration, characterising the minerals issue as a grave problem 'which could put an end to the Antarctic Treaty, bringing to light once more the troubles over sovereignty, creating problems of jurisdiction and opening up rivalry and discord' (ATCM, 1972g). Other parties cautioned *against* delay in addressing this issue. The UK argued at this meeting that 'the subject raises problems certain to multiply ... which cannot be evaded indefinitely without becoming increasingly difficult to solve' (ATCM, 1972e). There was some discussion of, but no consensus on establishing a moratorium on minerals activity until an agreement which had the support of all consultative parties could be negotiated. Moratorium proposals were motivated not by environmental concerns but were intended to prevent, in the short term, any activity which might prejudice the consultative parties' deliberations.[21]

In recommendation VII-6 the parties recited a number of general principles which seemed relevant to further discussion of the minerals issue (see ATS, 1989, p. 3303). Those principles recalled the use of the Antarctic for peaceful purposes only, the special responsibility of the consultative parties to protect the Antarctic environment and ensure that no-one engaged in activities contrary to the Treaty and included an oblique reference to the 'special situation' of the Antarctic regarding the territorial issue. The recommendation also drew attention to the need for further study on technological aspects of minerals activity in the Antarctic.

An informal conference on Antarctic resources was convened at the Nansen Foundation in Oslo in 1973. Experts from all twelve consultative parties attended in their personal capacities although at least ten of the 29 Nansen participants were on delegations to the following consultative meeting. The Nansen Meeting was not a formal part of the consultative process but its report was circulated to consultative party governments in advance of the 8th consultative meeting in 1975.

While there were differences of opinion on whether minerals activity was implicitly permitted or prohibited under the Antarctic Treaty, the political and legal working group agreed that minerals activity was a peaceful purpose under article I of the Antarctic Treaty and therefore a legitimate issue for the Treaty agenda.[22] This working group favoured a collective approach by the consultative parties over unilateral action but could not agree, from a political point of view, how urgent the need for regulation was. The second working group, on scientific and technical aspects, concluded that unless the Treaty parties decided to prohibit minerals exploration and exploitation altogether there was some urgency, on environmental grounds, in negotiating regulatory mechanisms. This working group suggested that the 'relatively undisturbed state of the Antarctic environment rendered its continued protection a matter of greater rather than lesser importance' (Nansen Foundation, 1973, p. 9) and questioned whether the severe environmental hazards of extracting Antarctic oil and gas were so great as to be out of proportion to the small gain in energy in world terms. Some participants expressed the 'philosophically attractive' view that if the prospect of a world energy crisis was as bleak as some portrayed it then the world's scientific capacity would be better applied to the development of alternative energy sources rather than a search for Antarctic oil (Nansen Foundation, 1973, pp. 13–14).

However, it was the potential for political instability, rather than environmental insecurity, which dominated debate at the next consultative meeting in 1975.[23] The consultative parties now accepted that a formal instrument was essential and that it was within their competence to make decisions on minerals activity although they had not yet decided whether to permit or prohibit minerals activities. It was at this meeting that New Zealand pressed for a permanent moratorium on minerals activity on environmental grounds and suggested that the Antarctic should be declared a World Park. New Zealand had at that time a Labour government which had taken a strong internationalist and independent foreign policy line. ASOC suggests (1989d, p. 6) that the New Zealand proposal also referred to the possibility of some UN involvement in the administration of the Antarctic, and that claimants might give up their claims. There was apparently some informal support for the New Zealand suggestion for a permanent moratorium from Argentina and Chile (Wallace, 1988, p. 6) although this may have been motivated as much by concerns about sovereignty as about environmental protection and would certainly have not included support for UN involvement or a relinquishing of claims. There was no support for the World Park proposal. The US perceived the moratorium suggestion as a way of avoiding dealing with the question altogether (Quigg, 1983, p. 194) and applied some pressure to its Treaty partners to make progress by suggesting that without an agreement 'those countries who do not recognise claims to sovereignty would surely have to assert the right to commence mineral resource activities at their will, subject only to the applicable provisions of the Antarctic Treaty' (Mitchell, 1977, p. 96). This was not so much the action of a hegemon as of one state (among many) with issue-specific power operating in a unanimity system to protect its interests.

No formal moratorium was adopted. However, in the final report of the meeting the parties urged

> states and persons to refrain from actions of commercial exploration and exploitation while, acting as Consultative Parties, they seek timely agreed solutions to the problems raised by the possible presence of valuable mineral resources in the Antarctic Treaty Area (ATCM, 1975a, pp. 9–10).[24]

The consultative parties also identified the need for further study of the technical and environmental aspects of minerals activity. SCAR was asked to provide an assessment of possible environmental impacts if minerals activity '*were* to occur' (recommendation VIII-14, ATS, 1989, p. 3303; emphasis added).[25]

A special preparatory meeting held in Paris in July 1976 set about clearly defining for the first time the substantive issues that would need to be addressed in a minerals regime. The question of whether minerals activity would be permitted was still not settled. A number of countries, including Australia, the Soviet Union, Argentina and Chile (see Mitchell, 1977, p. 97), raised concerns about the environmental impact of minerals activity.[26] The Soviet Union also continued to advocate a long-term moratorium as one way of preventing the US from asserting its technological advantage in Antarctic minerals exploitation. No state was openly talking about exploring for minerals. Nevertheless, as the problems were raised in earnest, delegations began to adopt negotiating positions based primarily on perceptions of their political and economic interests. The claimant states were adamant that they should benefit from any resource exploitation (Talboys, 1978, p. 10)[27] and that they should have some control over access to resources in their territory. The US, on the other hand, argued that they 'would be unwilling to take any steps which directly reduced the freedom of American companies to go where they wanted in the Antarctic and do whatever appeared to be commercially attractive' (cited in Mitchell, 1980, p. 17).[28] Mitchell (1977, p. 99) suggests that while the claimants displayed some willingness to compromise the US would not. Agreement was reached only on the most general of principles upon which to base a minerals agreement and those principles reflected the primary political concerns of the regime. The consultative parties would continue to play an active and responsible role in decision-making; the Antarctic Treaty would be maintained in its entirety; protection of the environment should be a basic consideration and any regime should not prejudice the interests of all humankind (ATS, 1976b). There was, at this stage, no direct reference to the sovereignty issue.

The Paris meeting also considered SCAR's initial response to recommendation VIII-14 (see SCAR/EAMREA, 1979, pp. 53–5). SCAR's report provided an assessment of short and long-term, localised and widespread environmental impacts of both on- and off-shore exploitation. Local impacts, it suggested, might be severe and irreversible, although possibly not significant in terms of the Antarctic ecosystem as a whole. Large-scale effects were thought likely to be more serious than the sum total of local effects. The report assumed that there would be an inevitable impact on the environment, but made no judgement about the desirability or otherwise of minerals activity. Those judgements, SCAR believed, were the responsibility of the consultative parties, not the scientific community.

At the 1977 consultative meeting the consultative parties endorsed the Paris principles as the basis for a minerals regime with the added provision that the principles of article IV of the Antarctic Treaty – that is the compromise on sovereignty – should not be affected. They also agreed to

> urge their nationals and other States to refrain from all exploration and exploitation of Antarctic mineral resources while making progress towards the timely adoption of an agreed regime ... to ensure that pending the timely adoption of agreed solutions pertaining to exploration and exploitation of mineral resources no activity shall be conducted to explore or exploit such resources (recommendation IX-1, ATS, 1989, pp. 3304–5).

This injunction is generally taken to enact a conditional moratorium through the principle of voluntary restraint although it is, as Triggs argues (1984, p. 535), 'vague and unclear [with] no mandatory injunction'. While the consultative parties cited it as an effective restraint upon their own activity and that of their nationals, even though they are only urged rather than required to enforce compliance, it is not clear what they could do to prevent non-Treaty states from undertaking minerals activity. It was also enacted as much for political reasons as for environmental ones.

The lack of information on the environmental impact of minerals activity and concerns about the inevitability of that impact were raised again at this meeting. A specially convened Group of Experts concluded that more knowledge was urgently needed about the Antarctic environment and geology so that adequate protection measures could be established in the event of minerals activity going ahead. This report was stronger in its warning of environmental damage and the unacceptability of such risks than SCAR's 1977 updated EAMREA report although several members of the SCAR group also contributed to this working group.[29] The consultative parties called for further study on the environmental implications of mineral resource activities and suggested convening yet another group of experts on this topic. However, in spite of recognising the problems raised by this lack of knowledge which had been noted by all the scientific and technical reports so far, and the consequent difficulties in monitoring the environmental impact of minerals activity, the consultative parties pressed ahead with their negotiations. In this case, then, scientific uncertainty on environmental concerns did not prevent action on minerals activity. Rather, as is often the case with environmental protection rules, it did not engender a precautionary approach to managing potential environmental damage. Two more meetings of experts were convened prior to ATCM-X in late 1979. Both identified potential environmental risks. In March 1979, the Rockefeller Foundation sponsored a workshop at the Bellagio Centre in Italy on oil and other minerals in the Antarctic with special attention to the environmental implications of possible mineral exploration and exploitation.[30] In conjunction with the preparatory meeting for ATCM-X, the awkwardly-named Group of Ecological, Technological and Other Related Experts on Minerals Exploration and Exploitation in Antarctica met in Washington in June 1979.

Themes that had been articulated at earlier meetings were raised again at the 10th consultative meeting in 1979 – the need to protect the Antarctic environment and avoid environmentally unacceptable activities, the need to deal with the question of jurisdiction and differences between consultative parties on territorial sovereignty, and the need to establish regulatory mechanisms. Several of the claimants argued that any regime should guarantee protection of their interests and an appropriate economic return. Argentina advised delegates that it would not 'give its sovereign consent' to any agreement which did not take the political realities of Antarctica into consideration (ATCM, 1979a, p. 70). While governments were reminded that the timely progress conditions for maintaining the voluntary moratorium were being met it is doubtful whether this *could* be called timely progress. Although the issue had been on the agenda for seven years the parties had not yet begun to negotiate a regime, only to debate how they might approach the task.

However, they did move a step further towards the formal negotiation of a minerals convention. As well as restating the principles upon which an agreement should be based, recommendation X-1 set out in general terms the kinds of rules and procedures that should be included (see ATS, 1989, pp. 3305–6). The parties agreed that a minerals agreement would have to include some means for assessing the possible environmental impact of mineral resource activities and determining whether such activities would be acceptable, as well as elaborating procedures to govern the ecological, technological, political, legal and economic aspects of mineral activities. Rules on protecting the environment and ensuring compliance would also be necessary. The preambular paragraphs of the recommendation focus particularly on protection of the Antarctic ecosystem and the need for further scientific information.[31] SCAR was asked to carry out further work on retrieving and analysing relevant information required to assess environmental impacts. SCAR was also asked to devise scientific programs which would assist in improving predictions of the environmental impact of activities and technologies associated with mineral resource exploration and exploitation. SCAR's Group of Specialists on Antarctic Environmental Implications of Possible Mineral Exploration and Exploitation (AEIMEE), which was established to undertake this work, produced two reports. The first, tabled at ATCM-XI in 1981, stressed the need for a scientific data base to assist in decision-making on minerals activity and for both basic and applied research to produce that data (see SCAR/AEIMEE, 1986). The second report, produced in 1983, focused on scientific programs related to off-shore hydrocarbon development (see SCAR/AEIMEE, 1986). However, these scientific programs were not being undertaken and the necessary information upon which rules should be based was, for the most part, still not available.

The 11th consultative meeting took place in 1981 against the background of the forthcoming conclusion of the Law of the Sea negotiations and the likelihood that Third World states would then turn their attention to the Antarctic. Divisions between the claimant and non-claimant states at this meeting were marked and the claimant delegations met together each morning in an effort to reach a common and unified position on the minerals issue (Quigg, 1983, p. 199). Recommendation XI-1 finally set in train the formal negotiations for a regime on Antarctic mineral resources by convening a special consultative meeting to determine the form of a regime and to establish a schedule for negotiations (see ATS, 1989, pp. 3306–7). As well as restating the Paris principles, this recommendation paid particular attention to the sovereignty issue. It required that the provisions and principles of article IV of the Antarctic Treaty remain unaffected by a minerals regime. Any agreement had to be 'acceptable and ... without prejudice to those states which have previously asserted rights of or claims to territorial sovereignty in Antarctica as well as to those states which neither recognise such rights of or claims to territorial sovereignty nor ... assert such rights or claims'. The regime was to apply to all minerals activity on the continent and offshore areas (without encroaching on the deep seabed) and had to include provisions to protect the consultative parties' responsibilities for the environment. A regime, the parties agreed, should also include procedures for adherence by parties other than consultative parties and provisions for cooperative arrangements with other organisations. The voluntary moratorium on minerals exploration would continue as long as timely progress was being made.

CONCLUSION

The early phases of coming to an agreement on minerals activity point to the broader application and importance of a 'pre-negotiation' phase, when parties are able to discuss the issue at hand without having to make formal commitment to negotiating texts. This may encourage, especially in a closed forum, a degree of openness before formal negotiations begin. However, with respect to environmental concerns, it is also a slow process which may not be advantageous if environmental degradation is already taking place. In the case of the minerals issue in the Antarctic this was not so. Nevertheless, the process here also indicates that the non-development option is unlikely to be seriously addressed. One reason for that may well be the compartmentalisation of issues. Scientific, environmental and technical concerns were dealt with separately from political and legal concerns. While this process has some advantages in taking discussions forward, it can also mean that environmental protection concerns are seen as separate, technical issues, rather than as political ones or as concerns which should drive resource or other agreements rather than being added as a supplementary concern. The experience of the early minerals discussions points again to the problem of scientific uncertainty and the inclination of states to proceed in spite of a lack of knowledge and in spite of warnings about environmental impact.

6 The Antarctic Minerals Convention (CRAMRA)

Once the in-principle decision was made that minerals activity in Antarctica was a legitimate activity and would be permitted, debates among the consultative parties in the 1970s had focused on how to define their collective interests. The principles adopted as the basis for a minerals regime were designed to ensure that those collective interests would be met, consultative party authority over Antarctic decision-making reinforced and the regime maintained. However, in spite of agreement on principles, there was little insight into the terms of the constitutional contract that would regulate minerals activity. In other words, as Young describes the process of institutional bargaining, 'the range of feasible options and the outcomes associated with these options [were not] well-defined at the outset' (1990b, p. 4). Nevertheless, as Bilder notes (1982, p. 199), the consultative parties had a 'strong mutual interest in reaching an agreed and orderly solution to the issue of mineral resources'. They were jointly agreed on the outcome they wished to avoid (conflict and the demise of the Treaty system) but there was no common agreement on the outcome they preferred. Individually, each state had different preferences set by its political (especially territorial) and economic interests.

SPECIAL CONSULTATIVE MEETING IV

The Convention on the Regulation of Antarctic Mineral Resource Activities was negotiated at the Fourth Special Consultative Meeting which met for its first session in 1982. Twelve formal negotiating sessions and three informal intersessional meetings were held over a period of six years. Fourteen consultative parties began the negotiations in July 1982. There were twenty consultative parties when the Convention was adopted in June 1988.[1] Non-consultative party participation was limited

– the NCPs were invited to attend negotiating sessions only from February 1985. This part of the chapter examines the parties' competing interests and briefly chronicles the formal negotiations. While the key points of difference among the parties were identified early there was no agreement, or even a clear idea, on how to overcome them. The second part of the chapter explores in more depth the outcome of those negotiations – that is, the Minerals Convention – and the ways in which it did, or did not meet those interests and the extent to which it reinforced the basic principles of the Antarctic regime.

States' Interests

States within the Treaty system were able to use the bargaining process to ensure that their interests were protected and at least the most important of their demands were adopted. There was no dominant coalition within the regime on this issue. Power-based explanations of the negotiations are therefore not especially insightful. Rather, the minerals negotiations are better understood through an analysis of the institutional bargaining process.The major split was between the claimants and non-claimants. It seemed doubtful that the ambiguity of article IV of the Antarctic Treaty, one of the mainstays of the Antartic regime, could be reproduced in a minerals convention which had to address directly the question of ownership of non-renewable resources. As Beeby noted (1982, p. 6), 'neither the claimant states nor the non-claimants could reasonably expect that the Special Consultative Meeting [would] endorse a solution either founded exclusively on the existence of the sovereignty claims ... or [one which] ignores the fact of those claims'. The claimants' relative numerical power continued to decline within the Treaty system as more states acceded to the Treaty and as the number of consultative parties grew. Yet they still maintained issue-specific power in a consensus-based system and were able to function as a relatively cohesive coalition. They argued that their sovereign rights in the Antarctic should ensure unilateral control over activities in their territory, including the exploitation of resources. While they would not achieve this in the convention, as they had not achieved it in the Treaty system generally, they were not prepared to forego some acknowledgement of their status especially in the institutions and decision-making procedures of a minerals regime. The non-claimants would not accept claimant control over resources and were reluctant to agree to any special privileges for the claimant states. It was more difficult for the non-claimants to maintain unity within the negotiations. Except for the specific issue of the exercise of sovereignty and jurisdiction, they were a diverse group with competing interests.

Throughout the negotiations the divisions between the claimants and the pro-development states became increasingly entrenched. The likely mining, non-claimant states (the US, West Germany and Japan) wanted non-discriminatory access to mineral resources and regulatory rules and procedures that would facilitate minerals activity. They were also keen to ensure security of ownership of any resources found or extracted, security of investment and protection against arbitrary interference with minerals activity. They were joined, to some extent, by the UK and France who were potential mining states but who also had claimant interests to protect. Countries with established mining industries, such as Australia and South Africa, were to some extent supportive of non-discriminatory access but were also concerned about competition with their own industries, although in Australia's case this was moderated by its claimant interests. They also sought provisions to prevent resource poor countries subsidising minerals activity.

The interests of the developing countries within the Treaty system had to be given some weight. Doing so would provide the consultative parties with an argument, although not an altogether water-tight one, to deflect criticism that they were a group of western industrialised states monopolising decision-making on, and the resources of, the Antarctic. The relative strength of the less-developed consultative parties grew as the negotiations progressed. Their ranks included Brazil and India from 1983 and China and Uruguay from 1985.[2] The developing countries established their own power base within the negotiations and eventually became a group to be reckoned with (Andersen, 1991, p. 106). Their interests primarily lay in participation in resource exploitation, access to technology, and a share in the benefits of exploitation.[3] These interests were in conflict with those of the technically advanced states, particularly the likely mining states, who were determined not to be required to share technology or to enter into mandatory joint ventures with less developed states.[4] The non-consultative parties were keen to ensure that they would be guaranteed some role in the convention's institutions and decision-making processes.

The consultative parties also took different positions on environmental protection under the proposed mineral regime. While all states were agreed that environmental protection rules were necessary they differed on the extent of those rules. Some states, such as Australia, New Zealand, Argentina and Chile, took a much stronger view on the need for stringent environmental protection provisions than the pro-development states. As Orrego Vicuña notes (1988, p. 186) the countries of the southern hemisphere 'have a particularly keen awareness of the environmental factor'. The British Foreign Office, on the other hand, argued that it did not accept 'that environmental risks of Antarctic mining are intrinsically unacceptable' (FCO, 1989b, p. 2) although it did suggest that they could be unacceptable in any particular case.

Divisions among the parties arose on legal and political grounds – on sovereignty and jurisdiction – and economic grounds. It was those divisions, not environmental concerns, which gave rise to the key issues which remained unresolved until the end of the negotiations. In this respect, the hierarchy of values that characterised the Antarctic regime generally also applied here. Environmental concerns, while not dismissed, were fitted into the solutions to political and economic concerns. Contentious issues arising from these cleavages included the nature of the institutional structure and relationships, financial provisions, enforcement and compliance rules, dispute resolution and liability and environmental regulations. Specific questions to be resolved included who would authorise minerals activity, under what authority appropriate control over resource activity would be exercised, and how payment of taxes and royalties would be settled. The task of devising an institutional compromise which would protect the collective and individual interests of the negotiating parties was difficult. Further, whatever compromise was acceptable to all parties also had to embody rules and procedures that would function satisfactorily once the Convention came into effect.

Institutional Bargaining

The negotiations, which were held in closed session, were characterised by informality and flexibility. Little time was spent in plenary session. The parties sought to maintain consensus and to avoid public dissension over the differences between them. Although never formally specified it was clear that the negotiations would have to proceed on the basis of consensus. Much of the discussion and debate, and working out of differences, took place in standing and ad hoc working groups. As the negotiations progressed, at least three standing working groups were established. The first, a legal issues group, dealt with questions on liability, dispute settlement, compliance, inspection and monitoring. The key issues group, from which non-consultative parties were often excluded, focused particularly on the institutional framework and decision-making procedures. The third formal group examined exploration and development issues. Informal groups were convened, as necessary, to address and reach agreement on specific topics. Environmental issues were, for the most part, also raised in informal groups often centred on a few key states – New Zealand, Australian, the US, Britain, Argentina and Chile.

Coffee break and corridor diplomacy was important. Andersen (1991, p. 104) describes the process thus:

> friendships, mutual respect and trust, as well as insight into one another's personalities, contributed much to the final outcome. Informal gatherings, talks over coffee, as well as working lunches and dinners, were often instrumental in breaking an apparent deadlock.

The bureaucratic elite which managed the consultative process, and who were as chapter 2 observed well socialised into the norms of the regime, also dominated the minerals negotiations. Foreign ministry officials were prominent. Many delegations included representatives from resource ministries, although only a few included mining industry representatives. Even fewer included officials from environment ministries although parties were urged to include on their delegations people with expertise in technical and environmental issues and people familiar both with administering environmental laws and with procedures for translating environmental principles into reality (Kimball, 1983, p. 15).[5] Continuity in negotiating teams was an asset in seeking to advance particular interests. Consultative party delegations which changed personnel frequently, as well as the new consultative parties, 'undoubtedly found the negotiations complicated' (Andersen, 1991, p. 96). No observers were invited and no interim reports were issued. While this secrecy within a limited membership forum attracted accusations of exclusivity, it nevertheless contributed to reaching an agreement. As Oxman observes (1986, p. 246), in the context of the Antarctic 'club', 'the states concerned may well make concessions to each other that they might never consider making in a universal negotiation'. While the scarcity and uncertainty of scientific knowledge had been a major theme in discussions on minerals activity in the 1970s, this difficulty was not addressed in detail in the 1980s. SCAR, for example, as the primary source of scientific knowledge in the Treaty system, had no formal role in the negotiations and, again, took little or no advocacy role. It was not even invited as an observer to the special consultative meeting and was not asked to provide further reports on the scientific, technical or environmental aspects of minerals activity or to comment on the proposed environmental protection rules in the convention.

The exercise of leadership is an important determinant of the success of institutional bargaining (Young, 1990b) in developing a regime. Entrepreneurial leadership was the key to the Antarctic minerals negotiation. This crucial role was filled by New Zealand diplomat Christopher Beeby who chaired the negotiations. As Chair, he sought to devise options to overcome bargaining difficulties and to line up support for those options in the way that Young (1990b, pp. 18–19) describes the role of an entrepreneurial leader. He represented a 'common memory base' (Andersen, 1991, p. 103) and provided continuity in the negotiations. He maintained contact with delegations through an informal key issues contact group and consulted regularly with heads of delegations and with convenors of working groups. Rather than adopt a formal single negotiating text, Beeby prepared subsequent versions of an informal draft text which was presented to delegates as his personal report.[6] These successive reports gave shape to the constitutional contract and assisted the search for compromise. The parties could move, however slowly, towards an agreement without making a formal commitment to a draft convention. Through his personal report, Beeby sought to ensure that the participants would be reluctant to abandon what had already been agreed upon when the tough issues were debated. Young suggests (1990b, p. 20) also that the work of an entrepreneurial leader is assisted if that person is the agent of a powerful state in the bargaining process although he does acknowledge that this is not a necessary condition for the successful exercise of this type of leadership. It is not clear that this proposition holds in Beeby's case. New Zealand possessed no more issue-specific power than a number of other states involved in the negotiations. Indeed, it is possible to argue that New Zealand's issue-specific power in the negotiations may have been augmented by Beeby's role as Chair of the negotiations rather than the other way around.

Non-governmental Environmental Organisations

In spite of its expertise, the scientific community, particularly SCAR, was not well incorporated into the formal process of negotiating the minerals regime, in contrast with its participation in working groups in the years before the special consultative meeting was convened, although individual scientists were members of national delegations. The environmental protection dimension of the negotiations was monitored closely by non-governmental organisations who, as described in chapter 3, were actively challenging and exposing the environmental record of the Treaty system. They criticised the assumptions upon which the negotiation of the minerals convention were based. They suggested that the principle of environmental protection was inherently incompatible with the objectives of regulating minerals activity and that *any* impact on the environment was unacceptable. As Schofield argued (1976,

p. 20) '[L]ost somewhere in all the rhetoric ... is the very first question that ought to be asked about the possible exploitation of Antarctica's mineral resources: not how, when or by whom, but *whether* to exploit them at all' (emphasis in original).

The NGOs adopted a dual strategy during the negotiations. Generally they opposed the negotiation of a minerals convention, arguing instead that the Antarctic should be permanently off-limits to minerals activity and should be declared a World Park. However, activists who were close to the meetings, or who were on national delegations, lobbied within the Treaty system for stringent environmental protection mechanisms in the event (which seemed most likely at that stage) that a convention was successfully negotiated and implemented. NGO activity was coordinated by the Antarctic and Southern Ocean Coalition and Greenpeace International. They monitored the formal sessions and liaised closely with NGO representatives on national delegations. Only three consultative party delegations included NGO representatives. The US did so from the first negotiating session in 1982. Australia adopted this practice from the July 1983 meeting and New Zealand from May 1984. The Danish delegation also included an NGO representative from February/March 1985 when non-consultative parties were first able to attend the SCM. ASOC and Greenpeace held press conferences, produced press releases and published the environmental newspaper *ECO* for distribution to delegates and the media.[7] They also produced sophisticated analyses of the Beeby texts and lengthy reports on the negotiating sessions which drew on their own legal and scientific expertise.[8] Those analyses highlighted environmental weaknesses in the texts and proposed changes which would improve the environmental protection mechanisms in the convention. While the draft conventions produced by Beeby were confidential, non-governmental organisations made them public 'in order that the world [could] participate in a more informed manner' (ASOC, 1984a, p. 1). In doing so they sought to impose transparency, to counter the secrecy of the negotiations, and accountability, especially with respect to the consultative parties' exercise of their responsibility for environmental protection. Despite their trenchant criticism of the negotiations and their publication of the Beeby drafts, NGO access to delegates was not withdrawn.

Negotiating Sessions

The first session of the Fourth Special Consultative Meeting convened in Wellington, New Zealand in June 1982 (see Table 6.1 which sets out the process in summary form). Delegates drew up a Schema for an Instrument on Antarctic Mineral Resources to guide their negotiations 'without prejudice to any decision on the format and structure of the regime' (in Bush, 1982, p. 450). That document set out a comprehensive listing of subject headings which included those matters which were likely to be most contentious: financial questions, applicable law and jurisdiction, and provisions to ensure the regime was acceptable and without prejudice to the claimants. The consultative parties agreed from the outset that any agreement had to establish a 'climate of confidence ... [ensuring] security of title for the operator, the protection of investment and certainty of regulation' (Beeby, 1983b, p. 194). They also accepted that, because of their differences over sovereignty in particular, and the lack of knowledge about mineral resources and future technologies and activities, it would be almost impossible to negotiate a detailed mining and environmental code which would set out rules and procedures to cover all possible problems and activities.

The consultative parties met again in Wellington in January 1983. According to Beeby (1985, pp. 22–3) disagreement at these first two sessions focused on where the powers required to regulate mineral activity should be located. Non-claimants favoured collective and non-discriminatory regulation. Claimants advocated regulation by individual states, preferably themselves over their own territory, even if that regulation was in accordance with general guidelines. In his first informal report – known colloquially as Beeby I – which was tabled at the conclusion of the second session, Beeby sought to avoid an early deadlock over regulation by switching attention to the institutions and procedures of the regime. There was some agreement on the institutional structure by this stage (a commission, an advisory committee and an executive secretariat) although the membership and functions of those institutions and the relationship between them proved to be contentious throughout the negotiations. The parties also agreed upon a three-stage approach to minerals activities – prospecting, exploration and

Table 6.1 Fourth Antarctic Treaty Special Consultative Meeting: minerals negotiations

Session		*Action and Outcome*
1	Wellington, New Zealand 14–25 June 1982	Schema for an Instrument on Antarctic Mineral Resources
2.	Wellington, New Zealand 17–28 January, 1983	Informal working groups: Area of applicability; Minerals to be covered and stages of activity; Environmental issues. **Beeby I** produced
3.	Bonn, West Germany 11–22 July 1983	Environmental Contact Group
4.	Washington, DC, USA 18–27 January 1984	**Beeby II** produced
5.	Tokyo, Japan 23–31 May 1984	
6.	Rio de Janiero, Brazil 26 February–8 March 1985	Working Group on objectives and principles of the regime; Working Group on prospecting; Working Group on legal issues and dispute settlement
7.	Paris, France 23 September–4 October 1985	Working Group on legal issues (liability and dispute settlement); Working group on guidelines for Operators; Working Group on Exploration and Exploitation; Informal Working Group on Key Issues
8.	Hobart, Australia 14–25 April 1986	Working Group on Legal Issues; Informal Heads of Delegations meeting on Key Issues; **Beeby III** produced
First Intersessional Meeting, Whangaroa, New Zealand, 1986 Claimant Caucus, San Diego, late 1986		
9.	Tokyo, Japan 27 October–12 November 1986	Legal Issues Working Group; Exploration and development Working Group; confidentiality of

		data Working Group; Key Issues Working Group; informal group on environmental issues

Second intersessional meeting, Whangaroa, New Zealand, March 1987. **Beeby IV** produced

10.	Montevideo, Uruguay 11–20 May, 1987	Working Group on Key Issues; Working Group on Legal Issues; Working Group on exploration and development; informal group on function of advisory committee; budgetary and financial provisions; confidentiality of data; area of application

Third intersessional meeting, Auckland, New Zealand, November 1987. **Beeby V** produced

11.	Wellington, New Zealand 18–29 January 1988	Working Group of the Whole (article by article review); Key Issues group; Drafting Committee; Informal working groups on area of application; financial matters; anti-subsidy provisions
12.1	Wellington, New Zealand 2–19 May 1988	Outstanding Key Issues Working Group; Working Group on matters of less-than-key importance; Drafting Committee. **Beeby VI** (Draft Convention) produced
12.2	Wellington, New Zealand 23–31 May 1988	Outstanding Key Issues Working Group; Working Group on matters of less-than-key importance; Drafting Committee. **Beeby VII** (Draft Convention) produced
	Final Conference 2 June 1988	**Convention on the Regulation of Antarctic Mineral Resource Activities** adopted along with Final Act of Special Consultative Meeting

development – although no definitions were yet settled upon. An indicative list of environmental principles was included.

Only modest progress was made at the third session in Bonn. The session Chair (who was not Beeby in this one case) reported that delegates 'discussed a number of very abstract and complicated models on which we were not in agreement' (*The Guardian* 23 July 1983).[9] The fourth session convened in Washington in January 1984. 'Sources close to the talks' were reported as saying that little of substance was agreed to (*The Guardian* 18 February 1984). All delegations except France were prepared to agree that all areas south of 60° south would remain *closed* to mineral development unless a decision was taken to open them. France (one of the likely mining states) wanted the reverse. Following this meeting Beeby prepared a second version of his personal report (Beeby II). Several delegations apparently did not receive it until shortly before the Tokyo negotiating session in May 1984 and a full reading could not be completed at Tokyo. The Tokyo meeting decided that non-consultative parties (NCPs) would be invited to future sessions as observers.[10] For the first time a working group was established to look at legal issues and dispute settlement, including questions of compliance and enforcement, liability, compensation for damages, and procedures for withdrawal from and amendment to the regime. Beeby reported (*Financial Times*, 13 March 1985) that progress had been made but that the toughest meetings were yet to come as the real issues separating the claimants from non-claimants were approached.

The Treaty parties met again in Rio de Janiero in February 1985 and in Paris from 23 September to 4 October 1985. Through discussions with Heads of Delegations and in small private meetings from which NCPs were generally excluded, Beeby continued to advance negotiations on the key issues – the composition and decision-making procedures of the institutions to be set up by the regime, international participation in benefits from the regime, and the internal accommodation between claimants and non-claimants (Kimball, 1985c, pp. 2–3). Following the eighth session in Hobart, Australia, in April 1986, he produced a third version of his personal report (Beeby III). Later in 1986 he convened an informal intersessional meeting in Whangaroa, New Zealand, to discuss

Beeby III. Only the US, USSR, UK, France, West Germany, Argentina and New Zealand were invited to participate.

A claimant caucus was held a few weeks later in San Diego and the parties met again formally in Tokyo in October/November 1986. After the Tokyo meeting Beeby called a second informal intersessional meeting in Whangaroa in March 1987.[11] Beeby IV was circulated soon afterwards. Negotiations continued in Montevideo in May 1987 where Beeby IV was modified considerably in the working groups. The Head of the Chilean Delegation, Fernando Zegers, suggested during an informal meeting that Antarctica could be designated an Antarctic Treaty Park if no more progress were possible in resolving certain key issues (Anon., 1988a, p. 1). While this was hailed by NGOs as a 'clever new development of the World Park idea' (*Antarctic News* #6, 1987, p. 2), it may have been a suggestion borne as much out of frustration with slow progress as out of environmental concern. A third intersessional meeting attended by 15 parties was held in November 1987, this time in Auckland, New Zealand. Beeby V was circulated prior to negotiations to be held in Wellington at the beginning of 1988. In December 1987, the US government made Beeby V public 'so that those interested in or potentially affected by the future agreement may review and analyse it' (Kimball, 1988a, p. 1).

The January 1988 session in Wellington convened a 'working group of the whole' to undertake an article-by-article review of Beeby V except for those articles being considered by the Key Issues group. The Key Issues group worked on the still contentious provisions on liability, dispute settlement, judicial review of decisions taken by institutions, inspection, compliance and approval procedures. Membership of this group changed depending on the issue under discussion. A number of small informal groups also met to consider the area of application, financial matters and anti-subsidy provisions. A drafting committee was also established. This committee, charged with ensuring that versions in all four official Treaty languages (and Chinese) were consistent with each other, was more contentious than originally anticipated. West Germany insisted on being represented. Other delegations argued that if there was a possibility of a German text, then their languages should also be represented. The Beeby VI draft left open the possibility that authentic texts might also be agreed to in German, Japanese,

Norwegian and Portuguese. At this session, delegates committed themselves to holding only one more negotiating session even though there were many issues still unresolved.

The final session convened at the beginning of May 1988, again in Wellington. All twenty consultative parties and thirteen of the 18 NCPs attended.[12] The first part of the session (2 to 19 May) divided into three groups: an Outstanding Key Issues group chaired by Beeby, a group to focus on matters of less-than-key importance, and the Drafting Committee. NCPs were generally excluded from the Key Issues group which met in small informal discussions followed by larger groups in the afternoons. After eleven formal sessions and three intersessional meetings, outstanding issues still included decision-making in the commission and regulatory committees, composition of the commission, the regulatory committees and the special meeting of parties, dispute settlement, liability, financial issues, participation, definitions of sponsoring state, operator and the area of applicability, compliance, entry into force, anti-subsidy provisions and languages of the text (ASOC, 1988k, p. 2). In a four day break between 19 and 22 May, Beeby consolidated a new negotiating text which was presented, for the first time, as a draft convention. Delegates met in plenary on 23 May to discuss the draft. Beeby stressed the necessity for delegations to forego their preferred positions in order to find 'tolerable common ground' if agreement was to be reached (ASOC, 1988k, p. 3). After the plenary the three working groups from the first part of the session met again. An all-night session was held on 30 May, attended by twelve consultative parties only (ASOC, 1988k, p. 3 – the twelve are not named). The following morning a second version of the draft convention – informally Beeby VII, officially MR/17 Rev V/Corr.1 – was distributed. A third plenary was convened with most delegations apparently satisfied with the Convention before them although there was still some concern over the decision-making and anti-subsidy provisions.

The Final Act of the Special Consultative Meeting, with the Convention appended, was distributed to delegations shortly before the final plenary on 2 June 1988. It was signed at a special ceremony after the formal close of the SCM.[13] Christopher Beeby hailed it as 'an historic occasion which I believe will go down in Antarctic history as the most important political development regarding the regulation of Antarctica since the Antarctic Treaty itself' (ATCM/SCM-IV, 1988b, p. 1).

THE CONVENTION ON THE REGULATION OF ANTARCTIC MINERAL RESOURCE ACTIVITIES[14]

The Minerals Convention is a pragmatic although complicated framework agreement which establishes principles, prescribes general rules and procedures and creates institutions for the purpose of regulating Antarctic mineral resource activities should they be considered acceptable. Much of the administrative machinery is to be established only when, and if, any party wishes to explore for minerals. Minerals activity – prospecting, exploration and development – may only be undertaken in accordance with the rules of the Convention. The Convention represents further change to the Antarctic regime, in the shape of single-loop learning, defined in chapter 1 as the pursuit of new rules and procedures for the same ends. There was no substantive change to the normative framework. Rather, new rules and procedures were developed to address a particular issue. More substantive compliance rules and a more soundly-based institutional framework reflected the necessities of such a complex agreement, rather than any departure from a reliance on sovereignty norms. There was also some acceptance of the importance of a precautionary approach, and of the need for knowledge-driven decision-making. However, for the most part, the sovereignty norms and political compromises which were a crucial determinant of the Antarctic regime also shaped the Minerals Convention. Given the tensions already identified in earlier chapters between those norms and the imperatives for environmental protection, the Minerals Convention could only further contribute to the internal inconsistency of a regime which purported to be capable of protecting the Antarctic environment.

The Convention manages skilfully to maintain the compromise between the claimants and non-claimants. Neither group attained its preferred outcome but that was unlikely from the beginning. Claimant concessions were, primarily, to the likely mining states, rather than to the non-claimants generally. The Convention leaves much of the detail of the compromise between claimants and non-claimants to be hammered out on a case-by-case basis should minerals activity begin. The decision-making procedures established under the Convention, especially in the regulatory committees, ensure that both claimants and non-claimants (especially mining non-claimants) have the opportunity to continue to press their particular interests. The Convention manages to step around the difficult sovereignty question. Like the Antarctic Treaty and subsequent instruments negotiated by the consultative parties, it does not solve the matter, but reproduces the creative ambiguity of article IV of the Treaty in asserting in the preamble that:

> a regime for Antarctic mineral resource activities must be consistent with article IV of the Antarctic Treaty and ... without prejudice and acceptable to those states which assert rights of or claims to territorial sovereignty in Antarctica, and those states which neither recognise nor assert such rights or claims, including those States which assert a basis of claim to territorial sovereignty in Antarctica.

However, as Triggs argues (1985, p. 204) 'it is doubtful that any such provision could protect the claimant states against a conclusion that their sovereignty in Antarctica, if it exists, [would be] diminished, were they to ratify a minerals regime which gives access to third states to resources within their claimant territory'.

The crux of the minerals negotiations and the Convention were the specific mechanisms of representation and process in which claimant and non-claimant interests would have to be reconciled. The generally held view among the Treaty parties was that the key to finding an accommodation between claimants and non-claimants would be found 'largely within the institutional structure of the regime' (Beeby, 1985, pp. 21–2). Yet, as chapter 2 noted, the consultative parties had also generally resisted any centralisation or institutionalisation of the Treaty regime.

Institutions

The Convention provides, in principle, that participation in mineral resource activities should be open to *all* states which have an interest in such activities and are prepared to subscribe to a regime governing them. In practice, only states which are already Antarctic Treaty parties may accede to the Convention and participate in its institutions. From the consultative party point of view, this was logical and acceptable given that the Minerals Convention was an integral part of the Antarctic Treaty system. It was also a way of maintaining 'ownership' of the Antarctic issue and reinforcing their collective power. Nevertheless, there is something contradictory in a convention that purports to be open to all but which puts restrictions on that openness. Further, any Antarctic Treaty party which had not signed the Minerals Convention would nevertheless be able to participate to some extent in the institutions of the Convention.

The Convention establishes a Commission, a Scientific Technical and Environmental Advisory Committee and an Executive Secretariat.[15] A regulatory committee is to be established for each area identified for exploration and development. The Convention also provides for a Special Meeting of Parties. Although the Minerals Convention is separate from the Antarctic Treaty, institutional and decision-making power resides directly with the consultative parties thus minimising any tendency towards centralisation. There was a general uncompromising opposition by the consultative parties to non-consultative party (NCP) wishes for a share in decision-making. The NCPs had no formal role in the minerals negotiations and thus had limited bargaining power. They were unsuccessful in their quest for a substantial role within the Convention. The authority of the institutions in which the NCPs participate as of right was also diluted during the negotiations.

Membership of the Commission is automatic for any party which was *also* an Antarctic Treaty consultative party on 25 November 1988, the date the Convention was opened for signature (article 18 (2)).[16] Other signatories may be Commission members *only* if they meet an activities criterion – undertaking research in the Treaty area which is relevant to decisions about minerals activities (most likely to be a new or aspiring consultative party) or sponsoring exploration and development – and only while that activity continues. If a party in either of these categories is a consultative party, then it is deemed to have met the requirements for Commission membership unless more than one-third of existing Commission members object. Otherwise, it may be a member *only* if its membership is accepted by *all* Commission members.[17] Any non-consultative party must, in applying for Commission membership, declare its intention to abide by recommendations made under the Antarctic Treaty.[18] This ties participation in the institutions of the Convention to the Antarctic Treaty and widens adherence to the Treaty's principles and rules. Any party which meets none of the criteria for Commission membership may become an observer. So too, however, may any Antarctic Treaty party which is not a party to CRAMRA.

All parties may be members of the Advisory Committee (article 23). Observer status is granted to any non-Convention state which is a party to the Antarctic Treaty or to CCAMLR. The Advisory Committee provides a forum for consultation and cooperation among the parties. It has no decision-making powers and thus its role as an important source of knowledge within the minerals regime is diluted, although the Commission and regulatory committees must take 'full account' of its advice (article 41). Developing country consultative parties, supported by the Netherlands and East Germany, joined the NCPs in pressing for a stronger role for the Advisory Committee which they saw as an important source of advice and assistance on areas in which they had little expertise. The Special Meeting of Parties (SMP), which also has no formal decision-making role, involves all parties to the Convention (article 28). While many ATCPs were completely opposed to it, the NCPs wanted a substantial role for the SMP and were supported by China and the USSR in this. Unless they are sponsoring minerals activity, NCPs are unlikely to be members of either of the decision-making institutions of the Convention. They therefore have little effective power in the minerals regime. The creation of the SMP was itself a concession to NCPs, but the only one they were successful in achieving. However the SMP has a role only in the threshold decision to open an area for potential minerals activity.

A ten-member regulatory committee is to be established for each area in which the Commission determines that minerals activity may take place (article 29).[19] The claimants aimed to retain as much power as possible on the important regulatory committees because it was here that detailed rules for minerals activity were set and individual applications decided upon. It was here that the claimants would be able to bargain for financial concessions and for influence over management schemes. Because of the smaller size of the regulatory committees, claimant power vis-a-vis the non-claimants was stronger than in the Commission.[20] The size of the regulatory committee was determined by political considerations. There was, as Andersen notes (1991, p. 100) a practical need for a body smaller than the Commission to undertake regulatory tasks and to establish contracts. However, this could easily have been accomplished by a small group of three or four states. The demands of the claimants meant that the regulatory committees would have to include at least three claimants.[21] The demand for political balance ensured that non-claimant numbers would be similar. Initial proposals for regulatory committee membership envisaged a split of four claimants and four non-claimants. The number was increased to ten to accommodate the demands of developing states and mining interests (Andersen, 1991, p. 100).

Membership of each regulatory committee is determined by the Commission according to a complex formula. Four committee members must be claimant states, including the claimant(s) to the area being regulated. Early Beeby drafts proposed that four of what was then to be an eight member regulatory committee would be claimants. There was some concern expressed that giving claimants equal weight to non-claimants could lead to side deals between claimants and mining companies when the management scheme was elaborated, on royalties for example, that could possibly give substance to their claims (see Antarctica Project, 1984). In the determination of regulatory committee membership, it was the superpowers rather than the claimants or non-claimants who gained most. Under the rules of article 29 they are members of every regulatory committee.[22] This was, in part, because both maintain a basis of claim to the Antarctic and because they were the two states with the largest presence in the Antarctic. In this context, they were able to argue for a 'permanent presence' on the regulatory committees.[23] Three of the ten must be developing countries.[24] Brazil, China, India and Uruguay tabled this proposal at the Montevideo meeting to secure some balanced representation on the regulatory committees and a role in decision-making.

The rule that the claimant to the area for which a regulatory committee has competence is able to *nominate* the other claimant members of the committee was adopted after the claimants submitted a written demand to this effect at the Montevideo meeting. By contrast, non-claimants members of the Commission may only 'present views' to the Commission chair on non-claimant membership (article 29(4)). Participation in regulatory activity, or specific minerals activity, cannot be used to support or deny a claim. While the claimants have a guaranteed position on the regulatory committees, this may not be 'interpreted as affecting article IV of the Antarctic Treaty' (article 29(7)). In other words, it does not amount to any recognition of claimant sovereignty over territory or resources and cannot be used by the claimants as evidence of such.

The potential mining states were successful in arguing that any regulatory committee should include states sponsoring or conducting minerals activity in the area of competence. Thus they had a guaranteed role in decision-making. Membership is also to include those parties which have contributed substantial scientific, technical and environmental information relevant to the identification of the area. The addition of extra members with a direct interest in the area – that is, states which sponsor minerals activity – was added at the suggestion of Japan and West Germany (Antarctica Project, 1987, p. 3). Any Party may attend regulatory committee meetings as an observer. NCPs are unlikely to sit on a regulatory committee unless they are sponsoring minerals activities.

Decision-Making

From a procedural perspective, a requirement for consensus in such a complex decision-making process as that established under CRAMRA might well have resulted in paralysis. Consensus, defined in the Convention as 'the absence of a formal objection' (article 22(5)), is required only in the Commission and only for the important threshold decision to identify an area for possible exploration and development, for the adoption of the budget and for decisions on budgetary and related matters. If the Chair of the Commission determines that an objection is likely with respect to the decision to open an area for minerals activity, then he or she is urged to consult with members of the Commission and convene those members most directly interested to seek to reconcile those differences and produce a generally acceptable proposal. A US Department of State official explained that the 'consensus requirement was particularly difficult for the US to accept since it ... has an interest in access to mineral resources (if they exist)' (Scully, 1989). In this context, then, consensus could well advantage environmental protection, rather than diminish it. Early Beeby drafts provided for a two-thirds majority within the Commission, even on the decision to open an area and on financial issues. At the Rio meeting (March 1985), Beeby proposed that the decision to open an area should be made by the Special Meeting of Parties on the basis of a consensus among consultative parties and a majority of NCPs. The ATCPs were apparently 'extremely hostile' (*Antarctic News*, #4, 1986, p. 3) to the idea and moved to block any power-sharing. A proposal was made at the final negotiating session for a qualified majority rule to apply to the identification of an area for exploration and development (Anon., 1988h, p. 1). Matters of substance in the Commission (including whether a matter is one of substance) require a two-thirds majority of those present and voting. A simple majority rule applies to procedural issues within the Commission.

The claimants would ideally have liked a veto over all regulatory committee decisions. This proposition was anathema to the non-claimants. The rules governing decision-making in the regulatory committees on issues of substance, such as approval of a management scheme or granting of a development permit, require a two-thirds majority which must include a clear majority of both claimants and non-claimants.[25] In the drafts up to and including Beeby V, regulatory committee decision-making was to be by a simple majority, but with a veto provision for both the claimant(s) and the sponsoring state. This was dropped in Beeby VI. Non-substantive issues, such as revising general guidelines, require approval by at least two claimants and three non-claimants. This chambered voting system means that neither claimants nor non-claimants have an absolute veto. However, neither can each group be 'out-voted'. In this respect they are treated equally. However, this provision modifies the accepted norm in the Treaty and consultative process where claimant (and non-claimant) interests are protected by consensus, to the potential disadvantage of the claimants. There is no rule that the claimant majority *must* include the claimant to the area over which the regulatory committee has competence and no procedure for a party to declare itself not bound by regulatory committee decisions against its wishes. It is therefore technically possible for the required majority of three of the four claimants not to include the relevant claimant. It is unlikely, however, that any claimant would refuse to support another claimant in respect of its territory. Such a refusal would carry the implication of interference in the exercise of territorial rights as well as raising the prospect of a reciprocal withdrawal of support at some time in the future. Thus claimant solidarity in this respect rests on diffuse reciprocity.

Managing Minerals Activity

The Antarctic is closed to all minerals activities (except prospecting) until the Commission decides, by consensus, to identify an area for possible exploration and development activities. This is an 'in principle' decision that minerals activity may, but not necessarily will proceed in a particular area. Any party may request such an identification which must include a detailed environmental impact assessment. In making this decision, the Commission has to be confident that minerals activities will not cause significant environmental impacts, that technology and procedures are available to provide for safe operations and limited environmental impact, and that the capacity exists to monitor key environmental parameters and to respond effectively to accidents (article 4). An area is identified only in relation to *specified* mineral resources.[26] In reaching its decision, the Commission must take into account the advice of the Advisory Committee and the conclusions of, and views expressed at, a Special Meeting of the Parties.

Because minerals activity is prohibited until specifically permitted, supporters of the Convention argue that the burden of proof is on potential operators to show that activities will not unduly damage the environment, rather than on those interested in protecting the environment to prove that it will. This reverses the major premiss of CCAMLR which permits harvesting of marine resources unless scientific evidence shows that conservation measures should be put in place. Unless otherwise specified, minerals activity is prohibited in any area designated as a Specially Protected Area (SPA), a Site of Special Scientific Interest (SSSI) or in any other protected area. The Commission may prohibit or restrict minerals activities in any area which it designates a protected area (article 13(2)). It may also designate a buffer zone around any area in which minerals activities are prohibited or restricted. The Commission is also responsible for promoting the collection and exchange of scientific, technical and other information so that environmental impact can be assessed. Measures for the protection of the environment are adopted by the Commission which also establishes rules on prospecting activities.[27]

The procedures in the Convention following identification of an area relate to exploration and development activities. No prior authorisation is required for prospecting, that is activities aimed at identifying areas of mineral resource potential, although any such activity must be conducted in accordance with the Convention. Nevertheless, nine months notification of an intention to prospect must be provided by any party which sponsors prospecting. Information on proposed prospecting must include an assessment of environmental and other impacts and the measures which will be taken to avoid harmful consequences or undue interference with other uses in the Antarctic. Prospecting confers no rights to any mineral resources.

The relevant regulatory committee is responsible for adapting the Commission's guidelines and identifying general requirements for exploration and development in its designated area. Any party can then apply to the committee for an exploration permit. Exploration activities are those aimed at identifying and evaluating specific mineral resource occurrences or deposits. The application must include an environmental impact assessment, a detailed description of the Operator, its proposed activities, its capacity to respond to accidents, to comply with general guidelines and its technical competence and financial capacity. The regulatory committee may decline the exploration application if it considers that the proposed activities are not consistent with the Convention. Specific terms and conditions for exploring are set out in a management scheme, drawn up by a sub-committee. An exploration permit will be granted if the management scheme – in effect a contract between the committee and the Operator – is approved by a two-thirds majority of the committee including a clear majority of both claimants and non-claimants. The claimant states wanted the right to prepare any management schemes for activity in their area and this provision was initially included in Beeby I.[28] The pro-mining states argued that any management scheme should be drawn up by the operator and the sponsoring state and then submitted to the regulatory committee with the application for an exploration permit. In an unsuccessful attempt to forestall a proposal that mining states should be guaranteed a position on regulatory committees, the claimants proposed that a technical subcommittee of the regulatory committee should be responsible for the management scheme. The final provision (article 46) gives responsibility for the management scheme to the sponsoring state, the relevant claimants and, if necessary, one or two additional members of the regulatory committee. Given that the management scheme has to be accepted by a majority of the claimant states it makes sense to have them involved in its preparation.

If an Operator wants to develop, that is, exploit specific mineral resources, a new application is required. This application must contain updated information including a further detailed environmental assessment. The regulatory committee considers whether the planned development would cause previously unforeseen impacts. If no modification to the management scheme is required, or if modifications are approved, then a development permit is issued. The decision to issue a development permit is made by the full regulatory committee (article 54). The consultative parties rejected a proposal at the Tokyo 1987 meeting that the Advisory Committee, to which NCPs belong, should be able to declare a management scheme unacceptable.[29] There is no specific procedure for declining a development application although any member of the Commission or regulatory committee can request a review of the decision to approve a management scheme or development permit. However, only a regulatory committee may suspend, modify or cancel a management scheme or permit.

Jurisdiction

As well as the specific questions of influence in the institutions of the Convention and authority over resource activity, the sovereignty issue was also an important factor in determining the area of applicability of the Convention. The general principles of the Convention apply to the Antarctic Treaty area (article 5(1)). However, in determining regulatory authority, the parties had to confront the disputed status of the continental shelf, where much of the hydrocarbon potential was thought to lie, and the disputed rights of the claimants, as coastal states, to declare Exclusive Economic Zones. The claimants viewed the Antarctic continental shelf as territory under their national jurisdiction. The continental shelf is contiguous with land and, in international law, is part of the territory of a coastal state rather than part of the deep seabed which lies beyond national jurisdiction. The non-claimants denied national or coastal state jurisdiction in the Antarctic. This position was complicated, however, because under the UN Convention on the Law of the Sea (UNCLOS), resources in the deep seabed and subsoil thereof *beyond national jurisdiction* are part of the global commons.[30] If no national jurisdiction was judged to exist in the Antarctic, then the continental shelf would share the legal status of the deep seabed. The logical outcome of this could have been that the authority of the International SeaBed Authority (ISBA) established under UNCLOS to manage deep seabed resources for the benefit of the international community would extend to the land margins of the Antarctic continent. Non-claimants and claimants were equally opposed to this proposition, but the non-claimants still opposed any provision that would appear to sanction coastal state jurisdiction in the Antarctic. The consultative parties adopted a legal cleverness to define the area over which regulatory authority was to be exercised. It avoids any direct mention of coastal state jurisdiction over the continental shelf but gives the parties authority to exercise joint regulatory control over off-shore areas. Decisions on regulating the continental shelf as part of an area stem from the authority of the Commission which comprises the consultative parties acting collectively.

The Convention distinguishes between offshore areas and the deep seabed in article 5(2) which says that

> the Convention shall regulate ... activities which take place on the continent of Antarctica and all Antarctic islands, including all ice shelves, south of 60°South latitude *and in the*

> *seabed and subsoil of adjacent offshore areas up to the deep seabed* (emphasis added).[31]

In other words, there are off-shore areas which are not deep seabed. The deep seabed is determined to be the seabed and subsoil *beyond* the geographic extent of the continental shelf as the term continental shelf is defined in accordance with international law (article 5(3)).[32] Thus without any reference to coastal state jurisdiction, the continental shelf (which is here defined to *not* be part of the deep seabed) can be equated with the adjacent offshore areas over which the parties assert regulatory control. In this way the continental shelf is not explicitly defined as part of the continent (in the way that ice-shelves are) but neither is it excluded from regulatory coverage. Nevertheless these articles do not provide any justification for the Treaty parties collective right to exercise jurisdiction or regulatory control over 'adjacent offshore areas'. Further, this emphasis on joint regulatory authority does undermine the claimant insistence on individual authority.

The status of coastal state declaration of EEZs posed a different problem. Under international law a coastal state must declare its jurisdiction over an EEZ. In the Antarctic, such a declaration by a claimant could well be interpreted as an enlargement of a claim and thus in contravention of article IV of the Antarctic Treaty. The Convention simply continues the regime practice of avoiding insurmountable jurisdictional questions and makes no reference to this matter.

The more general jurisdiction question centred on whose laws would apply to private enterprises and their employees engaged in minerals activity. The claimants obviously would have preferred that claimant law apply. West Germany (a likely mining state) argued that the law of the sponsoring state should apply. At the January 1988 meeting the USSR, Poland and the US suggested a compromise to the effect that if jurisdictional disputes arose, then the parties should consult. The Convention effectively adopts this approach and thus leaves the question unanswered.[33] The management scheme is required to include reference to the applicable law only 'to the extent necessary' (article 47(q)). Again, this follows the Treaty system practice of avoiding difficult questions or leaving them to be dealt with at a later date. It is likely that applicable jurisdiction would follow the unstated practice of the Treaty by which nationals are subject to the law of their own state, or to the law of the sponsoring state in the case of a joint venture or where the operator is not from the sponsoring state.

Economic Interests

Claimant demands for revenue-sharing and for the payment to them of royalties and taxes were unsuccessful although they do not specifically cede taxing rights to 'foreign' governments. In determining the disposition of surplus revenue the Commission must

> ensure that the interests of the members of Regulatory Committees having the *most direct interest* in the matter in relation to the areas in question are respected (article 35(7)(b); emphasis added).

The deliberate ambiguity in this provision means that 'interested states' can be interpreted as referring to either the claimants or the sponsoring state.[34] Nevertheless, 'surplus revenue' refers only to revenue raised by way of lodgement fees and other such administrative charges, not royalties or taxes. The Convention includes no specific anti-subsidy provisions. Australia was particularly active, but unsuccessful, in arguing for the Beeby III provisions requiring minerals activity to be conducted in an 'economically rational manner' to be retained. The Final Act contains a rather general, and probably unenforceable suggestion that 'unfair economic practices including certain forms of subsidies could cause adverse effects to the interests of Parties ... and ... should be addressed in the context of the relevant multilateral agreements' (ATCM/SCM–IV, 1988c, pp. 2–3). Those agreements are not specified but the reference is doubtless to the General Agreement on Tariffs and Trade.

There is some recognition of the problems faced by developing countries in undertaking minerals activity in the Antarctic, but the rules would probably have had little practical impact on this. In disposing of surplus revenue the Commission is required to promote participation in scientific research in Antarctica, especially that related to the environment and resources, by all parties but in particular by developing country parties (article 35 (7)(a)).[35] However, this does not specifically facilitate participation in minerals activities. The developing countries argued that priority should be given to those mining proposals which included opportunities for international participation and developing state involvement. Other states stressed that financial soundness, technological expertise and a commitment to environmental protection should be determining factors (Newman, 1987, p. 432). In Beeby III and IV, procedures for dealing with competing applications were to be decided on a first come, first served basis. The Convention makes some limited concession to developing country views on this. Regulatory committees are required, all other things being equal, to give priority to exploration applications with the broadest participation including developing countries (article 43(2)(e)). There was, therefore, some incentive for mining interests to include developing countries in some way. The Commission is also to elaborate joint ventures and other participatory opportunities with particular relevance to developing countries which are party to the regime, although what this means in practice is not spelled out. However, it was the guarantee of participation in regulatory committees which was most important in providing a platform for advancing the interests of developing countries.

In spite of claims to the contrary, the accommodation with non-Treaty parties was not high on the agenda of concerns. There were differences of opinion among the consultative parties on how to manage the external accommodation, in particular the questions of whether the convention should establish an open participation regime and whether joint ventures or a fund to assist with scientific work should be encouraged. Both the claimants and the potential mining states were opposed to any sharing of economic benefits of resource exploitation with other states inside the Treaty system, let alone outside it. The Convention pays some lipservice to the 'interests of all mankind [sic]' but makes no concrete moves in this direction. The preamble suggests that the effective regulation of Antarctic mineral resource activities is in the interests of the international community as a whole and one of the Convention's objectives is to take account of those interests (article 2(3)(g)). However, while participation in Antarctic mineral resource activities is technically open to all states, accession to the Convention is, as already noted, limited to signatories of the Antarctic Treaty (article 61). Participation by developing states was dealt with only as part of the internal accommodation. There is nothing in the Convention on the sharing of benefits with the international community, although Australian diplomat Richard Woolcott has argued (1985, p. 23) that the idea of revenue-sharing with the international community was considered premature simply because the development of mineral resources was so far away.

Compliance and Liability

As noted in chapter 1, compliance rules are crucial both to regime maintenance and to successful environmental protection. As chapters 2 and 3 have argued, the emphasis on sovereignty and decentralisation in the Antartic regime supported, for the most part, few compliance rules or binding dispute resolution procedures. The Convention builds on the rules established under the Antarctic Treaty and, for the most part, relies on national competence. It does, however, incoprorate some limited arrangements for centralised monitoring and enforcement. As a general rule, each party is required to take appropriate measures within its competence to ensure compliance with the provisions of the Convention (article 7). The Convention continues to rely on inspections (which were shown in chapters 3 and 4 to be of limited effectiveness) to identify violations of its environmental protection rules. Inspections may be conducted unilaterally in accordance with article VII of the Antarctic Treaty. All stations, installations and equipment relating to mineral resource activities must be opened for inspection (article 11). Inspectors may also be designated by any member of the Commission, by the Commission itself or the relevant regulatory committee. Procedures for inspection must be in place before exploration or development can proceed in a particular area. Reports from inspectors are to be transmitted to the Commission and to the appropriate regulatory committee (article 12).

The Convention includes detailed dispute settlement procedures (articles 56–59) and an Arbitral Tribunal established under an annex. Article 6(3) of the Annex to the Convention establishes emergency dispute settlement procedures for disputes relating to environmental protection measures, compliance, response action to environmental damage, inspection, non-discrimination, other uses of Antarctica, and prospecting. The Convention provides also that these matters may not be excluded from any of the disputes resolution procedures which are set out in the body of the Convention. Those procedures enable parties to choose either arbitration by the International Court of Justice or the Arbitral Tribunal, in such a way that gives preference to the Arbitral Tribunal if no choice is made or where disputant parties have made different choices.

The extent of liability and the question of residual state liability were contentious issues during the negotiations. Australia, Argentina and Chile had argued that no defences should be permissible (ASOC, 1988d) and were supported in this by France, Norway, the USSR and New Zealand. Japan opposed the suggestion that damage to the Antarctic environment and dependent ecosystems should be covered by the liability provisions. Operators are to be held strictly liable for damage to the environment and loss of or impairment to established uses in the Antarctic. The Convention also provides for residual *state* liability where damage, including environmental damage, occurs because a state has not fulfilled its obligations. However, liability is exempted in the case of unforeseen natural disasters – which could cover a wide range of events in the Antarctic – and acts of armed conflict including terrorism. More detailed liability rules were to be elaborated in a separate Protocol to the Convention, which was to be designed to enhance protection of the Antarctic environment. That Protocol had to be in force, for the party making the application, before an application for an exploration or development permit could be made. However the provisions also established that the Protocol could determine limits to liability, including environmental liability.

Transparency and Accountability

The rules on public availability of information under the Convention are more extensive than those adopted for the consultative process. Generally, however, transparency rules relate to the transmittal of information rather than procedures for public input. Further, the requirements are most limited with respect to the regulatory committees, which is where major decisions on exploration and development, as well as compliance, are taken. The Commission is required to give advance public notice of matters on which it has requested advice from the Advisory Committee including the decision to open an area for potential exploration and development. The Advisory Committee is also required to give advance public notice of meetings and matters before it. The Commission must maintain a publicly available record of its meetings and decisions (but not of documents provided to it unless specifically stated in the Convention) and of information submitted to it including Environmental Impact Assessments (EIAs) and regulatory committee monitoring reports. The Advisory Committee must make public reports of its meetings and decisions but only after they have been circulated to all parties and observers. Regulatory committees are similarly obliged to ensure that a record is available of all their decisions and management schemes in force. Commercially sensitive data, including the results of prospecting, which is often little different from scientific geological research, is exempted from this requirement.[36] Inspection reports are not automatically circulated to observer organisations or states, and the discussions on monitoring and compliance in the regulatory committees were not to be open to observer organisations either.

The Minerals Convention reflected the small steps that had been taken to open up the Treaty regime to international organisations although that participation was limited and discretionary. The Advisory Committee *may* consider the views of interested international organisations on matters before it and *may* establish procedures to transmit relevant information to those organisations. The Committee may also seek information and advice from other scientists and experts or scientific organisations on an ad hoc basis. The Commission may cooperate with the International Union for the Conservation of Nature and Natural Resources (IUCN) and 'other relevant international organisations, including non-governmental organisations'.[37] These organisations may be granted observer status on the Commission and the Advisory Committee (and, by default, to the Special Meeting of Parties).[38] These provisions in the Convention are vague and susceptible to interpretation.

Granting observer status is the prerogative of the consultative parties in accordance with the rules of the Convention. References to non-governmental organisations were intended, for the most part, to enable working relationships to be established with organisations such as the Scientific Committee on Oceanic Research (SCOR) and the IUCN rather than environmental interest groups. What is more surprising is that there was little formal role for SCAR in the institutions and procedures of the Convention, despite SCAR's importance as a source of information and knowledge on environmental impact and Antarctic geology. SCAR had argued it that it should be formally and specifically given observer status on the Commission and the Advisory Committee (ATCM, 1989j, p. 190). While the parties are generally encouraged to cooperate with SCAR, as a component of the Treaty system (article 34), what might be seen as a marginalisation of SCAR, at least in the formal rules, reinforces the state-centric nature of the minerals regime, which continues to restrict non-state access to decision-making and to diminish possible opportunities for the practical inclusion of, for the most part, independent scientific expertise and knowledge.

PROTECTION OF THE ENVIRONMENT: A CONVENTION FOR PROTECTION OR DEVELOPMENT?

The consultative parties justified their negotiation of the Minerals Convention in part on the need to ensure that the fragile Antarctic environment would be protected if minerals activity went ahead, citing their special ability and knowledge to undertake this task. Beeby emphasised that 'the objective of the Antarctic countries during the course of [the] negotiations ... had always been to preserve and strengthen the ATS *and* to ensure that the unique and fragile environment should not be damaged by mining activity' (ATCM/SCM-IV, 1988a, p. 1; emphasis added). The consultative parties argued that the growing interest in the mineral resources of the Antarctic would (if they took no action) result in an unregulated scramble which could have dire environmental consequences. The Convention, they suggested, was necessary to prevent this. NGOs disagreed with the logic of an 'unregulated scramble'.

They suggested that mining interests would be reluctant to undertake exploratory activities in the absence of any certainty of their rights to exploit any minerals found.[39] This view has some support. Crabbe argues (1985, p. 367) that 'no private or government enterprise would be willing to commit vast sums of money for exploitation or development prior to the establishment of international agreements on minerals rights'. Pat Quilty, a senior scientist with the Australian Antarctic Division, suggests (1986–87, p. 117) that the lack of any regime 'allowing for secure tenure of title ... [was a] severe hindrance to commercial minerals exploitation'. The oil industry took the view that 'the *sole hindrance* is jurisdiction – from whom do we get a drilling permit' (cited in Wallace, 1988, p. 9; emphasis added). Whether intended or not, the practical effect of the Convention would be to provide a secure tenure of title to operators over any resources discovered. If the Convention was based on the premise that minerals activity might occur at some stage, then it is doubtful that it could have avoided providing such rights for operators. Indeed, Beeby argued (1985, p. 21) that a convention would 'need to ensure that, if mining does take place ... those who undertake it have security of title [and] have their investment protected'. In this respect, NGOs argued that the Convention was a precondition for development and were concerned that, once in place, mining industry interests would want to put it to the test.

There are extensive environmental principles and rules in the Convention which impose obligations on the parties. However, two questions arise. First, were the environmental rules adequate? Second, were the provisions on compliance strong enough? On the first, the traditional Antarctic hierarchy of values which places political concerns, in the end, above environmental ones shaped the negotiations. Environmental protection mechanisms were weakened in the search for agreement on political and legal issues. In the second case, the dominance of sovereignty norms within the regime once again placed limits on what states would accept in terms of monitoring and enforcement institutions and rules. Thus their ability to ensure compliance must be questioned.

The Convention enshrines the fundamental principles of the Antarctic regime. Consultative party authority is guaranteed in the institutions of the Convention, especially in the decision-making bodies, the Commission and the regulatory committees. Regime stability, through a political balancing act, is the focus of the structure of the regulatory committees. As Joyner notes (1988, p. 141), both these principles 'occupy a higher place in the hypothetical hierarchy of principles than that of environmental protection'. Experience of the Treaty system had convinced the consultative parties, and the bureaucratic elites responsible for Antarctic policy, that environmental issues could only be addressed in the context of a wider agreement on political problems. The belief that the disputes over sovereignty and jurisdiction would limit what could be negotiated on environmental rules, especially with respect to monitoring and compliance, was strong.

Earlier chapters have demonstrated that the decision-making process within the Antarctic regime, dominated as it was by the need to settle political differences, was flawed from the perspective of negotiating good environmental rules. The minerals negotiations followed a similar path. The then Director of the British Antarctic Survey, Richard Laws, observed (1985, p. 54) that 'discussions on environmental matters have so far received less attention than the organisational structure of [the] regime'. Kimball (1983, p. 15) has also suggested that concern for the environment coexisted in 'tenuous harmony' with other imperatives in the negotiations (such as the sense of urgency to complete the negotiations and the internal accommodation between the claimants and non-claimants). NGOs similarly argued that 'providing a political solution to the Treaty partners' internal conflicts is an objective that has overridden protection of [the] environment' (cited in Shapley, 1985, p. 166). They suggested that the lawyers and diplomats who negotiated the convention had underestimated the natural and scientific values of the Antarctic and overestimated the ability of an unenforced set of rules to modify the hazardous behaviour of oil companies and other operators working at the limits of technology in extreme conditions (see ASOC, 1984a, p. 3).

Environmental rules adopted in the Minerals Convention *were* precautionary and, unlike CCAMLR, the Minerals Convention prohibits activity until judgements are made that it can be permitted to proceed. But monitoring and compliance is still based predominantly on sovereignty rules which favour national autonomy, little independent oversight and, more importantly, few sanctions. The Convention generally adopts a precautionary approach in principle. The preambular paragraphs of the Convention note that

> protection of the Antarctic environment and dependent and associated ecosystems must be a basic consideration in decisions taken on possible Antarctic mineral resource activities.

The general principles of the Convention (article 2) acknowledge the special responsibility of the consultative parties for the protection of the environment and the need to 'protect the Antarctic environment, respect Antarctica's significance for, and influence on, the global environment [and its] scientific value and aesthetic and wilderness qualities'.

Resource activity must not cause significant impact on air and water quality, or significant changes in atmospheric, terrestrial or marine environments or in the distribution, abundance or productivity of fauna or flora.[40] Endangered or threatened species must not be further jeopardised and there must be no degradation of, or substantial risk to, areas of special biological, scientific, historic, aesthetic or wilderness significance. Finally, no activity may take place until it is judged that it would not cause 'significant adverse effects on global or regional climate or weather patterns' (article 4(3)). Technology must be available to ensure safe operations. Operators must demonstrate a capacity to respond effectively to accidents and to monitor key environmental parameters and ecosystem components. In his June 1988 press statement Beeby called these the 'most stringent safeguards ever negotiated in an international treaty' (ATCM/SCM-IV, 1988b, p. 2). He argued that the approval process was a rigourous one. The consultative parties argued, in the words of a senior New Zealand diplomat, that the Convention was 'the best deal for Antarctica and the Antarctic environment that [was] attainable in an imperfect world. Tougher environmental provisions were just not negotiable' (C. Keating, 1988). However, the key question is whether, in the light of the parties record on other environmental protection issues, the best that could be achieved was good enough to ensure protection of the Antarctic environment, especially on a localised basis, if minerals activity went ahead.

In spite of this obligation to protect the Antarctic environment, the Convention also establishes the principle of acceptability of impact. Damage to the Antarctic environment or dependent or associated ecosystems is defined as 'any impact ... beyond that which is negligible or which has been assessed and judged to be acceptable pursuant to this Convention' (article 1(15)). Environmental protection obligations are also linked, in article 4, to a sufficiency of knowledge principle. Minerals activity may only occur if adequate information is available to make informed judgements about its possible environmental impact. Responsibility for collecting and generating this knowledge rests with the Commission. However, 'sufficiency of knowledge' is a problematic concept as the experience of CCAMLR shows. The procedures for increasing that knowledge, and for according a central role to environmental and scientific experts, are weak. Environmental factors are crucial in the decision to open an area, but once minerals activity begins the progression to development activities becomes harder to stop on environmental grounds.

Environmental organisations argued that the Convention was essentially a pro-mining document and that the environmental protection mechanisms were inadequate. A spokesperson for Greenpeace US argued that '[a]lthough the text's environmental provisions appear to be substantial, in practical terms there are major loopholes that will enable miners to escape the consequences of the inevitable environmental damage' (in Anon., 1988g, p. 12). NGOs argued that terms like 'adequate' and 'significant' with respect to environmental information were vague and open to interpretation. They suggested that the rules on EIAs gave insufficient detail on content or the procedures for implementing them. NGOs were also concerned that pressure could be brought to bear on any state attempting to exercise a veto on environmental grounds over the identification of an area for minerals activity, especially given the reconciliation process written into article 22(5).[41] The lack of prior authorisation for prospecting is a flaw from the perspective of protecting the environment, in spite of the requirement for an environmental impact assessment (a provision that the American Petroleum Institute considered 'wholly unnecessary' (API, [n.d])) and a description of measures which would be adopted to avoid harmful interference. Article 38 provides for steps to be taken if a member of the Commission is concerned about the environmental or other impact of prospecting, but this is a reactive approach and requires willingness to raise those concerns.

The powers of the Commission vis-a-vis the regulatory committees were progressively restricted during the negotiations and this undermined the strength of the environmental protection component of the Convention. Environmental protection interests may well have been better served if the Commission had been given relatively more functions and powers. In Beeby I and II, the Commission had responsibility for authorising permits for exploration and development, and approving or declining management schemes. However, these regulatory powers were transferred from the Commission to the regulatory committees during the negotiations. NGOs maintained that the complex regulatory committee membership formula, devised in response to political interests, could mean that '[n]ations in the regulatory committees will trade environmental protection for political and economic gain' (*Antarctic News*, #8, 1988, p. 1). They were worried that management schemes, which were to include specific terms and conditions for the protection of the environment and for timely response action, would be compromised by political tradeoffs in the regulatory committees. The regulatory process could also result in different rules being established for different management schemes in the same area. NGOs pointed to the difficulties of setting environmental rules for any specific area, even in accordance with Commission guidelines, when the ecosystem should be dealt with as a whole. A further potential weakness in environmental rules arose because once a management scheme had been established, and an exploration permit issued, progress to a development permit was almost automatic with few opportunities for independent reassessment of the operation or its environmental impact.[42]

The powers and responsibilities of the Advisory Committee, which is the locus for environmental and scientific expertise, were also limited during the negotiations to the extent that it was given an advisory role only with no power to initiate reports even on matters of concern. A stronger and more independent Advisory Committee would have been better placed to act as a balance to the politically-appointed regulatory committees. In Beeby IV, regulatory committees were required to refer any application for an exploration permit to the Advisory Committee for an assessment. If the Advisory Committee identified environmental risks the regulatory committee was *required* to reject the application, or to refer it to the Commission which could authorise the regulatory committee to proceed subject to such conditions as the Commission saw fit. Applications for development permits were also to be reviewed by the Advisory Committee in Beeby IV and the Advisory Committee could advise a regulatory committee to modify a management scheme if it deemed changes necessary. The regulatory committee was *required* to make those modifications. All these Advisory Committee functions were removed in Beeby V after the third intersessional meeting of the minerals negotiations. Beeby IV also provided that both the Advisory Committee and the regulatory committee would monitor compliance. In Beeby V this became the responsibility of the regulatory committee alone. The Advisory Committee's access to information was also increasingly limited in subsequent drafts. In Beeby V the Advisory Committee was to see only those parts of an application determined *by the regulatory committee* to be relevant. NGOs suggested, with some apparent justification, that Beeby V represented major concessions to mining interests (Anon., 1988b, p. 2).

Not only were the institutional relationships weakened from an environmental protection perspective but the rules on compliance and liability, which adhere to normal Treaty practice, are limited. In this regard alone, the ability of the Convention to protect the Antarctic environment, especially on a localised level, must be questioned. There is, as noted earlier, no provision for independent monitoring of minerals activity and environmental impact. There are few sanctions against non-compliance except after any non-compliance results in unacceptable environmental damage. While the initial decision to open an area is based on a precautionary and sufficiency of knowledge principle, subsequent action is most often reactive. For the most part, action can be taken only after environmental damage has occurred and only if any individual party is prepared to raise the issue.

As a compliance mechanism, inspection alone has proved an ineffective instrument to monitor or ensure compliance with environmental protection rules. There are no specific procedures in the Convention for action in instances of non-compliance identified by inspections. The Commission can consider regulatory committee monitoring reports but there is no specific guidance on what it may then do. Only parties to the Convention, not observers, may raise concerns about activities that affect the implementation of the Convention, including environmental impact issues. A management scheme or development permit can only be modified, suspended or cancelled, or a monetary penalty imposed, where activities have resulted in or are likely to result in environmental impacts *beyond* those already judged acceptable. The liability rules also opened the way for potential restrictions on liability for environmental damage to be established under a separate Protocol.

NGOs argued that there was not enough in the Convention to encourage transparency and accountability. They charged that the confidentiality of data rules established under the Convention with respect to information deemed commercially sensitive could prevent information necessary to make informed judgements on environmental impacts from being made public. 'Full availability of information', they argued, was 'vital to the success of all environmental protection measures ... openness is crucial to organisations outside the Convention wishing to comment on an application' (Anon., 1988c, p. 2). Thus restriction of knowledge was used by the parties to maintain power. There was in particular, NGOs argued, insufficient provision for adequate review of environmental impact statements and management schemes by observers and experienced NGOs. SCAR also supported early and timely release of data from minerals activity mainly because of the benefits to scientific knowledge (ATCM, 1989j, p. 190). As noted earlier in the chapter, the institutions of the Convention are required to maintain public records. But there are restrictions on this transparency. As Kimball notes (1988b, p. 19–20), incidents which potentially affect the implementation of the Convention are not automatically circulated to observer organisations. Neither are inspection reports. Discussions on monitoring and compliance in the regulatory committees are not open to observer organisations.

The environmental NGOs monitoring the minerals negotiations argued for changes to the Convention which they viewed as essential if its environmental protection mechanisms were to be improved (see ASOC, 1988e). They advocated a stronger role for the Advisory Committee and provision for the Commission to review regulatory committee decisions and argued that procedures were essential for a negative decision at the development permit stage. NGOs called for an independent Antarctic Environment Protection Agency which, if necessary, would be responsible for monitoring, enforcement and making judgements on environmental acceptability of any activity within a minerals regime. The criteria for environmental protection procedures should, they argued, be more clearly spelled out and better compliance, enforcement and liability rules were necessary. Finally, they demanded improved opportunities for public participation.

While NGOs argued that the Convention was a flawed environment protection agreement, some parts of the the mining industry styled it anti-development. The Director of the US Bureau of Mines commented that 'there is nothing in this Convention to encourage industry to expend any money to prospect, explore or develop resources. The Convention is anti-mineral development ... and will only serve as a deterrent ... [I]t is not a workable convention under which minerals will be developed' (Ary, [n.d.]). This view was shared, at least early on, by the American Petroleum Institute (API, [n.d.]) which referred to the 'excessively anti-development bent that currently permeates the draft regime'.

CONCLUSION

The negotiation of the Convention and the way in which the parties tried to manage a variety of competing interests points to the wider difficulty of reconciling resource use and environmental protection. It also points to potential problems when negotiation takes place without consideration of broader societal norms and values on environmental protection. International diplomacy is not, for the most part, a democratic process in that foreign policies and negotiating positions do not necessarily 'reflect the consensus or seek the consent of the majority of a country's citizens' (Kaiser, 1970, pp. 356–7) nor are community voices often involved in determining a state's foreign policy. The Convention rejects the non-development option. Wilson suggests (1986, p. 285) that this was because Antarctic policy is fashioned principally by civil servants in foreign ministries who are not directly answerable to the public. No public or political discussion which might have challenged conventional wisdom on the acceptability of minerals activity was sought. From an environmental perspective, the negotiation of the Convention points to a crucial dilemma. Is it possible for a regulatory mechanism (which is required if minerals activity is judged, in principle at least, to be an acceptable activity) to also function as a conservation and environmental protection regime? The IUCN argued (1987) that minerals activity 'will inevitably detract from the wilderness qualities of Antarctica'. The question is, then, how much detraction is acceptable? In the 1970s, minerals activity and environmental protection were thought compatible. In the 1980s, NGOs argued that this was an unacceptable premiss but the consultative parties, already eleven years into their discussions on the minerals question and now formally negotiating a convention, did not listen. They should have done so. The internal accommodation of the Minerals Convention did not hold, and it did not hold because the answer to the question of acceptability had changed.

7 Comprehensive Environmental Protection: from CRAMRA to Madrid

INTRODUCTION

The Minerals Convention was opened for signature on 25 November 1988 for one year by those consultative parties which had participated in the final negotiating session, plus any other state which had achieved consultative party status by that date. After that period any other signatory state to the Antarctic Treaty could accede to the Convention. The Final Act of the SCM provided that the voluntary moratorium on minerals activities, including prospecting, would continue pending the Convention's entry into force. The Convention would come into force when sixteen of the negotiating consultative parties had ratified it. That sixteen had to include all seven claimant states, the United States and the Soviet Union, and represent five developing and eleven developed countries. Ratification and entry into force seemed assured. The possibility that domestic support for the Convention might not be forthcoming – an involuntary defection in Putnam's term (1988, p. 438) – seemed unlikely. The Antarctic was not high on the agendas of parliaments or politicians. Nor had it been an issue of high public salience in spite of the activities of NGOs. Further, given the strong normative commitment to consensus within the Treaty system, the diplomats who negotiated the Minerals Convention had no reason to fear that any consultative government would *voluntarily* and unilaterally abandon the outcome of their deliberations. Thus the continued stability of the regime seemed assured.

They were mistaken. This agreement began to unravel quickly. The internal accommodation did not hold.[1] By May 1989 two key consultative parties, Australia and France, had withdrawn their support for the Convention. They sought to replace CRAMRA with a comprehensive environmental protection convention (which would include a ban on mining) and to have the Antarctic declared a wilderness reserve. The issue of environmental protection in the Antarctic took on a new importance as the consultative parties took sides on whether or not to ban minerals activity (an option they had rejected in the 1970s and the 1980s) and argued over how best to improve the complex and fragmented environmental sub-regime within the Treaty system. By October 1991 the consultative parties had negotiated a new, legally-binding agreement on environmental protection – the Madrid Protocol – which prohibited minerals activity and established comprehensive environmental principles and standards against which human activity in the Antarctic would be judged.

This change in the Antarctic regime was the proximate outcome of defection by key states and the construction by them of a coalition of support for new instruments. Continuing NGO activity and increased community interest were also key features of this process. When the legislation foreshadowing British ratification of the Convention was introduced into the House of Lords, Lord Alsa predicted that 'the conservationists would have the support of the vast majority of the public if the issue was put to them' (Parliament of the UK, 1989a, p. 934). In Britain, and in most other consultative states, the issue had *not* been put to public debate by Antarctic decision-makers. That task was taken up by non-governmental organisations and Lord Alsa's prediction was to prove correct, not just in Britain but elsewhere. Traditional Antarctic decision-makers – diplomats, lawyers and scientists – were marginalised by these events and their attempts to reclaim the issue were only partly successful. The debates and negotiations which followed the decisions by Australia and France to reject CRAMRA were characterised by a further and probably irreversible widening of participation in the regime.

Non-Governmental Organisations

After June 1988 and the adoption of CRAMRA, NGOs reassessed their strategy 'to try to prevent some of [the ATCPs] from signing and ratifying the Convention, or at least to severely delay them [to prevent] the Convention from coming into force' (ASOC, 1988l, p. 10). They targetted the key consultative states – the seven claimants and the US and USSR – as well as Italy and West Germany (see Greenpeace, 1988b). Non-ratification by any one of the claimants or the superpowers would result in CRAMRA not coming into force. In the case of Argentina and Chile, NGOs relied on arguments relating to loss of sovereignty. In both France and the UK, the campaign rested on environmental grounds although in both cases they recognised that the task would be difficult. In New Zealand and Australia, NGOs mobilised public awareness both of the Antarctic and of environmental concerns. They were also heartened by the fact that Australia had been the last state to hold out on signing the Final Act of the minerals negotiations and key governmental sources had indicated that the government was divided on whether to sign. In the US, NGOs hoped to use the lengthy congressional hearing process to delay ratification. The USSR, although it was likely to be difficult to influence, was concerned about the US gaining an upper-hand in minerals exploitation technology which the USSR could not match. With respect to Norway, the NGOs planned to use Prime Minister Brundtland's chairing of the World Commission on Environment and Development and its subsequent report as a point of leverage and linkage with Antarctic environmental issues. The Antarctic campaign was popular in West Germany where a strong public environmental awareness had translated into parliamentary representation. The Antarctic issue was also popular in Italy where the NGOs thought environmental arguments could be successful.

In support of their campaign for a ban on minerals activity and World Park status for the Antarctic, NGOs encouraged letter-writing campaigns and launched a number of successful international petitions against ratification to mobilise public awareness and put pressure on politicians.[2] They lobbied politicians, presented evidence to legislative committees and used the media to get their message to a wider audience. Information, advice and strategies were exchanged through the coordinating offices of ASOC and Greenpeace International. The anti-ratification campaign was actively supported by well-known individuals including French oceanographer Jacques Cousteau and British naturalist Sir Peter Scott. Cousteau lobbied politicians and state leaders, publicly and privately, in a way that ASOC representatives could not and attracted considerable media coverage in so doing.

Two events early in 1989 focused public attention on the environmental consequences of oil spills in polar regions. On 28 January, an Argentinian resupply vessel, the *Bahia Paraiso* went aground off the Antarctic Peninsula, spilling approximately 150 000 gallons of oil. US officials described the spill, which affected wildlife and scientific research at the nearby US Palmer station, as the first big environmental threat to the Antarctic (Darby, 1989).[3] On 24 March the *Exxon Valdez* went aground in Prince William Sound in Alaska, spilling 10.8 million gallons of oil. The world-wide media coverage of the devastating ecological consequences of the spill and the limited success of the lengthy and expensive clean-up attempt created a powerful visual and public image with which NGOs could challenge the adequacy of CRAMRA.[4] In this respect, the Antarctic regime and the environmental rules within it could not be insulated from other international environmental concerns.

CRACKS IN THE CONSENSUS

On 20 April 1989 the French government, one of the key ratifying states, called for the minerals negotiations to be reopened. Prime Minister Rocard argued that in the light of the *Exxon Valdez* accident, the environmental protection provisions in the Convention seemed inadequate (Reuters, 1989). He announced that France would 'not ratify the treaty [sic] in its present state' (AFP, 1989a). The future of CRAMRA was put further in doubt when, on 22 May, the Australian government announced that it was now committed to a position that no mining at all should take place in and around the continent and that it would therefore not sign the Minerals Convention. In a joint statement made by the Prime Minister and the Ministers for Foreign Affairs and the Environment, the Australian government argued that while

> the recently concluded Convention on the Regulation of Antarctic Mineral Resource Activities is very much better

> than no protective regime of any kind ... we believe that it is both desirable and possible to seek stronger protection for what remains the world's last great wilderness (Hawke et al. 1989, p. 1).

Australia, they announced, would pursue instead the 'urgent negotiation of a comprehensive environmental protection convention within the framework of the Antarctic Treaty system' and the 'prospects for the establishment of an Antarctic Wilderness Reserve'. They acknowledged that this was 'a significant change of approach in the management of Antarctica' and that achieving consensus was 'unlikely to be easy' (see Hawke et al, 1989). Australia's actions amounted to a voluntary defection from both the consensus on CRAMRA and from the fundamental norm in the Treaty system that a consensus, once reached, should be maintained. Whereas the French government had called only for the negotiations to be reopened, Australia had refused to sign or ratify the Convention and had also advocated that a different environmental policy path be followed. For that reason, much of the initial criticism from other consultative parties was directed at Australia. What is important, as chapter 1 suggested, is to identify the reason for such a change in policy which embodied, in some small degree, a normative commitment to a new priority of values within the Treaty regime and which challenged the orthodox views which underpinned that regime.

Both Australia and France cited environmental grounds for their opposition to the Minerals Convention and their support for wilderness status for Antarctica.[5] The Minerals Convention, the Australian Prime Minister suggested, was 'based on the clearly incorrect assumption – current in the 1970s – that mining in Antarctica could be consistent with the preservation of the continent's fragile environment' (Hawke, 1990a, p. 18). The most outspoken of their critics, particularly the US, UK and New Zealand, or more particularly, key figures in Antarctic policy-making in those states, sought actively to discredit this argument. They suggested that the real reason for Australia's decision, hidden behind the idealism of environmental arguments, could be found, first, in domestic and particularly electoral pressures and, second, in the reality of its sovereignty and economic interests.

New Zealand's Foreign Minister suggested that the Australian decision was 'dictated by political considerations rather than any fine feelings about Antarctica' (cited in Blay and Tsamenyi, 1990, p. 195), a reference to the forthcoming federal election and the Tasmanian state election early in May 1989 which gave Green Independents the balance of power in the state parliament. The government's decision was a response to community debate but it is simplistic and misleading to characterise it as a 'short term appeasement of the Green vote' (Fanshawe, 1990). Australia had not signed the Convention when it was opened for signature in November 1988. Instead, the Minister for Foreign Affairs, Senator Gareth Evans, tabled the Convention in Parliament for the express purpose of promoting community discussion. The tenor of that community discussion between November 1988, when the Convention was tabled in Parliament, and May 1989 indicated that public and parliamentary support for ratification was increasingly unlikely. There were already divisions within Cabinet and uncertainties about Australia's position on environmental as well as economic grounds even before the Convention was opened for signature and long before the Tasmanian election. That election was a minor factor which confirmed the new direction of Antarctic policy rather than informing it. As Blay and Tsamenyi suggest (1990, p. 198) 'the success of the Green Independents was ... an indicator of community concerns for the protection of the environment'.

There was also, as one foreign ministry officer has observed, 'a high level of public concern about Antarctica ... which had no precedent in earlier considerations in Australia of Antarctic issues' (Pascoe, 1990, p. 106). That public concern was mobilised by NGOs who conducted an intensive and well-organised campaign in seeking to convince the government not to sign or ratify the Convention. The Australian environment minister, Senator Graham Richardson, when presented with a plan for a World Park Antarctica by sixteen conservation groups early in May, conceded that public opinion in Australia was increasingly opposed to CRAMRA (Clark, 1989, p. 4). In explaining the Australian decision, Prime Minister Hawke referred to the

> changing world climate of opinion ... which is a function of the general concern and awareness of the responsibilities

> of governments in this generation to protect the planet generally and most particularly in regard to the Antarctic (cited in Bergin, 1991, p. 231).

Both the Federal Opposition and the Australian Democrats (who held the balance of power in the Senate) were opposed to ratification of CRAMRA. Labor government backbenchers also announced their opposition to mining in the Antarctic (Peake, 1989, p. 6). On 3 May 1989, Senator Evans advised the Senate, in response to a motion tabled the previous day by the Opposition which called on the government not to sign and ratify CRAMRA, that the 'Government is considering whether we should do just that' (Parliament of Australia, 1989, p. 1653). The Prime Minister's response to cabinet divisions, opposition in parliament and public concerns was to establish a committee to report to Cabinet on what steps would be necessary to designate the Antarctic as a World Park (Houweling, 1989, p. 2).[6] The Australian Treasurer, Paul Keating and the Minister for Resources, Senator Peter Cook, opposed Australian signature of CRAMRA in part because of their concerns that the absence of guaranteed royalty payments in the Minerals Convention could undermine Australia's sovereignty over its Antarctic territory and the absence of anti-subsidy provisions could result in unfair competition to Australia's mining industry.[7] Both interests had been important in defining Australia's position during the minerals negotiations. However, divisions within Cabinet arose over more than economic concerns. In a letter to the Prime Minister, Treasurer Keating suggested that 'Australia ... is in an excellent position to take a leading role in ensuring that the last wilderness is protected for future generations' (Keating, 1989, p. 2).[8]

New Zealand Prime Minister, David Lange's characterisation of the decision as a 'road-to-Damascus-like experience that tends to put Saul rather in the shade' (cited in Clifton, 1989, p. 2) ignores Australia's previous commitment to environmental protection in the Antarctic. Protection of the environment had long been one of Australia's Antarctic policy goals. However, as a senior Australian foreign ministry official has observed, 'those goals [were] set in another era ... Little if any critical consideration [was] given at the highest level of government to whether the individual goals [were] compatible with each other' (Brown,

1990a, p. 8). The Australian decision on the Minerals Convention represented a reassessment of those goals and gave, as Harris argues (1990, p. 17), 'explicit priority to environmental protection'.

Australia's support for an environmental protection agreement was already well established in Treaty meetings. Australia first tabled a proposal for an umbrella environmental protection agreement in 1983, an idea which 'proved too radical' for its Treaty partners (Rowland, 1988, p. 22).[9] At the preparatory meeting for the fifteenth consultative meeting, *before* the government's decision on the Convention, Australia argued that it saw 'merit in a catch-all convention ... subsuming over time existing instruments' (ATCM, 1989b, p. 1). Australia described the present system on environmental protection as 'piecemeal ... with many significant gaps'. The paper also referred to the enormous environmental consequences that could arise from mineral resource activities '*[i]f* CRAMRA enters into force...' (ATCM, 1989b, p. 3; emphasis added) thus indicating some doubt on the future of the minerals convention. Australia had also taken a lead in advocating improved rules on the Code of Conduct for expeditions and the protected area system and in pushing for better conservation measures to be adopted under CCAMLR. Throughout the minerals negotiations Australia had been one of the consultative parties arguing for strict environmental protection measures.

Australian support for wilderness reserve status (or something akin to a World Park) for Antarctica was new. Nevertheless, there is some evidence that Australian policy-makers were not as stridently opposed to this concept as were others. In the mid-1980s the Department of Foreign Affairs took the view that there were

> a number of problems in the proposal to declare Antarctica to be a World Park, not the least of which is uncertainty about the precise content and implications of that concept ... The Australian authorities therefore believe that it would be *premature* to raise any World Park concept in the Antarctic Treaty forum until there is an agreed, precise and feasible meaning of that concept which recognises the practical realities of Antarctica (Anon, 1984a, p. 97; emphasis added).

In other words, the possibility was not totally ruled out. In a speech reported in the Department of Foreign Affairs journal 18 months later, the Minister for Science suggested that it would be 'premature to *implement*' such a proposal until there was an agreed meaning (Jones, 1985, p. 841; emphasis added).[10]

The French position is more difficult to explain. Indeed, its commitment to environmental principles in the Antarctic was, with some justification, greeted with greater scepticism. Rowland (1988, p. 28) identifies France as one of the least conservationist of the Treaty parties. As a potential mining state it stood to benefit if CRAMRA entered into force and would have been less adversely affected (as a claimant) than Australia by the lack of specific provisions on royalties and subsidies. Its April announcement calling for the negotiations to be reopened can be located firmly in domestic political explanations rather than also building on a solid foundation of environmental concerns. In France, those domestic politics indicated a growing concern for environmental protection issues. Recent municipal elections had returned a 15 per cent green vote, and polls showed that the green vote in the forthcoming European Parliament elections was likely to be even higher (in the vicinity of 17 per cent) (Scott, 1989). NGOs conducted an active anti-ratification campaign with the support of Jacques Cousteau who was particularly influential in lobbying the French government, building on his close rapport with government leaders.[11]

Nevertheless there was something of a normative concern underpinning the French decision. The new Socialist government in France included members with strong environmental credentials. Environment Minister Brice Lalonde was a founder of Les Amis de la Terre, the French Friends of the Earth. The National Assembly also took a much more active role in the debate than previously. Indeed, the Parliamentary Committee which examined France's options on CRAMRA complained that the National Assembly had not been given a proper role in the development of Minerals Convention. The President of the National Assembly, M. Fabius, had announced early in April that France should not ratify CRAMRA (Iggulden, 1990, p. 103). The Committee supported the exploration of other options besides the minerals convention and recommended legislation to prohibit French nationals from participating in minerals activity (Anon., 1989i, pp. 2; 4).

While the French announcement had called for a reopening of the negotiations, the Australia announcement would, if carried through, mean the demise of the Minerals Convention. The Australian decision took everyone, including environmental organisations who had campaigned against ratification, by surprise. NGOs were overjoyed. The Director of the Australian Conservation Foundation called it a 'rare and visionary' action (Anon, 1989h, p. 1). Sir Peter Scott, son of the Antarctic explorer, told the Australian Prime Minister that

> Australia's magnificent stance ... may well at one brilliant stroke have ensured the future of the last virtually pristine continent on earth. This wonderful initiative gives us all new hope for the future of the planet (cited in Phillips, 1989, p. 8).[12]

Response to the announcement from other consultative governments was not so favourable. The New Zealand Minister of Foreign Affairs, Russell Marshall, argued that 'Australia will have a good deal to answer for if it fails' (cited in Anon, 1989g, p. 20). Hostility to the Australian decision and, to a lesser extent, the French was grounded in anger that both states had defected from the consensus rule which was so important to the Antarctic regime. Australian diplomat John Burgess (1990, p. 2) acknowledged that the concern which greeted the Australian and French announcements was as much about the *way* of doing things as the substance of the announcements. In particular, they had reneged on an agreement that had taken over six difficult years to negotiate. The British Parliament was told by the Parliamentary Under-secretary for Foreign and Commonwealth Affairs, Tim Eggar, that

> the Antarctic treaty has succeeded ... because ... no member has pushed its interest beyond the point where it can be tolerated by others. The treaty system has also depended on an implicit assumption that once an agreement has been reached by consensus, all parties to that consensus will confirm and implement the agreement. The Australian move has gone against both of those tenets... (Parliament of the UK, 1989b, pp. 217–18).

Australia (and France) had threatened the fundamental norms and practices of the Treaty system. For this reason, Eggar argued,

> Australia's Antarctic partners have a right to ask for its reasons for inflicting damage on the consensus machinery of the Antarctic treaty system. They have a right to ask why it has taken a path that could result in a collapse of that treaty (Parliament of the UK, 1989b, p. 218).

BUILDING A COALITION

Debates about the development of a comprehensive approach to environmental protection in the Antarctic had, for the most part, taken place outside the institutional boundaries of the Antarctic regime and the parties had been able to pay them limited attention. The proposals advocated by Australia and France meant that those challenges and debates were now internal to the regime and could not be ignored. While some hoped, indeed assumed, that Australia and France would back down in the face of stringent and apparently unshakeable opposition from other consultative parties, this was not to be the case. Australia and France now sought to have their initiative adopted as an institutional goal within the regime. However, the Australian and French defection from consensus was directed at a particular outcome – the Minerals Convention. It was not a rejection of the consensus norm nor of the value of the Treaty system. Prime Minister Hawke emphasised that his government was 'not challenging the Treaty System or the operation of the consensus principle that has underpinned its operation' (Hawke, 1990a, p. 19).[13] Both Australia and France stressed their continuing support for the Treaty system as fundamental to their interests in the Antarctic. They sought to find a new consensus among the Treaty parties on a ban on minerals activity and on strengthening environmental protection measures. However, as Blay and Tsamenyi have observed (1990, p. 201)

> [i]f CRAMRA is dead because of the absence of consensus, Australia can hardly expect to gain the consensus it will need

> to negotiate, let alone implement [their] ... proposed Convention ... It is a situation that could destroy the consensus approach to negotiations.

That Australian initiative sought to change the traditional ad hoc and sectoral approach to environmental decision-making in the Treaty system. In placing environmental concerns before economic ones and, indeed, before the values which supported the political stability of the Treaty system, Australia, soon to be joined formally by France, was advocating a reordering of the hierarchy of values in the regime. In a consensus-based regime such as the Antarctic Treaty system, that task could only be achieved through persuasion and bargaining. That process of regime change rests on the construction of coalitions of like-minded states through networks of influence rather than power hierarchies. It also involves isolating potential coalitions of states who could block a new consensus. Australia and France were able to translate their issue-specific power resources into bargaining leverage in support of their policy proposals. That is, they acted as structural as well as intellectual leaders. In Australia's case, the new initiative was taken up as part of the government's explicitly 'high activist, high profile foreign policy' (Evans and Grant, 1991, p. 312) which, since the late 1980s, had emphasised middle power leadership and good international citizenship.[14]

The first round of battles over the minerals convention and environmental protection had taken place behind closed doors, during the minerals negotiations, where the transgovernmental elite was most powerful. The second round was a more public debate. Traditional Antarctic decision-makers – lawyers, diplomats and scientists – were marginalised when they either ignored or failed to respond to critical voices and changing public opinion. NGOs were influential in this process. They mobilised public opinion and sought to 'present [their] case to the highest levels of government' (ASOC, 1989j). In that way, the process of change was characterised by a limited and enforced democratisation of the regime. Politicians in a number of Treaty states, including those most opposed to Australia and France, responded to the pressures of public campaigns and NGO lobbying, and urged a reassessment of national policy upon Antarctic bureaucrats. The Antarctic bureaucrats, who had dominated decision-making at a national and international level for so long, were reluctant to respond. Their resistance stemmed from their 'ownership' of the Antarctic issue and its decision-making processes and their continued belief that the regime could not function in any other way. Debates over the Minerals Convention involved a competition between knowledge-based actors and knowledges, which would determine the future shape of the regime and particularly of environmental protection in the Antarctic.

The French government now moved to align itself more closely with Australia. In June, French President Mitterand announced that he was 'very taken with the idea of turning [Antarctica] into a vast peaceful nature reserve' (cited in Ambassade de France, 1989b).[15] He said that he would ask the French Government to study a report he had commissioned from Jacques Cousteau to see 'whether along with countries who share our concern, this idea can be put into effect without delay' (Ambassade de France, 1989a). This alliance was further strengthened in a joint statement on environmental concerns issued by the French and Australian Prime Ministers in August 1989 when they announced that

> In the light of their shared concerns about the Antarctic Minerals Convention and other factors affecting the environment in the Antarctic [they had] agreed on a joint initiative to promote the protection of the environment in the Antarctic ... [M]ining in Antarctica is not compatible with protection of the fragile Antarctic environment ... the two countries will be proposing that the Treaty Parties negotiate a Comprehensive Environment Protection Convention which will turn the Antarctic into a wilderness reserve ... (Hawke and Rocard, 1989, p. 1).[16]

The Australian/French Initiative

There were three parts to the Australian and French initiative: first, there should be a ban on minerals activity; second, the Antarctic should be declared a wilderness reserve (although just what that meant was not yet spelled out); and third, a comprehensive approach to environmental protection could only be achieved in a wide-ranging new convention. There was strong opposition to all three components from other Treaty parties.

In advocating a ban on minerals activity, Australia and France effectively sought to replace CRAMRA which was based on the principle that minerals activity could be considered acceptable. The US and the UK were particularly opposed to a ban. While hastening to say that no-one was proposing to go to the Antarctic to mine in the short or even medium-term they were reluctant to rule out the possibility that they might want to do so at some stage in the future. The US argued that 'a total ban on mineral resource activity is not negotiable. The US will not agree to such action' (Scully, 1989). Supporters of the Minerals Convention argued that because a consensus was required before any area could be opened for minerals activity, Australia or France could enforce an effective prohibition on minerals activity using their power of veto. Australia responded that it was 'logical nonsense' to sign a convention which regulated and therefore accepted minerals activity, intending to use a veto to oppose mining permanently (Hope, 1989). To do so, the government suggested, would be an act of bad faith. Further, the effective veto applied only to the identification of an area for possible exploration and development: it did not apply to specific activities and neither did it cover prospecting. The veto could only be used if the proposed activities were inconsistent with the Convention. Given that the Minerals Convention accepts some environmental impact, *any* harm is therefore not 'significant' harm so it is doubtful whether Australia or France could use the argument of any environmental impact as a reason for exercising their veto. It is also debatable whether, as claimants, either Australia or France would have wanted to use their veto over activity in the territory of another claimant. Australia also asked why, if the other Treaty parties were prepared to advocate a de facto ban under the Minerals Convention through the use of the veto, they were not prepared to accept a ban in specific terms (Brown, 1990a, p. 7).

Opponents warned Australia and France that their new initiative could lead to the very situation they wished to avoid if a ban crumbled in the face of economic pressure to mine, leaving no environmental protection rules in place to regulate minerals activity. A further argument suggested that if

CRAMRA's entry into force was unlikely, the voluntary moratorium on minerals activity, which had been confirmed by the Final Act of the SCM, would become null and void. Eggar told the House of Commons that any state which did not ratify the Convention 'would, by its actions have destroyed the existing voluntary moratorium ... [and] would bear a heavy responsibility for the consequences of its decision' (Parliament of the UK, 1989b, p. 216). Australia and France argued that a ban on minerals activity under a comprehensive convention should not be inherently more unstable than the successful bans on military activity, nuclear explosions or disposal of radioactive waste under the Antarctic Treaty.

The Australian/French proposal for a wilderness reserve was dismissed by its critics as little different from the World Park proposal advocated by non-governmental organisations. This proposal dates to the 1972 Second World Conference on National Parks which called, at that stage, for the Antarctic to be designated a World Park under the auspices of the United Nations. The idea was canvassed in a series of international congresses over the following two decades including the Third World Parks Congress in 1982, UNEP's Session of Special Character in the same year and the Fourth World Wilderness Congress in 1987. In 1989, the UN General Assembly adopted a resolution calling, for the first time in that forum, for the protection of Antarctica as a nature reserve or World Park and urging a ban on prospecting and mining. The preservation of Antarctic environmental values was also debated at IUCN general assemblies in the 1980s. While the recommendations adopted here did not specifically call for the designation of a World Park, they did advocate full protection of the Antarctic environment.[17]

ASOC refined the World Park concept as an integral component of their campaign for a comprehensive approach to environmental protection in the Antarctic. They envisaged that the Antarctic, as a World Park, would remain a zone of peace and scientific investigation and cooperation. Wilderness values would be paramount. Environmentally damaging activities, including minerals exploitation, would be prohibited. Core elements of the NGO proposal included mandatory environmental impact assessment, the establishment of an Antarctic Environment Protection Agency, legally binding compliance mechanisms and participation by public interest groups.[18] The consultative parties had dismissed this proposal as the first unwelcome step in the internationalisation of the Antarctic and the erosion of their authority over Antarctic decision-making.[19] This proposal embodied values and practices that were at odds with those dominant within the Antarctic regime. It was not surprising, then, that the consultative parties argued that it was politically unrealistic and ill-defined and that it would restrict (or even prohibit) scientific research. The Australian/French proposal was opposed on similar grounds.

It was not clear just what Australia and France had in mind for defining a wilderness reserve. To some extent it was a piece of political rhetoric designed to reinforce the idea that the Antarctic would not be used for resource exploitation and that protection of the environment would be paramount. They later used the phrase 'nature reserve – land of science' to indicate, first, that scientific activity would not be proscribed and, second, to provide some political distance from the NGO concept of a World Park in order to make the proposal more acceptable to their Treaty partners.

The Australian and French initiative also demanded a new approach to environmental decision-making within the Treaty system. Australian diplomats argued that the 'incremental processes which [had] so far characterised the work of the Treaty [were] inadequate to meet current needs' of environmental protection in the Antarctic (Brown, 1990a, p. 12). They sought to replace the unwieldy collection of voluntary rules which were often poorly implemented with a legally binding agreement which would contain environmental principles, a set of clear and rigorous standards against which all activity in the Antarctic would be judged, compliance and monitoring mechanisms and an institutional framework which would include a new Antarctic Environment Commission.

The consultative parties had agreed at a preparatory meeting in early May that the issue of comprehensive environmental protection measures would be on the agenda for ATCM-XV which was scheduled for October 1989. While there was some agreement that environmental protection measures should be improved, there was no agreement on how that should be done and very little support for the approach now being advocated by Australia and France. Few consultative parties were convinced that the existing system of rules should be replaced, rather than simply upgraded. There was strong opposition to any independent decision-making institution with responsibility for monitoring and compliance. The British also argued that too strict rules might unacceptably restrict the pursuit of scientific knowledge.[20] The suggestion that a comprehensive convention would also incorporate a ban on minerals activity meant, in the context of the rejection of CRAMRA, that it was received with more hostility than might otherwise have been the case. It was characterised as a politically expedient alternative to the Minerals Convention, rather than an initiative on improving environmental protection its own right.

Support for Australia and France continued to grow prior to the fifteenth Consultative Meeting. India had announced support for the Australian initiative in June (Murdoch, 1989, p. 6). In the same month, the Belgian Parliament passed legislation prohibiting Belgian nationals and companies from undertaking any minerals activity in the Antarctic on the grounds, the Foreign Affairs Minister Mark Eyskens advised, that 'the ... Convention is dangerous and it is inappropriate for us to approve it. The best solution is to make Antarctica a world nature reserve' (Pallemaerts, 1989). In Britain, bipartisan support for the Antarctic Minerals Bill, introduced first into the House of Lords, broke down in the House of Commons when the Opposition Labour Party moved that the House decline the second reading of the bill (Parliament of the UK, 1989b, p. 220). While the motion was unsuccessful, it did show that the legislation and the Convention would now be subject to parliamentary debate and scrutiny. In September, the Chamber of Deputies of the Italian Parliament passed a resolution opposing the Minerals Convention and supporting world nature park status for Antarctica. The Under-Secretary for Foreign Affairs told Parliament that the Italian Government would support the World Park option at the forthcoming consultative meeting. In the US, Senator Al Gore introduced a resolution (co-sponsored by Senator Claiborne Pell, Chair of the Senate Foreign Relations Committee) into Congress on 26 September calling for full protection for the Antarctic and a ban on minerals activities. A similar resolution (HJR 418) opposing the minerals treaty and calling for strong environmental protection of the Antarctic was introduced into the House of Representatives by

Congressman Owens. In the light of this, the US Environmental Protection Agency and the Council on Environmental Quality called for US Antarctic policy to be reviewed before the Minerals Convention was submitted to the Senate for advice and consent on ratification. The State Department, however, maintained its support for the Convention in the belief that Australia would back down.

There was also further public support from within the non-governmental community. On 9 October, WWF-International came out in support of the Australian/French initiative. At the 1989 Annual General Meeting of Friends of the Earth, held in Washington DC, the 38 member groups unanimously agreed to a resolution calling on the Treaty parties not to ratify CRAMRA. FOE's position advocated the negotiation of a comprehensive Environmental Protection Convention for Antarctica and a permanent ban on minerals activity which together would maintain the continent as a wilderness reserve or Antarctic Treaty Park (Anon., 1989j, p. 4).

ATCM-XV, 9–20 October 1989

Four-and-a-half months after the Australian announcement, the consultative parties met in Paris for the 15th consultative meeting. By this time fifteen Treaty parties, including the other seven key states, had signed the Minerals Convention although none had yet ratified it.[21] Opening statements were, on the whole, restrained. Some referred to the problems of a lack of consensus or stated their support for the Minerals Convention, but none openly censured Australia or France. Nevertheless, the head of the West German delegation summed up the feelings of many in noting that they

> share[d] the concern expressed over the fact that doubts have now been raised from among our ranks … [we] believe that a divided stand of the Antarctic Treaty States on the Minerals Convention will lead to risks with implications well beyond our circle (ATCM, 1989j, p. 137).[22]

The agenda was dominated by environmental issues. As well as comprehensive protection measures, delegates were scheduled to discuss waste disposal, marine pollution, environmental impact procedures, the protected area system, the siting of stations, the effects of tourism and non-governmental expeditions and environmental monitoring. Thirteen of the twenty-two recommendations adopted at the consultative meeting addressed these issues and the meeting also approved a declaration on the ozone layer and climate change.

Australia and France, Chile, New Zealand, the US and Sweden submitted proposals on the issue of improved protection measures. In a joint paper, Australia and France called, as expected, for a new convention on comprehensive environmental protection but did not table a draft convention nor mention a ban on minerals activity (see ATCM, 1989e). They did set out suggestions on appropriate principles and procedural and institutional requirements, including an environmental commission with decision-making powers, a scientific and technical committee, an arbitration body and an inspection and monitoring corps. The New Zealand paper was the most detailed of those tabled and drew attention to 'significant gaps in the system' noting that 'implementation of some of the instruments previously adopted has not always been up to the required or desired standards' (ATCM, 1989j, p. 214). It proposed monitoring and compliance mechanisms and an improved institutional framework which would build on existing components of the Treaty system.

Australia and France received support from a number of Treaty parties, including Belgium, Italy, India, Poland and Spain (all consultative parties) as well as Austria, Bulgaria and Greece. None of these were key Treaty powers although India had some sway as a developing country within the Treaty system. With the exception of Belgium and Poland, all were recent signatories. However, that may well indicate that the newer parties were not prepared to be led on the question of minerals activity or environmental protection by the more established powers such as the US, Britain, New Zealand, Argentina or Chile whose diplomats were particularly influential in the regime. Indeed, they may have harboured some resentment at this predominant role and the way in which the interests of some of those parties had dominated the minerals negotiations.

The consultative parties agreed to convene a special consultative meeting late in 1990 to 'explore and discuss all proposals relating to the comprehensive protection of the Antarctic environment' (recommendation XV-1, ATCM, 1989j, pp. 39–42). This purpose of this meeting would be to 'undertake as a priority objective the further elaboration, maintenance and effective implementation of a comprehensive system for the protection of the Antarctic environment'. This would involve a review of existing measures to consider if and to what extent new institutional arrangements were required. The recommendation made no concession to the Australian/French initiative but neither did it reject it. At the very least it kept the issue alive. Nevertheless agreement to convene a special meeting was reached only on the understanding that there would also be a meeting to discuss proposals for a liability protocol under the Minerals Convention indicating again that some parties were still committed to CRAMRA.

As mentioned in earlier chapters, the environmental regime within the Antarctic Treaty system was further developed at the fifteenth consultative meeting. Delegates adopted several recommendations on other environmental issues which strengthened a number of the rules that were to be reviewed at the special consultative meeting, possibly in the hope that such a review would find no need for a new and comprehensive convention. A detailed recommendation on waste disposal included quite specific and, in comparison with earlier recommendations (see chapter 3), rigorous guidelines on acceptable practices. Governments had to establish rules at least as stringent as those applying to their nationals and vessels outside the Antarctic and were required to ensure compliance with those rules by *all* those subject to their jurisdiction, including private operations. The new rules emphasised the need to retrograde wastes – that is, return to country of origin – and encouraged the use of new technology. The final report included a detailed statement of further work to be undertaken by the consultative parties with respect to improving waste disposal practices (ATCM, 1989j, pp. 17–19).[23] A detailed recommendation on marine pollution (XV-4), the first to be adopted on this issue, required governments to take measures to prohibit discharges from vessels and to ensure that vessels complied with the relevant provisions of specified international conventions.[24] There was lengthy debate on how best to implement the EIA procedures adopted in 1987. The general view was that these procedures should apply to all activities, including tourism and non-governmental expeditions, but there was no agreement on how to extend those obligations beyond official expeditions. Thus the rules existed but were not yet being put into practice. The debate on tourism was, as usual, unresolved and the parties decided that this was a matter that could be addressed by the special consultative meeting.

The protected area system was modified considerably to improve its coverage (recommendations XV 8–11, ATCM, 1989j, pp. 69–77). Management plans were now to be required for Specially Protected Areas and two new categories of protected area were adopted. Specially Reserved Areas were intended to protect values not already covered by existing criteria including scenic, aesthetic and wilderness values. Multiple-Use Planning Areas were designed to minimise the risk of activities interfering with each other and of cumulative impacts by encouraging cooperative planning and coordination of activities in areas so designated. Although the categories were established, no specific designations were made.

From Paris to Madrid

In the months following the consultative meeting, Australia and France continued to lobby for support for their proposals as part of their coalition-building strategy designed to effect change within the regime. The prime ministers of both countries took up the issue as a personal crusade. The Australian Prime Minister and his Cabinet colleagues lobbied hard on this issue.[25] Non-governmental organisations also maintained their pressure on governments opposed to Australia and France. In Britain, where the Foreign Office was still adamantly opposed to a minerals ban and any new instrument for environmental protection, Greenpeace and the WorldWide Fund for Nature mobilised a public campaign in a somewhat unusual coalition with the National Federation of Women's Institutes. In Germany, a petition circulated by members of the WorldWide Fund for Nature demanding full protection for the Antarctic and a permanent ban on minerals activity collected over 300 000 signatures (Anon., 1990e, p. 2). US NGOs were active in lobbying government agencies, congressional leaders and the President in support of the anti-ratification campaign.[26] The US-based Antarctica Project spearheaded a collective endeavour by ASOC members to research and draft a model Antarctic protection convention. French oceanographer and environmentalist Jacques Cousteau met privately with New Zealand Prime Minister Palmer in February 1990 and with US President Bush in October. Australian NGOs put their efforts into supporting colleagues in other countries. Through these NGO networks, which were less formal and more speedy than diplomatic communications, they were also able to provide information to the Australian and French governments on behind-the-scenes manoeuvres in other countries.

There was also growing opposition to the Minerals Convention from within the scientific community, primarily from scientists who were not also part of the Treaty system's decision-making network. For example, eighty-four per cent of Antarctic scientists who responded to a poll in Germany opposed ratification of CRAMRA and favoured a World Park for Antarctica (Anon., 1989k, p. 4). In the UK almost 250 scientists, including some with the British Antarctic Survey, signed a Greenpeace petition urging support for the Australian/French initiative (Anon., 1990i, p. 4).[27] New Zealand scientists also expressed their support for a ban on minerals activity and protected status for the Antarctic. Others, particularly those in executive positions within SCAR or with close ties to the diplomatic processes of the Treaty system, adopted a similar line to the diplomats and bureaucrats who argued against Australia and France. They were also strident in their condemnation of NGOs. The then President of SCAR, Richard Laws (1991b, p. 231) accused ASOC of 'engineer[ing] the wrecking of CRAMRA'. He referred to them elsewhere as 'vociferous [and] well-financed' (Laws, 1991a, p. 4).

When the year-long signature period for the Minerals Convention passed in November 1989, just after the 15th consultative meeting, only nineteen Treaty states had signed. With the exception of the UK, none of the signatory states had begun ratification processes. Significantly West Germany, which had been one of the strongest pro-mining nations during the negotiations, allowed the signature date to pass without making a decision. Two committees of the Bundestag had before them for consideration World Park motions put down by the Greens Party. It was likely that all opposition parties would support a ban on mining and the negotiation of a comprehensive convention.

There were increasingly strong indications of support for Australia and France from within the Treaty system although the US and Britain showed no signs of policy change. On 19 January 1990, President Gorbachev told a World Forum on Environment and Development in Moscow that the Soviet Union was 'ready to participate in the survival of the Antarctic, of this world reserve, which is our common natural laboratory' (Anon, 1990b, p. 4). 'Our children', he said, 'will never forgive us for not having protected this extraordinary ecosystem' (Barnes, 1990d, p. 8). Soviet Prime Minister, Nikolai Ryzhkov told a parliamentary dinner during his visit to Australia in February that the Soviet Union was 'prepared to collaborate with Australia and other countries in implementing ... initiatives that cover the survival of the Antarctic' (Ryzhkov, 1990). The Soviets had long favoured a moratorium on mining in the Antarctic in part because of a lack of funds and polar minerals technology vis-à-vis the US. They were also keen to establish their environmental credentials in the post-cold war international order. The Spanish Minister of Education and Science told that country's Senate in February that the Spanish government 'cannot help but view with sympathy ... the declaration of Antarctica as a World Park' (Spain, 1990). Later in the year, on 20 September, the Senate unanimously adopted a motion which had been tabled in July by the Parliamentary Socialist Group, that the government should support measures which would give the greatest possible protection to the Antarctic, and should not adhere to CRAMRA but support Antarctica becoming a nature reserve and land of science.

Australia and France received a significant boost in support when, on 26 February 1990, New Zealand Prime Minister Geoffrey Palmer announced that his government would set aside the ratification of CRAMRA to explore the possibility of a 'long-term legally binding moratorium on mining activity in Antarctica' (Palmer, 1990, p. 2). When pressed the following day by the Leader of the Opposition on whether that meant that the government might reinstate the Minerals Convention as its preferred option if it did not make progress in some other form, Palmer replied 'No, it does not' (Parliament of New Zealand, 1990a, p. 147). On 6 July Palmer announced that his government would promote a new protocol to the Antarctic Treaty and a *moratorium* on mining and on 4 August he told a conference of environment and conservation organisations that New Zealand would now support a *permanent* ban on mining. New Zealand's decision was important. As a claimant state its ratification was required to effect entry into force and it had also invested considerable diplomatic energy in the negotiation of CRAMRA through Christopher Beeby.

Three of the nine key states had now put the Convention aside. It was a further blow to the US and UK arguments that Australia and France were isolated within the Treaty system. Indeed, Australia and France were increasingly successful in isolating this potential blocking coalition. It was an important victory for Australia and France, and for NGOs who had conducted an active anti-Convention campaign in New Zealand. In the US, State Department support for CRAMRA was being further challenged in Congress amid a growing concern over the administration's unwillingness to consider seriously any alternatives to the Minerals Convention. On 9 November 1989 senior members of the House of Representatives Committee on Interior and Insular Affairs wrote to President Bush asking the President to 'direct the Department of State to consider Australia's proposal for an Antarctic Environment Protection Convention in lieu of the CRAMRA' (Udall et al., 1989, p. 2).[28] Between February and May 1990, five legislative initiatives on the Antarctic were introduced into the House of Representatives and the Senate.[29] These draft bills sought, first, to prevent US nationals and corporations from participating in Antarctic minerals activity and second, to require the Bush administration to enter negotiations with other consultative parties for an international agreement banning all minerals activities in the Antarctic. They also advocated World Park status for the Antarctic.

The Congressional resolutions introduced earlier in the year by Senator Gore and Congressman Owens were adopted in October 1990. In a further blow to State Department support for the Minerals Convention, legislation based on the Conte and Kerry bills passed Congress and was signed into law as the Antarctic Protection Act by President Bush on 16 November. US citizens or companies were prohibited from engaging in, financing or providing assistance to any Antarctic minerals activities, pending Congressional approval of a new agreement to ban minerals activities and protect the Antarctic environment (Anon., 1990f, p. 1). Any treaty or agreement on the Antarctic submitted to the Senate for advice and consent had to be consistent with this legislation. CRAMRA clearly was not. The Environmental Protection Agency asked the State Department to change instructions to the US delegation to the forthcoming special consultative meeting to reflect the substance of the new legislation. The State Department indicated only that it was 'prepared to consider means of addressing the concerns which have been raised over the Convention' (Bohlen, 1990, p. 11) but would make no further commitment.

The growth in support for the Australian/French initiative was also influenced by debates and events outside the Treaty states. At its 18th General Assembly, held in Australia in early December 1990, the IUCN called for the consultative parties to work towards the adoption of a comprehensive and legally binding environmental protection regime for Antarctica which would incorporate a 'permanent exclusion of mineral prospecting, exploration and exploitation' (IUCN, 1990, p. 2).[30] This resolution, which gave further effective support to the Australian/French initiative, was important. A number of consultative parties were state members of IUCN. It was perceived by them as a legitimate and credible conservation organisation which represented international opinion and it had been given observer status at various institutions within the Treaty system. The IUCN had worked closely with SCAR and the consultative parties on Antarctic conservation issues.

The UN General Assembly passed a resolution, sponsored again by Malaysia, which welcomed the increased support for the establishment of a World Park and for the banning of minerals activities in the Antarctic. It also, predictably, demanded that negotiations for a comprehensive convention should be done with the full participation of the international community. The UN resolution was a further signal that the protection of the Antarctic environment, and consultative party stewardship of that environment, would be the new focus for debate and challenge to the Treaty system. While the Treaty parties did not respond directly to the UN resolution, these debates and the Malaysian intention to raise the issue at the forthcoming UN Conference on Environment and Development, provided some incentive for the consultative parties to be seen to be dealing with the internal dispute about how best to protect the Antarctic environment and to be resolving it in a way that acknowledged broader environmental concerns.

THE MAKING OF THE MADRID PROTOCOL: SPECIAL CONSULTATIVE MEETING XI

The special consultative meeting on comprehensive environmental protection was the first time in the thirty-year history of the Antarctic regime that the consultative parties had met in a session devoted exclusively to environmental issues and it marked a redirecting of the regime. Four sessions were held between November 1990 and October 1991. Over this period of less than a year the consultative parties moved from division to consensus on a new environmental agreement, the Madrid Protocol, which included an effective prohibition on minerals activity. This institutional bargaining process was somewhat different from those on earlier issues. The key issues were known, positions were clear, and there was a body of informal legislation through existing recommendations to structure the negotiations.

By the time of the first session, Australia and France had been joined by Belgium and Italy in a formal coalition known as the Group of Four. By the second session at least eight 'like-minded' countries had been identified – India, New Zealand, Finland, Denmark, Greece, Austria, Romania, and Ecuador. This kind of formal coalition activity, which identified a pro-environmental caucus within the Treaty system, was new although the claimants had caucused informally during the minerals negotiations. It demonstrated that Australia and France were no longer isolated and it also provided NGOs with a weapon in their public campaigns in states still opposed to the prospect of a ban on minerals and a new environmental protection agreement. Formal NGO participation in this special consultative meeting was also more extensive than it had been in the past. The Antarctic and Southern Ocean Coalition was granted official observer status and sent three representatives to act in this capacity.[31] NGO representatives were included on several national delegations – the usual ones of Australia, US, Denmark and New Zealand as well as those of Chile and France. Another twenty people from ten countries worked as lobbyists outside the meeting. In contrast, SCAR's involvement was limited. SCAR was present as an observer but it was not asked to provide advice on the issues under discussion, a factor which exacerbated some of the tensions between scientific and NGO communities.

The first session was held in Viña del Mar, Chile, from 19 November to 6 December 1990. Debate on the form and content of comprehensive measures focused on four proposals. In March 1990, Australia and France had circulated a revised proposal for a possible instrument on environmental protection. Just before the SCM, the Four circulated a draft convention. New Zealand tabled a comprehensive draft protocol to the Treaty. An outline proposal for a protocol was tabled jointly by Argentina, Norway, the UK, the US and Uruguay, and was supplemented by a draft UK protocol and a draft US protocol and annexes. India also tabled an outline working paper. There was still some resistance to the Group of Four proposals. The first area of dispute was the *form* of any comprehensive measures. There were differences over whether a new agreement was needed or whether existing measures could be updated and expanded. The second area of contention was the *content* of any comprehensive measures. There was no agreement on whether a ban on minerals activity should be adopted and, if so, in what form – a moratorium (implying that minerals activity might occur at some stage) or a prohibition. The UK and the US had not abandoned their support for the Minerals Convention and remained opposed to any ban. As might be expected, there was dispute over the need for new institutional arrangements, such as an independent inspectorate and a standing committee for environmental protection, and over what the powers of any new institution would be. A further area of dispute was how to give effect to any requirement that all activities should be subject to prior environmental impact assessment and whether this should incorporate provisions for collective review.

The Group of Four (G4) proposal sought to establish clear environmental principles and a system of regulation, monitoring and enforcement which would be applicable to all activities in the Antarctic. All activities would be subject to environmental impact procedures and would be categorised according to the risk of environmental harm as low risk, high risk or prohibited (which would include minerals activity). The draft convention proposed that collective authorisation would be required for certain high risk categories and the consensus process would be used to identify which activities would be exempt from collective authorisation and subject only to assessment by national authorities in accordance with the convention (Kimball, 1991b, p. 6). The institutional framework would include a committee for the protection of the environment, a secretariat, a scientific and technical advisory committee and an inspectorate.

New Zealand's proposed legally-binding environment protection protocol to the Antarctic Treaty also included a prohibition on all minerals activities and updated existing recommendations. Environmental assessment procedures would be required in some form for all activities and all non-negligible impacts on the Antarctic environment would be avoided. A secretariat, advisory committee for environmental protection and an inspectorate would be established although decision-making would remain in the hands of the consultative parties. The proposal co-sponsored by Argentina, Norway, the UK, US and Uruguay, and the supporting draft protocol and annexes circulated by the UK and the US, envisaged little change to existing rules. In particular, there was no provision for inspection or enforcement procedures or a ban on minerals activity. Several draft annexes were also circulated which, in the main, reproduced existing measures. As expected, the United Kingdom and the US delegations were not prepared to accept a permanent ban on minerals activity or wilderness reserve status for the Antarctic. They argued that the question of minerals activity should be dealt with separately from a comprehensive environmental protection convention.[32] British diplomat John Heap told the meeting that

> the minerals convention ... provides the only mechanism on offer that has the capacity to defuse an otherwise explosive political issue and regulate minerals activities in a way that meets all reasonable environmental demands (Anon., 1990g),

thus suggesting,1 by implication, that the environmental demands of the Four and the NGOs were unreasonable. The UK argued that there would be no consensus on a World Park (Hamer, 1990, p. 5).

The atmosphere in Working Group 1, which considered these proposals, was tense. The head of the Belgian delegation, describing the climate as 'not good' (Anon, 1990h) suggested that the G-4 proposal had 'not been seriously discussed at all ... We have the impression we must buy the so-called American protocol' (Anon, 1990h). Working Group 2, on the other hand, made steady progress in its review of existing measures on environmental protection. Nevertheless, this group's efforts ultimately relied on agreement in Working Group 1 on the form of a comprehensive agreement and that seemed increasingly unlikely.

There was some agreement on the form of an environmental protection instrument, but still considerable disagreement on its content. The deadlock was broken by the efforts of senior Norwegian diplomat Rolf Anderson. After extensive consultation with heads of delegations he drafted an informal and incomplete text for a protocol to the Treaty. The draft articles in the Andersen protocol covered environmental principles, scientific cooperation, environmental impact assessment, compliance, inspection, response action, liability, requirements for annual reporting and provisions for dispute settlement. It proposed a Committee for Environmental Protection and provided for a series of annexes on the specific environmental issues under consideration by Working Group 2. It also declared the Antarctic a 'natural reserve, devoted to science' (article 1). Article 6 proclaimed a ban on minerals activity although it did so in a way that made it clear there was still considerable disagreement.[33] A draft article establishing a secretariat was removed at the insistence of the Argentinian delegation which argued that the SCM did not have the mandate to make decisions that would affect the full consultative process, although it agreed that the question of a secretariat and more frequent meetings of the consultative parties could be discussed at the Bonn 16th consultative meeting. The meeting agreed that the current restraint on mineral resource activities would continue, thereby delinking the voluntary restraint on minerals activity from entry into force of CRAMRA.

Delegates were prepared to use the Andersen draft as the basis for further discussions, although it was not accorded the formal status of a negotiating text.[34] The US and UK, as the core of a blocking coalition were prepared to make few concessions on the unfinished articles of the Andersen text. Early in 1991 the US government circulated a diplomatic note to all consultative parties proposing a fixed term moratorium, with consensus required to *extend* it, a position based on its long-standing opposition to anything that closed off the Antarctic to future minerals activity.[35] The British government announced in March 1991 that it would support a temporary moratorium on mining so that consensus could be achieved, but that it would leave the elaboration of further details for debate at the April session of the SCM. This was a limited concession. Opposition leaders in the UK criticised the government for not going far enough.[36] Neither the US nor the UK had abandoned CRAMRA or willingly accepted an alternative agreement. Rather they had been forced into this negotiating position in the face of growing support for Australia and France from key states within the Treaty system.

The Australian/French coalition grew when two more influential Treaty states announced their support for the Group of Four, leaving the US and the UK further isolated. The Japanese delegation to the November meeting in Chile had told the US that it had not realised how strong opposition to Antarctic mining had become and that it would have to review its position in the light of this (MacKenzie, 1990, p. 4). On 23 April, Japan announced that it would now support a prohibition on minerals activity. In response to public and parliamentary pressure the German Minister of Industry Juergen Moellemann also announced that his government would now support nothing less than a permanent ban on mining in Antarctica.[37]

At the end of January 1991 the Four met in Paris to consider their response to the Anderson draft. They were joined on the second day of their meeting by the eight likeminded states. All twelve met again in Rome at the end of March 1991 to consolidate amendments to the Anderson draft. In particular, they wanted the provisions on compliance to be strengthened and proposed further changes to the rules on inspection, response action and liability. The key elements of environmental impact assessment procedures, they argued, should be included in the body of the agreement rather than set out in an annex.

The second session of SCM XI was held in Madrid from 22 to 30 April 1991. There was still no final agreement on a number of key issues – minerals, impact assessment, the environmental committee, compliance and the liability and response action provisions. Several delegations opposed anything more than an advisory role for the Environmental Protection Committee suggested in the Anderson draft. They preferred the traditional Antarctic approach of voluntary compliance on a national basis, with few formalised institutions, even though this had proved inadequate as a mechanism for ensuring implementation of environmental rules. Working Group 2 had almost finished the draft annexes on marine pollution, waste disposal and protection of fauna and flora. The draft annex on EIA was problematic because it raised questions about the role of the proposed Committee for Environmental Protection, especially whether it would be advisory only or have a substantive decision-making role. Some of the newer consultative parties, such as India, were worried also that EIA procedures would increase the already high costs of Antarctic research.

Andersen's diplomatic skills and consultations – in effect his entrepreneurial leadership – resulted in a second version of the draft protocol upon which there was at least consensus that it be referred to governments. The contentious issue was still the prohibition on minerals activity. Anderson's protocol proposed a two-stage procedure to lift the ban on mining. This provision could be amended after fifty years by a majority of all parties including a majority of all consultative parties but would then only enter into force when ratified by three-quarters of the consultative parties, including all 26 parties who were consultative parties at the time the protocol was adopted.[38] Delegates agreed to return to Madrid for a final meeting from 17 June and it was anticipated that the Protocol would be opened for signature on 23 June 1991, the 30th anniversary of the entry into force of the Antarctic Treaty.[39] However, the need for consensus on the agreement meant that the US and the UK could still block the protocol. They had agreed only reluctantly to the consensus arrangement, even with its modifications, for lifting a minerals ban. On 10 May, the British Prime Minister, John Major, in answer to a written question in the House of Commons, indicated that his government *would* support the Madrid agreement. In doing so he overrode the objections of the Foreign Office, and the Departments of Energy, and Trade and Industry and removed a potential obstacle to adoption of the new agreement. Early in June the US advised that it still had difficulties with the provisions for lifting the prohibition on minerals activity. It proposed that only a three-quarters consultative party ratification majority be required and that any state could withdraw from the ban if that required majority was not forthcoming within three years.[40] Later in June, the US delegation formally advised that it could not sign the protocol in the form agreed to at the April meeting.

The Madrid session reconvened on 17 June. On 19 June, Australia led a group of 15 nations in a closed session in expressing their opposition to the US decision (Hunt, 1991, p. 6). They were hopeful, although not overly optimistic, that an agreement could still be reached. A compromise plan, which seemed acceptable to the US, proposed a three-quarter majority both to adopt an amendment lifting the ban and for its subsequent entry into force. The withdrawal provisions would apply to the protocol as a whole, not just the minerals prohibition as proposed by the US. In spite of anger that the earlier agreement had itself been a package deal and that some states had made concessions to reach a consensus, the compromise was accepted by other parties. Delegations were asked to forward it to their respective governments for consideration. On the last day of the meeting, however, the US delegation again advised that its government wanted more time to consider alternative proposals. As a result, the signing ceremony scheduled for 23 June was cancelled.

This latest announcement was not well received. Diplomats privately acknowledged that they were disappointed at the outcome. The head of the Australian delegation was reported as saying that 'it would be inappropriate for me to display irritation at this moment' (Anon., 1991f, p. 6). New Zealand Prime Minister, Jim Bolger said that he would be taking the issue up with the US ambassador to New Zealand. Australian Prime Minister Hawke wrote to President Bush urging him to sign the agreement. Legislators from Japan, the European Community, and the Soviet Union (as well as from the US) sent Bush a letter from a meeting in Tokyo (see Austin, 1991). The strength of the coalition in favour of the Protocol was by now such that the US could not translate its preferences into institutional action. In response to this pressure President Bush announced on 3 July that the US would sign the compromise protocol agreed to at the June meeting.

The Final Act of the special consultative meeting was adopted in Madrid on 4 October 1991 and the Protocol on Environmental Protection to the Antarctic Treaty was opened for signature until 3 October 1992. Twenty-three of the twenty-six consultative parties and eight of the non-consultative parties signed in Madrid.[41] The voluntary moratorium on minerals activity, which dates to 1977, will continue until the Protocol formally comes into effect upon ratification by all twenty-six signatory consultative parties.

INSTITUTIONAL BARGAINING

The formal negotiations were characterised by Andersen's exercise of entrepreneurial leadership in his brokerage of the various interests of the consultative parties in subsequent drafts of the Protocol. Intellectual leadership – that is, the generation of new ideas and 'intellectual capital' (Young, 1991, pp. 298–302) – was also important in this process. Those new ideas and values came from NGOs. They were taken up first by the defecting states, Australia and France and then, through their coalition-building in which NGOs were also important, by the regime. Particular individuals were also able to carry forward that leadership in ideas. In this respect we can look to Prime Ministers Hawke and Rocard and to non-governmental representatives such as Jacques Cousteau and Sir Peter Scott. Cousteau 'played an instrumental role in lobbying world leaders' to accept the Australian and French proposals (Anon., 1990a, pp. 4–5).

The impact of NGOs on the institutional bargaining process was explicitly recognised by the formal actors within the regime.[42] Australian Prime Minister Hawke argued (1990, p. 2) that the 'world-wide environment movement … has played a large part in this success'. The impact of public debate was also recognised. In opening the 16th consultative meeting, the German Minister of State at the Foreign Office noted that 'one factor proved to be particularly helpful in ensuring the continuation of work on the Protocol ... public opinion' (ATCM, 1991b, p. 150).[43] Australian Foreign Minister Senator Evans suggests (1990, p. 4) that 'entrenched bureaucratic opposition' was overwhelmed by the 'weight of public opinion'.

SCAR was not supportive of the Madrid Protocol nor of the process by which it had come about. Indeed, it had almost no involvement in these debates in contrast with its input into the development of the environmental rules described in chapter three. In a wide-ranging speech at the sixteenth consultative meeting, Richard Laws (then President of SCAR and formerly director of the British Antarctic Survey) argued against strict environmental assessment rules (on the grounds that they threatened freedom of scientific investigation)[44] and warned the consultative parties, in effect, that they could no longer take SCAR for granted.[45] Laws recognised NGOs as a new factor in the Treaty system, but one which he did not welcome. He suggested that it was their 'disproportionate influence' which was responsible for the diminution of SCAR's role.[46] Whereas NGOs were influential actors (as a source of new ideas and in generating support for those ideas), the scientific community did not function in this process as an epistemic community (in policy advocacy) nor even successfully in support of its own interests in the regime. In this respect, the scientific community and NGOs were in competition for influence and power within the Treaty system.

8 The Madrid Protocol ... and After

THE PROTOCOL ON ENVIRONMENTAL PROTECTION TO THE ANTARCTIC TREATY

The Protocol represents changing ideas and values within the Antarctic regime on how protection of the environment can best be achieved. In this context it represents a re-ordering of the hierarchy of values within the Treaty system. The move from CRAMRA to the Protocol required economic interests to be put to one side, albeit not always willingly. Political interests were still important in determining the final shape of the agreement (although the initial defection showed that the collective political values could be questioned), but the acceptance of an overarching instrument, imposing 'top-down' principles and rules, signals that environmental values will also be important determinants of the regime. The Protocol differs from the environmental rules examined in chapters 3 and 4 in other ways. There is a move away from voluntary compliance, a potential for limited centralisation, and a greater emphasis on the importance of monitoring and the need to respond to changing environmental knowledge. The transparency rules also improve on previous practice, ensuring that the consultative parties and the bureaucratic elites no longer have total control over knowledge and information generated by processes within the regime.

The Protocol establishes a comprehensive framework which provides the basis for more effective environmental protection in the Antarctic. Its environmental principles are much more rigorous and extensive than those adopted in the 1970s. The Protocol requires that environmental protection values, which includes aesthetic and wilderness values, are to be *fundamental* considerations in the planning and conduct of *all* activities in the Antarctic. Article 2 commits the parties to the 'comprehensive protection of the Antarctic environment and dependent and associated ecosystems' and designates the Antarctic a 'natural reserve, devoted to peace and science' (see ATCM/SCM

XI, 1991b). Any activity relating to mineral resources, except for scientific research, is effectively prohibited for fifty years with stringent requirements to lift that ban or any other provisions of the Protocol.[1] A conference to review the Protocol, including the prohibition on minerals activity, may be called after fifty years. Any amendment tabled at such a conference must be adopted by a majority of all parties, but that majority must include three-quarters of the twenty-six consultative parties at the time the Protocol was adopted. Any amendment agreed to in this way then comes into effect only when it has been ratified by three-quarters of the consultative parties at that time, including all twenty six original consultative signatories to the Protocol. The prohibition also cannot be lifted unless there is in place 'a binding legal regime' on minerals activities (article 25 (5)(a)). While this is not a permanent ban, nor is the Antarctic formally declared a World Park, both provisions reflect more closely the preferences of Australia and France than they do of either the US or the UK.

The Protocol is a legally binding agreement, subject to compulsory dispute settlement procedures whereas the recommendations it effectively replaces were not. Those procedures are based, in great measure, on those agreed to under the Minerals Convention. Stronger compliance provisions replace the system of voluntary compliance and non-consultative parties now also have the right to instigate dispute action in cases of violation of the rules. While there is still no independent inspectorate which can undertake monitoring or enforce compliance the way is open for improved inspection by individual parties and the potential exists, in principle at least, for some form of centralised inspection procedures to be adopted at a later stage. Inspectors can be appointed by consultative meetings as well as by individual parties. There are no provisions on liability in the Protocol (in contrast with the Minerals Convention, for example). Liability rules must be elaborated in an annex to the Protocol. This has not yet been done. While steps have been taken towards this, the absence of such rules at this stage remains a weakness in the agreement and therefore in the environmental regime.[2]

The Protocol also moves to improve the institutional basis of environmental protection in the Antarctic. It establishes a new regime institution, the Committee for Environmental

Protection, to provide advice and formulate recommendations on the implementation of the Protocol. The task of formulating recommendations was a concession to those parties which wanted a more activist role for the Committee. However, the Committee has no formal decision-making powers. Decisions remain the responsibility of consultative meetings although the consensus rule is modified to allow for flexibility in decision-making. Any amendment to the annexes will be deemed effective one year after the close of the consultative meeting at which that amendment was adopted unless one (or more) of the parties notifies that it wants an extension of time or is unable to approve the measure. This ensures that amendments will be legally-binding within a year if there is no objection.

The transparency requirements of the Protocol are a considerable improvement on existing practice. Reports of Committee meetings, inspection reports, annual reports on steps taken to implement the Protocol, waste management plans, comprehensive environmental evaluations and annual lists of initial environmental evaluations must be made publicly available. While publicly available does not always mean easily available (especially in the absence of a permanent secretariat) improved transparency may well contribute to improved accountability and compliance. NGOs will continue to have an important role in creating an atmosphere which encourages and, if necessary, imposes accountability.

As adopted at Madrid, the Protocol included four annexes which set out rules and guidelines on specific activities – environmental impact assessment, conservation of fauna and flora, waste disposal and management and marine pollution. While those annexes built on existing measures they did so in a way that produced an integrated and, as noted above, legally binding environmental regime. The annexes also emphasise the importance of monitoring environmental impact and reviewing and updating rules in response to that monitoring and to new information. In that respect, the Protocol may engender an environmental regime that is more explicitly knowledge-driven.

Requirements for EIA are entrenched in the body of the Protocol and elaborated further in an annex. Environmental impact assessment is mandatory for all activities (and any change in activity) conducted in connection with scientific research programmes, tourism, expeditionary and station activity and logistic support. Significant or adverse environmental impacts must be avoided although neither of these terms is defined. The parties are required to establish contingency plans for responding to incidents with potentially adverse environmental impacts. The Protocol also requires regular and effective monitoring so that the impact of ongoing activities can be assessed and possible unforeseen effects can be detected, but it does not establish any specific procedures to achieve this. There is, also, still no oversight of the preliminary stages of the evaluation process – the judgement as to whether an impact is likely to be minor or transitory, or otherwise, is the responsibility of the individual national operator. Nor are there any mechanisms to ensure consistency of assessment procedures between parties although there is some guidance on what an evaluation should contain. The annex on environmental impact assessment reproduces, for the most part, the rules adopted in 1987. It also requires draft Comprehensive Environmental Evaluations (CEEs) to be circulated to the Committee on Environmental Protection and all parties to the Protocol. The final decision on any activity is still made at the national level but it cannot be made until the CEE has been considered by a consultative meeting. Final CEEs must be made publicly available at least 60 days prior to the commencement of the proposed activity and must include a summary of any comments made on the draft evaluation. While there is still no formal centralised authorisation for any activity (as there was over minerals activity in the Minerals Convention, for example) these administrative and transparency requirements relating to environmental impact of human activity in Antarctica are a welcome advance on earlier rules in the regime.

The annex on fauna and flora is based on the relevant provisions of the Agreed Measures but streamlines the rules and procedures to make them less confusing. It includes more detailed guidelines which require that the diversity of species, the habitats essential to their existence and the balance of ecosystems must be maintained.[3] However permits for taking native fauna and flora are still to be issued by national authorities. The waste disposal annex (which is to apply to non-governmental activities as well) improves upon existing management provisions and expands the list of materials which must be removed from the Antarctic. The Committee on Environmental Protection may review national waste management plans and offer comments and advice on their content. There is also a formal requirement for the annex to be reviewed regularly to take into account new waste management technology. Nevertheless, the rules still permit incineration as a method of waste disposal and the discharge of sewage and domestic liquid waste into the sea. The marine pollution annex sets out much more specific detail than the recommendation adopted at ATCM XV in 1989 and includes details on prohibited practices although its sovereign immunity provisions are a potential flaw.[4] Parties must, however, ensure that ships covered by this provision act in a manner consistent with the annex and the provisions of the annex are to be taken into account when new ships are constructed.

In spite of its flaws, the Protocol marks a qualitative change in the approach to environmental issues in the Antarctic regime and replaces the ad hoc and unwieldy network of measures which had characterised environmental protection under the Treaty. It is comprehensive in that it establishes general environmental principles which *must* be applied to all activities in the Antarctic. The provisions for prior environmental assessment, the dispute settlement procedures and the improved transparency rules contribute to the integrated and comprehensive nature of the agreement. The influence of interdependence norms in the Protocol and some move away from the decentralisation that had characterised the framework of the regime can be identified in the rules that require prior environmental impact assessment, the potential for a greater role for the CEP and the potential for inspection, which remains the main compliance mechanism, on a collective basis. The acceptance of legally binding environmental rules and principles, and submission to a compulsory dispute settlement procedure, also represent the slow incursion of interdependence norms into the Treaty system. The way is opened for a further move towards interdependence norms, especially if a secretariat is established and the CEP is politically supported by the consultative parties. This in itself identifies a problem – in spite of strengthened compliance rules, political will still determines the success of the Protocol and the future of environmental protection in the Antartic. Nevertheless, without this political will on the part of states, it is likely that the future stability of the regime will be undermined, in particular because of the participation of non-governmental organisations who will, no doubt, take steps to identify violation of the rules and flaws in those rules, and increasing interest in the effectiveness of the Protocol on the part of the international community outside the regime. The balance of participation and knowledge, which in the past has favoured the consultative parties and bureaucratic actors, has changed to include non-governmental organisations. While, as noted in chapter 2, violations of environmental rules have in the past thought to be less destabilising to the regime than violations of the fundamental norms of the regime, it is not clear that this will continue to be the case.

... AND AFTER!

Immediately following the final session of SCM-XI and the signing of the Protocol the consultative parties met for the 16th consultative meeting in Bonn. ASOC was included in the list of invited expert organisations and seven delegations included NGO representatives.[5] The list of invited observer organisations was extensive and, for the first time, included the UN Environment Programme, whose participation had been blocked in the past.[6] Opening speeches were self-congratulatory on the adoption of the Madrid Protocol. Only Britain referred openly to the dissension caused over CRAMRA.[7] Eleven of the thirteen recommendations adopted related to the agenda of environmental issues under the Protocol – ten on the protected area system and one on tourism and non-governmental activities. Delegates also agreed to convene meetings of experts on other Protocol issues, including marine pollution and environmental monitoring.[8] Delegates agreed that, as far as possible, consultative parties would apply the provisions of the Protocol in the interim and stressed the importance of early ratification and entry into force.[9] The parties agreed to conduct a review of consultative recommendations, which remain in force at least until the Protocol comes into effect, to identify those which have been superseded or affected by the Protocol and its annexes. The meeting also decided that, to facilitate decision-making under the Protocol, consultative meetings would be held annually.[10]

The Swedish delegation noted that the Protocol 'is not an end in itself, and it must not become an end to our efforts'

(ATCM, 1991f, p. 196). Nevertheless, some of the momentum of Madrid was lost. Efforts to establish the Committee for Environmental Protection prior to entry into force of the Protocol were blocked. The meeting did agree that the following consultative meeting would begin drafting the Rules of Procedure for the CEP. There was little advance on negotiating a liability annex to the Protocol in the face of US concern that to do so could delay ratification, although it is not clear how it would do so. The question of how to address liability matters was deferred until the next consultative meeting. There was still no consensus on establishing a secretariat, an essential move if the CEP is to carry out its responsibilities and if the Protocol is to function successfully. It was blocked again by Argentina.[11]

The meeting adopted an extensive annex to the Protocol on the Antarctic Protected Area system.[12] It rationalises the system of protected areas and elaborates specific guidelines for conduct in such areas. It establishes two categories of protected areas – Antarctic Specially Protected Areas which can only be entered with a permit and Antarctic Specially Managed Areas.[13] Management plans must accompany proposals for area designation, and a code of conduct is required for all SMAs. The annex sets out detailed guidelines on what must be included in a management plan. Details about management plans and entry permits issued for entry into protected areas must be made publicly available. Designations can be proposed not only by any party but also by the CEP, SCAR or CCAMLR. This annex was essential to improve the protection of the Antarctic because, as chapter 3 argued, the existing system of protected areas had become complicated and unwieldy.

There was a 'tense and somewhat bitter' debate (Greenpeace, 1991d, p. 11) on the tourism question and whether a separate annex on non-governmental activities was necessary, given that the Protocol and its annexes specifically apply to all human activities in the Antarctic including 'non-official' activities. Opposing views were taken by France, who remained strongly committed to a new annex, and the US, who argued that no further regulation was necessary. NGOs have also opposed a separate annex, fearing that this could give consultative parties an effective veto over any NGO activity in the Antarctic.[14] The meeting agreed, following a suggestion from

Japan, to convene an informal group of experts to consider this question once more.

Between the sixteenth and seventeenth consultative meetings several workshops were held on issues relating to the implementation of the Protocol. A Group of Experts meeting on environmental monitoring was convened in June 1992. SCAR and the IUCN sponsored a joint workshop on the Antarctic protected area system in July. The consultative parties met for an informal meeting on tourism and non-governmental activities just prior to the seventeenth consultative meeting in Venice in November 1992.

While further steps were taken on Protocol-related issues at the 17th consultative meeting, which was held in Venice from 11 to 20 November 1992, there is still a need for further organisational and practical aspects to be addressed.[15] The view taken by the parties is that as much as possible should be agreed so that procedures can be implemented as soon as the Protocol comes into force. However, they are not prepared to establish interim procedures prior to that event. By the time of the 17th meeting, thirty-six of the Treaty parties, including all twenty-six consultative parties had signed the Protocol, but ratification was proving a slow process, especially for those parties who required legislative action as part of the ratification process.[16] The meeting agreed, however, that it was important to ensure uniformity of interpretation of the Protocol's provisions, although they agreed to do this only through diplomatic channels rather than the institutions of the Protocol. Some small steps were taken on key issues of environmental impact assessment, the protected area system and tourism, all of which are vital to the successful management of human activity in the Antarctic. COMNAP distributed draft guidelines on EIAs, although ASOC has argued that these are inconsistent with the Protocol and therefore inadequate as a guideline (Anon., 1992). Parties were urged to submit revised or new management plans for protected areas for consideration when the Protocol enters into force, to institute permit systems, and to ensure that information on location and restrictions is available to all persons in the Antarctic. There was no mention of a future role for the CEP in this. The tourism issue remains unsettled. The key point of difference continues to be whether a specific annex is required. According to the draft final report

(ATCM, 1992, p. 17), some parties feel that even though the Protocol applies to all activities, 'more precise regulation having legally binding force was required for such activities'. While some favoured calling yet another meeting to continue examination of this issue, others were prepared to do so only if such a meeting would give prospect of progress on this issue.

Although the parties were not prepared to establish an interim Committee on Environmental Protection, they did agree on some of the organisational details for the CEP. Committee meetings will be held in the same location and just prior to the annual consultative meetings, but there will be provision for intersessional meetings. The CEP will not have its own secretariat but will be assisted by the Antarctic Treaty secretariat. An informal contact group was established at the 17th meeting to prepare draft rules of procedure. The negotiations for a secretariat, upon which there is now consensus, were advanced but not completed. In particular, there were tensions over where the secretariat would be located. Both Argentina and the US have offered their capitals as sites. While the US has the advantage of also being the depositary government for the Treaty, several delegations expressed views 'favouring the insertion of Latin America in the institutional framework of the Antarctic Treaty system' (ATCM, 1992, p. 8). An informal meeting is scheduled before the 1994 consultative meeting to reach overall agreement on several outstanding issues relating to the establishment of a secretariat. The parties also agreed to put the question of jurisdiction on the agenda of the next consultative meeting, for the first time in thirty years.

No liability annex was agreed to. The parties agreed only to convene a meeting of legal experts to elaborate liability rules and acknowledged, in effect, that further meetings may be required.[17] Some steps were taken to improve the basis of inspections, in particular their emphasis on environmental matters, but with continued resistance to any form of centralised inspectorate. SCAR and COMNAP have been asked to prepare an indicative checklist for inspections. However, the parties rejected ASOC's view that a permanent technical and scientific team should be established to carry out inspections relating to the Antarctic environment (ATCM, 1992, p. 11). This resistance may have arisen, in part, as a reaction to ASOC's suggestion that it be designated an inspector under the Protocol. Some steps were also taken on the question of environmental monitoring but with something of a minimalist emphasis on programmes designed to answer specific questions or test particular hypotheses, and only what is considered essential to meet the requirements of the Protocol.

The Antarctic regime is thus in a transition period. The changed regime under the Protocol exists in principle, but it has not yet been put into practice, in spite of urgings that its rules be observed, as far as practicable, in the interim. There is, then, still not an integrated, legally binding environmental regime in place. Political concerns have, it seems, slowed efforts to give effect to the Protocol's provisions. The next meeting has been deferred until 1994, in spite of an agreement that meetings would be held annually. There is some hope that the Protocol will be in force by then, but uncertainty surrounds this. In any case, the process of ratification means that it is taking longer for the Protocol to come into force than it took to negotiate it and somewhat undermines the sense of urgency that surrounded those negotiations. It is also uncertain whether the institutional framework will be sufficiently developed to enable the rules of the Protocol to be given effect as soon as the final instrument of ratification is deposited. The Protocol's future, therefore, is still speculative.

Conclusion

Three themes were identified in the introduction to this book. The primary focus of this study has been the international politics of environmental protection in the Antarctic. A second theme was to consider whether this study might provide some insights into the practice and prospects of international environmental politics more generally. In other words, what lessons if any might be learned from the Antarctic experience that are relevant to our attempts to negotiate and implement better environmental regimes. A third, less explicit purpose was to interrogate the adequacy of international relations paradigms for exploring environmental issue-areas and politics.

Environmental protection in the Antarctic was managed within the constraints of a regime structured by the Antarctic Treaty of 1959. That Treaty was not an environmental treaty. It had, as chapter 2 explained, limited objectives. It was designed primarily to prevent conflict among states over territorial claims and to manage potential tensions between the superpowers. Notwithstanding that the Antarctic Treaty represented a commitment to cooperation, sovereignty rather than interdependence norms prevailed and the hierarchy of values favoured political concerns such as the importance of consensus and the imperative for maintaining the compromise on sovereignty over other values including environmental protection. These characteristics ensured that the Antarctic regime was a successful political and security regime. The measure of success was that consensus was maintained and that the political differences over sovereignty were managed within the confines of the Treaty. However, this normative framework did not support an effective environmental protection regime. Indeed, in the long term it could not do so. Although the Treaty was not designed as an environmental protection regime, protection of the Antarctic environment became one of the main focuses of decision-making. In adopting environmental protection measures the consultative parties extended the principles and objectives of the Treaty and their obligations under it. They claimed for themselves the right to manage the Antarctic environment and justified their continuing authority over the Antarctic in part on their stewardship of that environment.

The environmental agenda grew in a fairly ad hoc fashion based on the only explicit reference to these issues in the Treaty – that of the conservation and preservation of living resources. In developing the environmental regime, the Treaty parties established rules and procedures to manage protected areas, scientific activity, waste disposal, environmental impact assessment, tourism and non-governmental activities. There was an in-principle commitment to protecting the environment of the Antarctic, but that commitment was not translated effectively into rules that could ensure that commitment. As chapters 3 and 4 have shown, the rules adopted did not go as far as they might have, or as quickly as they might in protecting the Antarctic environment or in minimising localised environmental impact as a result of human activity. Although environmental principles were adopted, they established only a general statement of intent and were not translated into rigorous guidelines or mandatory rules to manage human activity in the Antarctic. The environmental protection rules were not integrated into a comprehensive protection regime. The limitations of those rules were compounded by a poor compliance record. Indeed, the Treaty parties themselves were forced to accept that these were justified charges. That is not to suggest that the rules were ignored completely, but there was a sufficient degree of non-compliance to question the consultative parties' claimed stewardship of the Antarctic environment. Further, non-compliance with environmental rules was not considered destabilising to the Antarctic regime, whereas disputes over that non-compliance might well have been.

The limitations to the environmental regime were a direct result of the features that had guaranteed the regime's political success. The need for consensus engendered a slow and piecemeal decision-making process that proved increasingly unable to respond to the demands of an increase in human activity in the Antarctic or to new information and knowledge about the environmental impact of that activity. The application of sovereignty norms supported a permissive approach to compliance, resisted any centralised monitoring, compliance or inspection agency and did not encourage the parties to adopt legally-binding measures. The recommendations adopted were hortatory rather than mandatory. The unilateral inspection provisions in the Treaty, which were not underpinned by sanctions, were ineffective as compliance mechanisms. The interpretation and implementation of environmental rules remained in the hands of national governments with nothing to ensure consistency of application among consultative parties. In spite of concerns to protect the Antarctic environment, the environmental regime was driven and shaped by political concerns rather than by environmental ones. It was not primarily a knowledge-driven regime. Decision-making lagged increasingly behind knowledge and the scientific knowledge and advice which the parties requested from SCAR, while crucial to the development of an effective environmental protection regime, was rarely accepted as the primary determinant of environmental rules. The scientific community, through SCAR, did not function as an epistemic community. While the advice and information was often consensual, and scientists were sometimes participants in decision-making, there was no conscious and active policy advocacy. The hierarchy of values ensured that, in spite of the recognised importance of protecting the Antarctic environment, rules that would effectively achieve that goal were compromised by political and economic concerns. The bureaucratic elite which was so important to the management of the political compromises of the Treaty were convinced, for the most part, that environmental issues could only be managed in the context of those compromises. It was only if those political bargains were maintained that other issues could be addressed.

In the 1960s and for most of the 1970s, there was little public interest in Antarctic issues and only a nascent public environmental consciousness. In that context, the adequacy of the environmental regime was not questioned. However, that changed in the 1980s. Non-governmental organisations focused attention on the Antarctic Treaty system and, in particular, the adequacy of environmental protection rules and the adequacy of compliance. Their criticisms of environmental rules and their identification of the impact of non-compliance imposed accountability on the Treaty parties in both the policy and practice of Antarctic environmental protection and there was a gradual improvement in the rules adopted and in compliance with those rules. In this way, it was non-governmental organisations which engendered a change in the environmental regime, not the scientific community.

The Minerals Convention proved to be the catalyst for a more comprehensive change in the Antarctic regime, although neither the Treaty parties nor the NGOs could have anticipated this. While the Treaty parties accepted their responsibility to establish strict environmental protection rules in the Convention, the negotiating process and the final agreement was, as chapter 6 shows, still shaped by political and economic concerns. Given that the non-development option had been rejected, it is unlikely that it could have been otherwise. The non-governmental challenge to those negotiations and to Convention itself was shaped by demands for normative change – an emphasis on interdependence norms and a rethinking of the hierarchy of values to ensure that decisions on environmental protection were guided by environmental and not political knowledge.

The rejection of the Minerals Convention and the negotiation of the Madrid Protocol described in chapter 7 arose out of more than state action and cannot be understood simply in terms of conventional power explanations. The coalitions that arose to mobilise change in the Treaty regime were not just between states. They arose between states and other actors. Defection by Australia and France was an effective strategy for change and for generating an institutional response because, in the first instance, Australia and France possessed issue-specific power in the Treaty system as well as over the entry into force of CRAMRA. Their success in mobilising change in the regime rested on the exercise of structural leadership and coalition building strategies. NGOs had clearly lobbied for the Minerals Convention to be rejected on the grounds that minerals activity was incompatible with the goals of environmental protection. NGO success in overturning CRAMRA when they had not been able to prevent its negotiation was in part due to their ability to mobilise public debate and exploit domestic politics to pressure political leaders on a specific agreement. It was also due to their increased experience and credibility in the regime and to the transnational organisation of the campaign. Yet, without the policy decisions made by Australia and France, that is without the support of states, they could not have been successful. Yet it is equally the case that without the support of NGOs, Australia and France could not have successfully built a coalition of support in favour of their new initiative. The successful challenges to the Minerals Convention also affected power relationships in the regime, not between states but between actors. The traditional bureaucratic elite and the scientific community were marginalised. NGO values, and their participation in the regime were enhanced.

In spite of the effective demise of the Minerals Convention and the adoption of a comprehensive agreement on environmental protection, we have not observed a paradigm shift in the Antarctic regime. The Protocol represents the outcome not of paradigmatic learning but of cognitive learning. Institutional policy goals have changed in response to new knowledge in the form of ideas and values about the acceptability of environmental impact and about how environmental protection can best be attained. Sovereignty norms have not been replaced by interdependence norms, but they have given way somewhat. The Madrid Protocol also represents something of a change in the hierarchy of values although again this should not be overstated. Environmental concerns have been given greater priority and emphasis than they were before, in terms of the kinds of rules adopted. The Protocol is legally-binding and it establishes a top-down comprehensive set of principles and guidelines which must be taken into account in the planning and management of all activities in the Antarctic. In this respect it does reflect a change of direction in the regime. Nevertheless, the implementation of those rules is still constrained by political concerns, in particular continuing resistance to centralised and independent monitoring and inspection authorities. The Protocol has, however, changed the terms of the debate about the Antarctic continent. In this respect, it is possible to assert that the Antarctic Treaty governments are now debating the NGO agenda.

The Antarctic case does not stand in isolation. The experience there has some relevance for our understanding of international environmental politics. In particular it offers some contributions, which are considered here only briefly, to the processes of institutional bargaining, the problems of managing environmental protection in a decentralised world of sovereign states, and for the processes by which that environmental protection might be improved.

The Antarctic environmental regime was developed through the process of institutional bargaining. The lessons of the Antarctic confirm the propositions that leadership is an important component of successful negotiation although it is difficult to determine the causal nature of this relationship or to assert confidently that it was both a necessary and sufficient condition to the successful completion of the bargaining process. While it is easy to identify and describe leadership, it is less easy to identify the conditions under which that leadership can be successfully exercised. The Antarctic case study may suggest, however, that on environmental issues at least intellectual and entrepreneurial leadership are as important as structural leadership (especially for the process of change), or that structural leadership to bring about improved environmental protection is difficult to exercise in the absence of other types of leadership.

At a more practical level, the process of institutional bargaining in the Antarctic regime confirms the value of informal negotiating texts as a way of moving towards a contractarian agreement. It would also suggest that the use of working groups is an important contributor to joint problem-solving. Nevertheless, there can be problems with this, as the minerals negotiations in particular showed. The compartmentalisation of issues usually involves environmental concerns being dealt with independently of political or legal issues. This can result in the bargaining process being politically driven rather than environmentally driven. Success in bargaining is often measured as much by the reaching of an agreement as it is on the basis of the content of that agreement. It also means that the process is not always primarily knowledge-driven which can make it difficult for an epistemic community to be effective during the negotiation process.

These problems, I would suggest, do not arise from any fundamental flaw in institutional bargaining as the basis for negotiating environmental regimes. Rather they are the outcome of processes which are constructed to protect political and economic interests and shaped by continuing reliance on sovereignty norms. This suggests, then, that there is a limit to what can be achieved when sovereignty rather than interdependence norms dominate and identifies but does not answer the difficult question of how a reliance on and acceptance of interdependence norms can be brought about.[1]

The stance taken by the Treaty parties on the implementation of the Madrid Protocol also provides some positive and negative lessons. The parties have moved, although not as far as they might, to ensure that procedures and organisational details are agreed upon so that they can be put into practice as soon as the Protocol comes into force. However, what they have not been prepared to do is to negotiate or put in place interim procedures in advance. Where ratification processes can take a long time, this means that no action is likely to be taken under environmental protection agreements. While there is some merit in the Treaty parties' policy of urging voluntary compliance 'in the interim', this simply compounds the difficulties of voluntary restraint.

The Antarctic case highlights the difficulty of securing environmental protection in a decentralised world where sovereignty norms prevail. It raises the question of whether voluntary restraint and compliance can be effective in achieving environmental protection. It confirms the need for centralised and independent monitoring and inspection institutions and for the scientific community and scientific knowledge, along with NGOs as representatives of community voices, to be more strongly incorporated into environmental decision-making. However it also points to the difficulties of moving towards interdependence norms.

The paradigm shift is not to be easily achieved. If what is needed are improved protection rules and improved compliance in the first instance, and in the longer-term a change in values, does the Antarctic case provide any insights into how that might be brought about? The judgement of what is the best that can be negotiated in an imperfect world can be changed as the experience of the Antarctic shows. The case study suggests at least that coalition building is an important strategy for achieving institutional change. There are conditions attached to the successful use of coalition-building strategies by middle-powers (see Grant and Evans, 1991, p. 324). However international or transnational environmental issues are one area in which these strategies can be successfully employed, especially when conventional economic or military capabilities and relationships are less salient to the task of finding and managing solutions to environmental problems.

It may not be enough, however, for coalitions in international environmental politics to be constructed simply around states. The overturning of the Minerals Convention and the negotiation of the Madrid Protocol suggest that coalitions between states and non-state actors are likely to be more effective especially in negotiated regimes. The case study is, however, less insightful on normative change or a paradigm shift except to confirm that the process, if it occurs, is likely to arise from incremental processes rather than a radical change and that is unlikely to be motivated by states alone.

The Antarctic case study contributes not only to the research agenda of international environmental politics. It also reveals a more theoretical research agenda, or at least confirms the direction of an existing research agenda which seeks to understand and to theorise two connected areas of study which have implications for international relations paradigms. Those areas of study focus on the place of non-state actors in international politics, and the interpretation of power as a basic concept in international relations. Theoretical propositions arising from the Antarctic case study are not explored in any depth here. Rather this conclusion simply makes a few brief observations. Non-state actors have been confirmed as important actors in international environmental politics. With the exception of the work on epistemic communities, there has been little attempt to theorise non-state actors even within theories of regimes or institutions. If non-state actors, particularly the scientific community and non-governmental environment and conservation organisations are recognised as important transnational participants in international regimes, then international relations paradigms need to take account of them in a way that goes beyond simple description. State-centric analysis cannot do this and is, therefore, limited in its ability to understand and explain these dimensions of international environmental politics. Those dimensions encourage further work on the exercise of epistemic power. They also suggest that some aspects at least of international relations can be interrogated through an identification of relationships between knowledge-based actors. This study confirms also the importance of continuing to explore the connection between knowledge and power as it relates to international relations. The Antarctic experience supports the argument that power over outcomes, for both states and non-state actors, can be based on the exercise and possession of knowledge as well as, or perhaps even instead of the more conventionally understood power attributes of military and economic capabilities.

While the Antarctic is an isolated continent, it is nevertheless an integral part of the global ecosystem and of the world of international environmental politics. Debates about environmental protection in the Antarctic have matched changing international views on how best to protect the environment and about what we now consider acceptable and unacceptable in that context. The Antarctic, as a scientific laboratory and for the most part pristine environment, can reveal valuable information for the management of global environmental insecurities such as the depletion of the ozone layer and the problems of climate change arising from the accumulation of greenhouse gases. The steps taken by the Antarctic Treaty consultative parties to improve the protection of this continent and to better manage their own activities there, and the processes by which they have taken those steps, also provide us with valuable political lessons for the future of environmental security.

Appendices

APPENDIX 1 THE ANTARCTIC TREATY

The Governments of Argentina, Australia, Belgium, Chile, the French Republic, Japan, New Zealand, Norway, the Union of South Africa, the Union of Soviet Socialist Republics, the United Kingdom of Great Britain and Northern Ireland, and the United States of America,

Recognising that it is in the interest of all mankind that Antarctica shall continue for ever to be used exclusively for peaceful purposes and shall not become the scene or object of international discord;

Acknowledging the substantial contributions to scientific knowledge resulting from international cooperation in scientific investigation in Antarctica;

Convinced that the establishment of a firm foundation for the continuation and development of such cooperation on the basis of freedom of scientific investigation in Antarctica as applied during the International Geophysical Year accords with the interests of science and the progress of all mankind;

Convinced also that a treaty ensuring the use of Antarctica for peaceful purposes only and the continuance of international harmony in Antarctica will further the purposes and principles embodied in the Charter of the United Nations;

Have agreed as follows:

Article I

1. Antarctica shall be used for peaceful purposes only. There shall be prohibited, *inter alia*, any measure of a military nature, such as the establishment of military bases and fortifications, the carrying out of military manoeuvres, as well as the testing of any type of weapon.

2. The present Treaty shall not prevent the use of military personnel for scientific research or for any other peaceful purpose.

Article II

Freedom of scientific investigation in Antarctica and co-operation toward that end, as applied during the International Geophysical Year, shall continue, subject to the provisions of the present Treaty.

Article III

1. In order to promote international cooperation in scientific investigation in Antarctica, as provided for in Article II of the present Treaty, the Contracting Parties agree that, to the greatest extent feasible and practicable:
(a) information regarding plans for scientific programs in Antarctica shall be exchanged to permit maximum economy of and efficiency of operations;
(b) scientific personnel shall be exchanged in Antarctica between expeditions and stations;
(c) scientific observations and results from Antarctica shall be exchanged and made freely available.

2. In implementing this Article, every encouragement shall be given to the establishment of co-operative working relations with those Specialized Agencies of the United Nations and other international organisations having a scientific or technical interest in Antarctica.

Article IV

1. Nothing in the present Treaty shall be interpreted as:
(a) a renunciation by any Contracting Party of previously asserted rights of or claims to territorial sovereignty in Antarctica;
(b) a renunciation or diminution by any Contracting Party of any basis of claim to territorial sovereignty in Antarctica which it may have whether as a result of its activities or those of its nationals in Antarctica, or otherwise;
(c) prejudicing the position of any Contracting Party as regards its recognition or non-recognition of any other State's rights of or claim or basis of claim to territorial sovereignty in Antarctica.

2. No acts or activities taking place while the present Treaty is in force shall constitute a basis for asserting, supporting or denying a claim to territorial sovereignty in Antarctica or create any rights of sovereignty in Antarctica. No new claim, or enlargement of an existing claim, to territorial sovereignty in Antarctica shall be asserted while the present Treaty is in force.

Article V

1. Any nuclear explosions in Antarctica and the disposal there of radioactive waste material shall be prohibited.

2. In the event of the conclusion of international agreements concerning the use of nuclear energy, including nuclear explosions and the disposal of radioactive waste material, to which all of the Contracting Parties whose representatives are entitled to participate in the meetings provided for under Article IX are parties, the rules established under such agreements shall apply in Antarctica.

Article VI

The provisions of the present Treaty shall apply to the area south of 60° South Latitude, including all ice shelves, but nothing in the present Treaty shall prejudice or in any way affect the rights, or the exercise of the rights, of any State under international law with regard to the high seas within that area.

Article VII

1. In order to promote the objectives and ensure the observance of the provisions of the present Treaty, each Contracting Party whose representatives are entitled to participate in the meetings referred to in Article IX of the Treaty shall have the right to designate observers to carry out any inspection provided for by the present Article. Observers shall be nationals of the Contracting Parties which designate them. The names of observers shall be communicated to every other Contracting Party having the right to designate observers, and like notice shall be given of the termination of their appointment.

2. Each observer designated in accordance with the provisions of paragraph 1 of this Article shall have complete freedom of access at any time to any or all areas of Antarctica.

3. All areas of Antarctica, including all stations, installations and equipment within those areas, and all ships and aircraft at points of discharging or embarking cargoes or personnel in Antarctica, shall be open at all times to inspection by any observers designated in accordance with paragraph 1 of this Article.

4. Aerial observation may be carried out at any time over any or all areas of Antarctica by any of the Contracting Parties having the right to designate observers.

5. Each Contracting Party shall, at the time when the present Treaty enters into force for it, inform the other Contracting Parties, and thereafter shall give them notice in advance, of

(a) all expeditions to and within Antarctica, on the part of its ships or nationals, and all expeditions to Antarctica organised in or proceeding from its territory;

(b) all stations in Antarctica occupied by its nationals; and

(c) any military personnel or equipment intended to be introduced by it into Antarctica subject to the conditions prescribed in paragraph 2 of Article 1 of the present Treaty.

Article VIII

1. In order to facilitate the exercise of their functions under the present Treaty, and without prejudice to the respective positions of the Contracting

Parties relating to jurisdiction over all other persons in Antarctica, observers designated under paragraph 1 of Article VII and scientific personnel exchanged under sub-paragraph 1(b) of Article III of the Treaty, and members of the staffs accompanying any such persons, shall be subject only to the jurisdiction of the Contracting Party of which they are nationals in respect of all acts or omissions occurring while they are in Antarctica for the purposes of exercising their functions.

2. Without prejudice to the provisions of paragraph 1 of this Article, and pending the adoption of measures in pursuance of sub-paragraph 1(e) of Article IX, the Contracting Parties concerned in any case of dispute with regard to the exercise of jurisdiction in Antarctica shall immediately consult together with a view to reaching a mutually acceptable solution.

Article IX

1. Representatives of the Contracting Parties named in the preamble to the present Treaty shall meet at the City of Canberra within two months after the date of entry into force of the Treaty, and thereafter at suitable intervals and places, for the purpose of exchanging information, consulting together on matters of common interest pertaining to Antarctica, and formulating and considering, and recommending to their Governments, measures in furtherance of the principles and objectives of the Treaty, including measures regarding:–

(a) use of Antarctica for peaceful purposes only;
(b) facilitation of scientific research in Antarctica;
(c) facilitation of international scientific co-operation in Antarctica;
(d) facilitation of the exercise of the rights of inspection provided for in Article VII of the Treaty;
(e) questions relating to the exercise of jurisdiction in Antarctica;
(f) preservation and conservation of living resources in Antarctica.

2. Each Contracting Party which has become a party to the present Treaty by accession under Article XIII shall be entitled to appoint representatives to participate in the meetings referred to in paragraph 1 of the present Article, during such times as that Contracting Party demonstrates an interest in Antarctica by conducting substantial research activity there, such as the establishment of a scientific station or the despatch of a scientific expedition.

3. Reports from the observers referred to in Article VII of the present Treaty shall be transmitted to the representatives of the Contracting Parties participating in the meetings referred to in paragraph 1 of the present Article.

4. The measures referred to in paragraph 1 of this Article shall become effective when approved by all the Contracting Parties whose representatives were entitled to participate in the meetings held to consider those measures.

5. Any or all of the rights established in the present Treaty may be exercised as from date of entry into force of the Treaty whether or not any measures facilitating the exercise of such rights have been proposed, considered or approved as provided in this Article.

Article X

Each of the Contracting Parties undertakes to exert appropriate efforts, consistent with the Charter of the United Nations, to the end that no one engages in any activity in Antarctica contrary to the principles or purposes of the present Treaty.

Article XI

1. If any dispute arises between two or more of the Contracting Parties concerning the interpretation or application of the present Treaty, those Contracting Parties shall consult among themselves with a view to having the dispute resolved by negotiation, inquiry, mediation, conciliation, arbitration, judicial settlement or other peaceful means of their own choice.

2. Any dispute of this character not so resolved shall, with the consent, in each case, of all parties to the dispute, be referred to the International Court of Justice for settlement; but failure to reach agreement on reference to the International Court of Justice shall not absolve the parties to the dispute from the responsibility of continuing to seek to resolve it by any of the various peaceful means referred to in paragraph 1 of this Article.

Article XII

1(a) The present Treaty may be modified or amended at any time by unanimous agreement of the Contracting Parties whose representatives are entitled to participate in the meetings provided for under Article IX. Any such modification or amendment shall enter into force when the depositary Government has received notice from all such Contracting Parties that they have ratified it.

(b) Such modification or amendment shall thereafter enter into force as to any other Contracting Party when notice of ratification by it has been received by the depositary Government. Any such Contracting Party from which no notice of ratification is received within a period of two years from the date of entry into force of the modification or amendment in accordance with the provision of sub-paragraph 1(a) of this Article shall be deemed to have withdrawn from the present Treaty on the date of the expiration of such period.

2(a) If after the expiration of thirty years from the date of entry into force of the present Treaty, any of the Contracting Parties whose representatives are entitled to participate in the meetings provided for under Article IX so requests by a communication to the depositary Government, a Conference of all the Contracting Parties shall be held as soon as practicable to review the operation of the Treaty.

(b) Any modification or amendment to the present Treaty which is approved at such a Conference by a majority of the Contracting Parties there represented, including a majority of those whose representatives are entitled to participate in the meetings provided for under Article IX, shall be communicated by the depositary Government to all Contracting Parties immediately after the termination of the Conference and shall enter into force in accordance with the provisions of paragraph 1 of the present Article.

(c) If any such modification or amendment has not entered into force in accordance with the provisions of sub-paragraph 1(a) of this Article within a period of two years after the date of its communication to all the Contracting Parties, any Contracting Party may at any time after the expiration of that period give notice to the depositary Government of its withdrawal from the present Treaty; and such withdrawal shall take effect two years after the receipt of the notice by the depositary Government.

Article XIII

1. The present Treaty shall be subject to ratification by the signatory States. It shall be open for accession by any State which is a Member of the United Nations, or any other State which may be invited to accede to the Treaty with the consent of all the Contracting Parties whose representatives are entitled to participate in the meetings provided for under Article IX of the Treaty.

2. Ratification of or accession to the present Treaty shall be effected by each State in accordance with its constitutional processes.

3. Instruments of ratification and instruments of accession shall be deposited with the Government of the United States of America, hereby designated as the depositary Government.

4. The depositary Government shall inform all signatory and acceding States of the date of each deposit of an instrument of ratification or accession, and the date of entry into force of the Treaty and of any modification or amendment thereto.

5. Upon the deposit of instruments of ratification by all the signatory States, the present Treaty shall enter into force for those States and for States which have deposited instruments of accession. Thereafter the Treaty shall enter into force for any acceding State upon the deposit of its instruments of accession.

6. The present Treaty shall be registered by the depositary Government pursuant to Article 102 of the Charter of the United Nations.

Article XIV

The present Treaty, done in the English, French, Russian and Spanish languages, each version being equally authentic, shall be deposited in the archives of the Government of the United States of America, which shall transmit duly certified copies thereof to the Governments of the signatory and acceding States.

APPENDIX 2 CONTRACTING PARTIES TO THE ANTARCTIC TREATY

(as at October 1993)

State	*Status**	*Date of Ratification*	*Date of CP Status*[1]
United Kingdom	OS/CP/CL	31 May 1960	
South Africa	OS/CP	21 June 1960	
Belgium	OS/CP	26 July 1960	
Japan	OS/CP	4 August 1960	
United States of America	OS/CP	18 August 1960	
Norway	OS/CP/CL	24 August 1960	
France	OS/CP/CL	16 September 1960	
New Zealand	OS/CP/CL	1 November 1960	
Soviet Union[2]	OS/CP	2 November 1960	
Argentina	OS/CP/CL	23 June 1961	
Australia	OS/CP/CL	23 June 1961	
Chile	OS/CP/CL	23 June 1961	
Poland	AS/CP	8 June 1961	29 July 1977
Czechoslovakia	AS/NCP	14 June 1962	
Denmark	AS/NCP	20 May 1965	
Netherlands	AS/CP	30 March 1967	19 November 1990
Romania	AS/NCP	15 September 1971	
Germany			
German Democratic Republic	AS/CP	19 November 1974	5 October 1987
Federal Republic of Germany	AS/CP	5 February 1979	3 March 1981
Brazil	AS/CP	16 May 1975	12 September 1983
Bulgaria	AS/NCP	11 September 1978	
Uruguay	AS/CP	11 January 1980	7 October 1985
Papua New Guinea	AS/NCP	16 March 1981[3]	
Italy	AS/CP	18 March 1981	5 October 1987
Peru	AS/CP	10 April 1981	9 October 1989
Spain	AS/CP	31 March 1982	21 September 1988
China, People's Republic of	AS/CP	8 June 1983	7 October 1985
India	AS/CP	19 August 1983	12 September 1983
Hungary	AS/NCP	27 January 1984	
Sweden	AS/CP	24 April 1984	21 September 1988
Finland	AS/CP	15 May 1984	9 October 1989
Cuba	AS/NCP	16 August 1984	
Republic of Korea	AS/CP	28 November 1986	9 October 1989

Greece	AS/NCP	8 January 1987	
Democratic People's Republic of Korea	AS/NCP	21 January 1987	
Austria	AS/NCP	25 August 1987	
Ecuador	AS/CP	15 September 1987	19 November 1990
Canada	AS/NCP	4 May 1988	
Columbia	AS/NCP	31 January 1989	
Switzerland	AS/NCP	15 November 1990	
Guatemala	AS/NCP	31 July 1991	
Ukraine	AS/NCP	28 October 1992	

* OS: Original Signatory CL: Claimant State AS: Acceding State
CP: Consultative Party NCP: Non-consultative Party

1. Original Signatories have automatic consultative party status.
2. The Soviet Union's party status is now held by the Russian Federation.
3. Papua New Guinea's succession to the Antarctic Treaty followed its indepedence from Australia.

APPENDIX 3 PROTOCOL ON ENVIRONMENTAL PROTECTION TO THE ANTARCTIC TREATY

Preamble

The States Parties to this Protocol to the Antarctic Treaty, hereinafter referred to as the Parties,

Convinced of the need to enhance the protection of the Antarctic environment and dependent and associated ecosystems;

Convinced of the need to strengthen the Antarctic Treaty so as to ensure that Antarctica shall continue forever to be used exclusively for peaceful purposes and shall not become the scene or object of international discord;

Bearing in mind the special legal and political status of Antarctica and the special responsibility of the Antarctic Treaty Consultative Parties to ensure that all activities in Antarctica are consistent with the purposes and principles of the Antarctic Treaty;

Recalling the designation of Antarctica as a Special Conservation Area and other measures adopted under the Antarctic Treaty system to protect the Antarctic environment and dependent and associated ecosystems;

Acknowledging further the unique opportunities Antarctica offers for scientific monitoring of and research on processes of global as well as regional importance;

Reaffirming the conservation principles of the Convention on the Conservation of Antarctic Marine Living Resources;

Convinced that the development of a comprehensive regime for the protection of the Antarctic environment and dependent and associated ecosystems is in the interest of all mankind as a whole;

Desiring to supplement the Antarctic Treaty to this end;

Have agreed as follows:

Article 1 Definitions

For the purposes of this Protocol:

(a) 'The Antarctic Treaty' means the Antarctic Treaty done at Washington on 1 December 1959;
(b) 'Antarctic Treaty area' means the area to which the provisions of the Antarctic Treaty apply in accordance with Article VI of that Treaty;
(c) 'Antarctic Treaty Consultative Meetings' means the meetings referred to in Article IX of the Antarctic Treaty;
(d) 'Antarctic Treaty Consultative Parties' means the Contracting Parties to the Antarctic Treaty entitled to appoint representatives to participate in the meetings referred to in Article IX of that Treaty;
(e) 'Antarctic Treaty system' means the Antarctic Treaty, the measures in effect under that Treaty, its associated separate international instruments in force and the measures in effect under those instruments;
(f) 'Arbitral Tribunal' means the Arbitral Tribunal established in accordance with the Schedule to this Protocol, which forms an integral part thereof;
(g) 'Committee' means the Committee for Environmental Protection established in accordance with Article 11.

Article 2 Objective and Designation

The Parties commit themselves to the comprehensive protection of the Antarctic environment and dependent and associated ecosystems and hereby designate Antarctica as a natural reserve, devoted to peace and science.

Article 3 Environmental Principles

1. The protection of the Antarctic environment and dependent and associated ecosystems and the intrinsic value of Antarctica, including its wilderness and aesthetic values and its value as an area for the conduct of scientific research, in particular research essential to understanding the global environment, shall be fundamental considerations in the planning and conduct of all activities in the Antarctic Treaty area.

2. To this end:

(a) activities in the Antarctic Treaty area shall be planned and conducted so as to limit adverse impacts on the Antarctic environment and dependent and associated ecosystems;

(b) activities in the Antarctic Treaty area shall be planned and conducted so as to avoid:
   (i) adverse effects on climate or weather patterns;
   (ii) significant adverse effects on air or water quality;
   (iii) significant changes in the atmospheric, terrestrial (including aquatic), glacial or marine environments;
   (iv) detrimental changes in the distribution, abundance or productivity of species or populations of species of fauna and flora;
   (v) further jeopardy to endangered or threatened species or populations of such species; or
   (vi) degradation of, or substantial risk to, areas of biological, scientific, historic, aesthetic or wilderness significance;

(c) activities in the Antarctic Treaty area shall be planned and conducted on the basis of information sufficient to allow prior assessments of, and informed judgments about, their possible impacts on the Antarctic environment and dependent and associated ecosystems and on the value of Antarctica for the conduct of scientific research; such judgments shall take full account of:
   (i) the scope of the activity, including its area, duration and intensity;
   (ii) the cumulative impacts of the activity, both by itself and in combination with other activities in the Antarctic Treaty area;
   (iii) whether the activity will detrimentally affect any other activity in the Antarctic Treaty area;
   (iv) whether technology and procedures are available to provide for environmentally safe operations;
   (v) whether there exists the capacity to monitor key environmental parameters and ecosystem components so as to identify and provide early warning of any adverse effects of the activity and to provide for such modification of operating procedures as may be necessary in the light of the results of monitoring or increased knowledge of the Antarctic environment and dependent and associated ecosystems; and
   (vi) whether there exists the capacity to respond promptly and effectively to accidents, particularly those with potential environmental effects;

(d) regular and effective monitoring shall take place to allow assessment of the impacts of ongoing activities, including the verification of predicted impacts;

(e) regular and effective monitoring shall take place to facilitate early detection of the possible unforeseen effects of activities carried on both within and outside the Antarctic Treaty area on the Antarctic environment and dependent and associated ecosystems.

3. Activities shall be planned and conducted in the Antarctic Treaty area so as to accord priority to scientific research and to preserve the value of Antarctica as an area for the conduct of such research, including research essential to understanding the global environment.

4. Activities undertaken in the Antarctic Treaty area pursuant to scientific research programmes, tourism and all other governmental and non-

governmental activities in the Antarctic Treaty area for which advance notice is required in accordance with Article VII(5) of the Antarctic Treaty, including associated logistic support activities, shall:
(a) take place in a manner consistent with the principles in this Article; and
(b) be modified, suspended or cancelled if they result in or threaten to result in impacts upon the Antarctic environment or dependent or associated ecosystems inconsistent with those principles.

Article 4 Relationship with other components of the Antarctic Treaty System

1. This Protocol shall supplement the Antarctic Treaty and shall neither modify nor amend that Treaty.

2. Nothing in this Protocol shall derogate from the rights and obligations of the Parties to this Protocol under the other international instruments in force within the Antarctic Treaty system.

Article 5 Consistency with the other components of the Antarctic Treaty system

The Parties shall consult and co-operate with the Contracting Parties to the other international instruments in force within the Antarctic Treaty system and their respective institutions with a view to ensuring the achievement of the objectives and principles of this Protocol and avoiding any interference with the achievement of the objectives and principles of those instruments or any inconsistency between the implementation of those instruments and of this Protocol.

Article 6 Co-operation

1. The Parties shall co-operate in the planning and conduct of activities in the Antarctic Treaty area. To this end, each Party shall endeavour to:
(a) promote co-operative programmes of scientific, technical and educational value, concerning the protection of the Antarctic environment and dependent and associated ecosystems;
(b) provide appropriate assistance to other Parties in the preparation of environmental impact assessments;
(c) provide to other Parties upon request information relevant to any potential environmental risk and assistance to minimise the effects of accidents which may damage the Antarctic environment or dependent and associated ecosystems;
(d) consult with other Parties with regard to the choice of sites for prospective stations and other facilities so as to avoid the cumulative impacts caused by their excessive concentration in any location;
(e) where appropriate, undertake joint expeditions and share the use of stations and other facilities; and

(f) carry out such steps as may be agreed upon at Antarctic Treaty Consultative Meetings.

2. Each Party undertakes, to the extent possible, to share information that may be helpful to other Parties in planning and conducting their activities in the Antarctic Treaty area, with a view to the protection of the Antarctic environment and dependent and associated ecosystems.

3. The Parties shall co-operate with those Parties which may exercise jurisdiction in areas adjacent to the Antarctic Treaty area with a view to ensuring that activities in the Antarctic Treaty area do not have adverse environmental impacts on those areas.

Article 7 Prohibition of Mineral Resource Activities

Any activity relating to mineral resources, other than scientific research, shall be prohibited.

Article 8 Environmental Impact Assessment

1. Proposed activities referred to in paragraph 2 below shall be subject to the procedures set out in Annex I for prior assessment of the impacts of those activities on the Antarctic environment or on dependent or associated ecosystems according to whether those activities are identified as having:
(a) less than a minor or transitory impact;
(b) a minor or transitory impact; or
(c) more than a minor or transitory impact.

2. Each Party shall ensure that the assessment procedures set out in Annex I are applied in the planning processes leading to decisions about any activities undertaken in the Antarctic Treaty area pursuant to scientific research programmes, tourism and all other governmental and non-governmental activities in the Antarctic Treaty area for which advance notice is required under Article VII(5) of the Antarctic Treaty, including associated logistic support activities.

3. The assessment procedures set out in Annex I shall apply to any change in an activity whether the change arises from an increase or decrease in the intensity of an existing activity, from the addition of an activity, the decommissioning of a facility, or otherwise.

4. Where activities are planned jointly by more than one Party, the Parties involved shall nominate one of their number to coordinate the implementation of the environmental impact assessment procedures set out in Annex I.

Article 9 Annexes

1. The Annexes to this Protocol shall form an integral part thereof.

2. Annexes, additional to Annexes I-IV, may be adopted and become effective in accordance with Article IX of the Antarctic Treaty.

3. Amendments and modifications to Annexes may be adopted and become effective in accordance with Article IX of the Antarctic Treaty, provided that any Annex may itself make provision for amendments and modifications to become effective on an accelerated basis.

4. Annexes and any amendments and modifications thereto which have become effective in accordance with paragraphs 2 and 3 above shall, unless an Annex itself provides otherwise in respect of the entry into effect of any amendment or modification thereto, become effective for a Contracting Party to the Antarctic Treaty which is not an Antarctic Treaty Consultative Party, or which was not an Antarctic Treaty Consultative Party at the time of the adoption, when notice of approval of that Contracting Party has been received by the Depositary.

5. Annexes shall, except to the extent that an Annex provides otherwise, be subject to the procedures for dispute settlement set out in Articles 18 to 20.

Article 10 Antarctic Treaty Consultative Meetings

1. Antarctic Treaty Consultative Meetings shall, drawing upon the best scientific and technical advice available:
(a) define, in accordance with the provisions of this Protocol, the general policy for the comprehensive protection of the Antarctic environment and dependent and associated ecosystems; and
(b) adopt measures under Article IX of the Antarctic Treaty for the implementation of this Protocol.

2. Antarctic Treaty Consultative Meetings shall review the work of the Committee and shall draw fully upon its advice and recommendations in carrying out the tasks referred to in paragraph 1 above, as well as upon the advice of the Scientific Committee on Antarctic Research.

Article 11 Committee for Environmental Protection

1. There is hereby established the Committee for Environmental Protection.

2. Each Party shall be entitled to be a member of the Committee and to appoint a representative who may be accompanied by experts and advisers.

3. Observer status in the Committee shall be open to any Contracting Party to the Antarctic Treaty which is not a Party to this Protocol.

4. The Committee shall invite the President of the Scientific Committee on Antarctic Research and the Chairman of the Scientific Committee for the Conservation of Antarctic Marine Living Resources to participate as observers at its sessions. The Committee may also, with the approval of the Antarctic Treaty Consultative Meeting, invite such other relevant scientific, environmental and technical organisations which can contribute to its work to participate as observers at its sessions.

5. The Committee shall present a report on each of its sessions to the Antarctic Treaty Consultative Meeting. The report shall cover all matters considered at the session and shall reflect the views expressed. The report shall be circulated to the Parties and to observers attending the session, and shall thereupon be made publicly available.

6. The Committee shall adopt its rules of procedure which shall be subject to approval by the Antarctic Treaty Consultative Meeting.

Article 12 Functions of the Committee

1. The functions of the Committee shall be to provide advice and formulate recommendations to the Parties in connection with the implementation of this Protocol, including the operation of its Annexes, for consideration at Antarctic Treaty Consultative Meetings, and to perform such other functions as may be referred to it by the Antarctic Treaty Consultative Meetings. In particular, it shall provide advice on:

(a) the effectiveness of measures taken pursuant to this Protocol;
(b) the need to update, strengthen or otherwise improve such measures;
(c) the need for additional measures, including the need for additional Annexes, where appropriate;
(d) the application and implementation of the environmental impact assessment procedures set out in Article 8 and Annex I;
(e) means of minimising or mitigating environmental impacts of activities in the Antarctic Treaty area;
(f) procedures for situations requiring urgent action, including response action in environmental emergencies;
(g) the operation and further elaboration of the Antarctic Protected Area system;
(h) inspection procedures, including formats for inspection reports and checklists for the conduct of inspections;
(i) the collection, archiving, exchange and evaluation of information related to environmental protection;
(j) the state of the Antarctic environment; and
(k) the need for scientific research, including environmental monitoring, related to the implementation of this Protocol.

2. In carrying out its functions, the Committee shall, as appropriate, consult with the Scientific Committee on Antarctic Research, the Scientific Committee for the Conservation of Antarctic Marine Living Resources and other relevant scientific, environmental and technical organisations.

Article 13 Compliance with this Protocol

1. Each Party shall take appropriate measures within its competence, including the adoption of laws and regulations, administrative actions and enforcement measures, to ensure compliance with this Protocol.

2. Each Party shall exert appropriate efforts, consistent with the Charter of the United Nations, to the end that no one engages in any activity contrary to this Protocol.

3. Each Party shall notify all other Parties of the measures it takes pursuant to paragraphs 1 and 2 above.

4. Each Party shall draw the attention of all other Parties to any activity which in its opinion affects the implementation of the objectives and principles of this Protocol.

5. The Antarctic Treaty Consultative Meetings shall draw the attention of any State which is not a Party to this Protocol to any activity undertaken by that State, its agencies, instrumentalities, natural or juridical persons, ships, aircraft or other means of transport which affects the implementation of the objectives and principles of this Protocol.

Article 14 Inspection

1. In order to promote the protection of the Antarctic environment and dependent and associated ecosystems, and to ensure compliance with this Protocol, the Antarctic Treaty Consultative Parties shall arrange, individually or collectively, for inspections by observers to be made in accordance with Article VII of the Antarctic Treaty.

2. Observers are:
(a) observers designated by any Antarctic Treaty Consultative Party who shall be nationals of that Party; and
(b) any observers designated at Antarctic Treaty Consultative Meetings to carry out inspections under procedures to be established by an Antarctic Treaty Consultative Meeting.

3. Parties shall co-operate fully with observers undertaking inspections, and shall ensure that during inspections, observers are given access to all parts of stations, installations, equipment, ships and aircraft open to inspection under Article VII(3) of the Antarctic Treaty, as well as to all records maintained thereon which are called for pursuant to this Protocol.

4. Reports of inspections shall be sent to the Parties whose stations, installations, equipment, ships or aircraft are covered by the reports. After those Parties have been given the opportunity to comment, the reports and any comments thereon shall be circulated to all the Parties and to the Committee, considered at the next Antarctic Treaty Antarctic Treaty Consultative Meeting, and thereafter made publicly available.

Article 15 Emergency Response Action

1. In order to respond to environmental emergencies in the Antarctic Treaty area, each Party agrees to:

(a) provide for prompt and effective response action to such emergencies which might arise in the performance of scientific research programmes, tourism and all other governmental and non-governmental activities in the Antarctic Treaty area for which advance notice is required under Article VII(5) of the Antarctic Treaty, including associated logistic support activities;
(b) establish contingency plans for response to incidents with potential adverse effects on the Antarctic environment or dependent or associated ecosystems.

2. To this end, the Parties shall:
(a) co-operate in the formulation and implementation of such contingency plans; and
(b) establish procedures for immediate notification of, and co-operative response to, environmental emergencies.

3. In the implementation of this Article, the Parties shall draw upon the advice of the appropriate international organisations.

Article 16 **Liability**

Consistent with the objectives of this Protocol for the comprehensive protection of the Antarctic environment and dependent and associated ecosystems, the Parties undertake to elaborate rules and procedures relating to liability for damage arising from activities taking place in the Antarctic Treaty area and covered by this Protocol. Those rules and procedures shall be included in one or more Annexes to be adopted in accordance with Article 9(2).

Article 17 **Annual Report by Parties**

1. Each Party shall report annually on the steps taken to implement this Protocol. Such reports shall include notifications made in accordance with Article 13(3), contingency plans established in accordance with Article 15 and any other notifications and information called for pursuant to this Protocol for which there is no other provision concerning the circulation and exchange of information.

2. Reports made in accordance with paragraph 1 above shall be circulated to all Parties and to the Committee, considered at the next Antarctic Treaty Consultative Meeting, and made publicly available.

Article 18 **Dispute Settlement**

If a dispute arises concerning the interpretation or application of this Protocol, the parties to the dispute shall, at the request of any one of them, consult among themselves as soon as possible with a view to having the dispute resolved by negotiation, inquiry, mediation, conciliation, arbitration, judicial settlement or other peaceful means to which the parties to the dispute agree.

Article 19 Choice of Dispute Settlement Procedure

1. Each Party, when signing, ratifying, accepting, approving or acceding to this Protocol, or at any time thereafter, may choose, by written declaration, one or both of the following means for the settlement of disputes concerning the interpretation or application of Articles 7, 8 and 15 and, expect to the extent that an Annex provides otherwise, the provisions of any Annex and, insofar as it relates to these Articles and provisions, Article 13:
(a) the International Court of Justice;
(b) the Arbitral Tribunal.

2. A declaration made under paragraph 1 above shall not affect the operation of Article 18 and Article 20(2).

3. A Party that has not made a declaration under paragraph 1 above or in respect of which a declaration is no longer in force shall be deemed to have accepted the competence of the Arbitral Tribunal.

4. If the parties to a dispute have accepted the same means for the settlement of a dispute, the dispute may be submitted only to that procedure, unless the parties otherwise agree.

5. If the parties to a dispute have not accepted the same means for the settlement of a dispute, or if they have both accepted both means, the dispute may be submitted only to the Arbitral Tribunal, unless the parties otherwise agree.

6. A declaration made under paragraph 1 above shall remain in force until it expires in accordance with its terms or until three months after written notice of revocation has been deposited with the Depositary.

7. A new declaration, a notice of revocation or the expiry of a declaration shall not in any way affect proceedings pending before the International Court of Justice or the Arbitral Tribunal, unless the parties to the dispute otherwise agree.

8. Declarations and notices referred to in this Article shall be deposited with the Depositary who shall transmit copies thereof to all Parties.

Article 20 Dispute Settlement Procedure

1. If the parties to a dispute concerning the interpretation or application of Articles 7, 8 or 15 or, except to the extent that an Annex provides otherwise, the provisions of any Annex or, insofar as it relates to these Articles and provisions, Article 13, have not agreed on a means for resolving it within 12 months of the request for consultation pursuant to Article 18, the dispute shall be referred, at the request of any party to the dispute, for settlement in accordance with the procedure determined by Article 19(4) and (5).

2. The Arbitral Tribunal shall not be competent to decide or rule upon any matter within the scope of Article IV of the Antarctic Treaty. In addition,

nothing in this Protocol shall be interpreted as conferring competence of jurisdiction on the International Court of Justice or any other tribunal established for the purpose of settling disputes between Parties to decide or otherwise rule upon any matter within the scope of Article IV of the Antarctic Treaty.

Article 21 Signature

This Protocol shall be open for signature at Madrid on the 4th of October 1991 and thereafter at Washington until the 3rd of October 1992 by any State which is a Contracting Party to the Antarctic Treaty.

Article 22 Ratification, Acceptance, Approval or Accession

1. This Protocol is subject to ratification, acceptance or approval by signatory States.

2. After the 3rd of October 1992 this Protocol shall be open for accession by any State which is a Contracting Party to the Antarctic Treaty.

3. Instruments of ratification, acceptance, approval or accession shall be deposited with the Government of the United States of America, hereby designated as the Depositary.

4. After the date on which this Protocol has entered into force, the Antarctic Treaty Consultative Parties shall not act upon a notification regarding the entitlement of a Contracting Party to the Antarctic Treaty to appoint representatives to participate in Antarctic Treaty Consultative Meetings in accordance with Article IX(2) of the Antarctic Treaty unless that Contracting Party has first ratified, accepted, approved or acceded to this Protocol.

Article 23 Entry into Force

1. This Protocol shall enter into force on the thirtieth day following the date of deposit of instruments of ratification, acceptance, approval or accession by all States which are Antarctic Treaty Consultative Parties at the date on which this Protocol is adopted.

2. For each Contracting Party to the Antarctic Treaty which, subsequent to the date of entry into force of this Protocol, deposits an instrument of ratification, acceptance, approval or accession, this Protocol shall enter into force on the thirtieth day following such deposit.

Article 24 Reservations

Reservations to this Protocol shall not be permitted.

Article 25 Modification or Amendment

1. Without prejudice to the provisions of Article 9, this Protocol may be modified or amended at any time in accordance with the procedures set forth in Article XII (1)(a) and (b) of the Antarctic Treaty.

2. If, after the expiration of 50 years from the date of entry into force of this Protocol, any of the Antarctic Treaty Consultative Parties so requests by a communication addressed to the Depositary, a conference shall be held as soon as practicable to review the operation of this Protocol.

3. A modification or amendment proposed at any Review Conference called pursuant to paragraph 2 above shall be adopted by a majority of the Parties, including 3/4 of the States which are Antarctic Treaty Consultative Parties at the time of adoption of this Protocol.

4. Any modification or amendment adopted pursuant to paragraph 3 above shall enter into force upon ratification by 3/4 of the Consultative Parties, including ratification, acceptance, approval or accession by all States which are Consultative Parties at the time of the adoption of this Protocol.

5(a) With respect to Article 7, the prohibition on Antarctic mineral resource activities contained therein shall continue unless there is in force a binding legal regime on Antarctic mineral resource activities that includes an agreed means for determining whether, and, if so, under which conditions, any such activities would be acceptable. This regime shall fully safeguard the interests of all States referred to in Article IV of the Antarctic Treaty and apply the principles thereof. Therefore, if a modification or amendments to Article 7 is proposed at a Review Conference referred to in paragraph 2 above, it shall include such a binding legal regime.

(b) If any such modification or amendment has not entered into force within 3 years of the date of its adoption, any Party may at any time thereafter notify to the Depositary of its withdrawal from this Protocol, and such withdrawal shall take effect 2 years after receipt of the notice by the Depositary.

Article 26 Notifications by the Depositary

The Depositary shall notify all Contracting Parties to the Antarctic Treaty of the following:

(a) signatures of this Protocol and the deposits of instruments of ratification, acceptance, approval or accession;
(b) the date of entry into force of this Protocol and any additional Annex thereto;
(c) the date of entry into force of any amendment or modification to this Protocol; and
(d) the deposit of declarations and notices pursuant to Article 19; and
(e) any notification received pursuant to Article 25(5)(b).

Article 27 Authentic Texts and Registration with the United Nations

1. This Protocol, done in the English, French, Russian and Spanish languages, each version being equally authentic, shall be deposited in the archives of the Government of the United States of America, which shall transmit duly certified copies thereof to all Contracting Parties to the Antarctic Treaty.

2. This Protocol shall be registered by the Depositary pursuant to Article 102 of the Charter of the United Nations.

SCHEDULE TO THE PROTOCOL: ARBITRATION

Article 1

1. The Arbitral Tribxunal shall be constituted and shall function in accordance with this Protocol, including this Schedule.

2. The Secretary referred to in this Schedule is the Secretary General of the Permanent Court of Arbitration.

Article 2

1. Each Party shall be entitled to designate up to three Arbitrators, at least one of whom shall be designated within three months of the entry into force of this Protocol for that Party. Each Arbitrator shall be experienced in Antarctic affairs, have thorough knowledge of international law and enjoying the highest reputation for fairness, competence and integrity. The names of the persons so designated shall constitute the list of Arbitrators. Each Party shall at all times maintain the name of at least one Arbitrator on the list.

2. Subject to paragraph 3 below, an Arbitrator designated by a Party shall remain on the list for a period of five years and shall be eligible for redesignation by that Party for additional five year periods.

3. A Party which designated an Arbitrator may withdraw the name of that Arbitrator from the list. If an Arbitrator dies or if a Party for any reason withdraws from the list the name of an Arbitrator designated by it, the Party which designated the Arbitrator in question shall notify the Secretary promptly. An Arbitrator whose name is withdrawn from the list shall continue to serve on any Arbitral Tribunal to which that Arbitrator has been appointed until the completion of proceedings before the Arbitral Tribunal.

4. The Secretary shall ensure that an up-to-date list is maintained of the Arbitrators designated pursuant to this Article.

Article 3

1. The Arbitral Tribunal shall be composed of three Arbitrators who shall be appointed as follows:

(a) the party to the dispute commencing the proceedings shall appoint one Arbitrator, who may be its national, from the list referred to in Article 2. This appointment shall be included in the notification referred to in Article 4.

(b) Within 40 days of the receipt of that notification, the other party to the dispute shall appoint the second Arbitrator, who may be its national, from the list referred to in Article 2.

(c) Within 60 days of the appointment of the second Arbitrator, the parties to the dispute shall appoint by agreement the third Arbitrator from the list referred to in Article 2. The third Arbitrator shall not be either a national of, or a person designated by, a party to the dispute, or of the same nationality as either of the first two Arbitrators. The third Arbitrator shall be the Chairperson of the Arbitral Tribunal.

(d) If the second Arbitrator has not been appointed within the prescribed period, or if the parties to the dispute have not reached agreement within the prescribed period on the appointment of the third Arbitrator, the Arbitrator or Arbitrators shall be appointed, at the request of any party to the dispute and within 30 days of the receipt of such request, by the President of the International Court of Justice from the list referred to in Article 2 and subject to the conditions prescribed in subparagraphs (b) and (c) above. In performing the functions accorded him or her in this subparagraph, the President of the Court shall consult the parties to the dispute.

(e) If the President of the International Court of Justice is unable to perform the functions accorded him or her in subparagraph (d) above or is a national of a party to the dispute, the functions shall be performed by the Vice-President of the Court, except that if the Vice-President is unable to perform the functions or is a national of a party to the dispute the functions shall be performed by the next most senior member of the Court who is available and is not a national of a party to the dispute.

2. Any vacancy shall be filled in the manner prescribed for the initial appointment.

3. In any dispute involving more than two Parties, those Parties having the same interest shall appoint one Arbitrator by agreement within the period specified in paragraph 1(b) above.

Article 4

The party to the dispute commencing proceedings shall so notify the other party or parties to the dispute and the Secretary in writing. Such notification shall include a statement of the claim and the grounds on which it is based. The notification shall be transmitted by the Secretary to all Parties.

Article 5

1. Unless the parties to the dispute agree otherwise, arbitration shall take place at The Hague, where the records of the Arbitral Tribunal shall be kept. The Arbitral Tribunal shall adopt its own rules of procedure. Such rules shall ensure that each party to the dispute has a full opportunity to be heard and to present its case and shall also ensure that the proceedings are conducted expeditiously.

2. The Arbitral Tribunal may hear and decide counterclaims arising out of the dispute.

Article 6

1. The Arbitral Tribunal, where it considers that *prima facie* it has jurisdiction under this Protocol, may:
(a) at the request of any party to a dispute, indicate such provisional measures as it considers necessary to preserve the respective rights of the parties to the dispute;
(b) prescribe any provisional measures which it considers appropriate under the circumstances to prevent serious harm to the Antarctic environment or dependent or associated ecosystems.

2. The parties to a dispute shall comply promptly with any provisional measures prescribed under paragraph 1(b) above pending an award under Article 9.

3. Notwithstanding Article 20(1) and (2) of this Protocol, a party to a dispute may at any time, by notification to the other party or parties to the dispute and to the Secretary in accordance with Article 4, request that the Arbitral Tribunal be constituted as a matter of exceptional urgency to indicate or prescribe emergency provisional measures in accordance with this Article. In such case, the Arbitral Tribunal shall be constituted as soon as possible in accordance with Article 3, except that the time periods in Article 3(1)(b), (c) and (d) shall be reduced to 14 days in each case. The Arbitral Tribunal shall decide upon the request for emergency provisional measures within two months of the appointment of its Chairperson.

4. Following a decision by the Arbitral Tribunal upon a request for emergency provisional measures in accordance with paragraph 3 above, settlement of the dispute shall proceed in accordance with Articles 18 and 19 of the Protocol.

Article 7

Any Party which believes it has a legal interest, whether general or individual, which may be substantially affected by the award of an Arbitral Tribunal, may, unless the Arbitral Tribunal decides otherwise, intervene in the proceedings.

Article 8

The parties to the dispute shall facilitate the work of the Arbitral Tribunal and, in particular, in accordance with their law and using all means at their disposal, shall provide it with all relevant documents and information, and enable it, when necessary, to call witnesses or experts and receive their evidence.

Article 9

If one of the parties to the dispute does not appear before the Arbitral Tribunal or fails to defend its case, any other party to the dispute may request the Arbitral Tribunal to continue the proceedings and make its award.

Article 10

1. The Arbitral Tribunal shall, on the basis of this Protocol and other applicable rules of law that are not incompatible with such provisions, decide such disputes as are submitted to it.

2. The Arbitral Tribunal may decide, *ex aequo et bono*, a dispute submitted to it, if the parties to the dispute so agree.

Article 11

1. Before making its award, the Arbitral Tribunal shall satisfy itself that it has competence in respect of the dispute and that the claim or counterclaim is well founded in fact and law.

2. The award shall be accompanied by a statement of reasons for the decision and shall be communicated to the Secretary who shall transmit it to all Parties.

3. The award shall be final and binding on the parties to the dispute and on any Party which intervened in the proceedings and shall be complied with without delay. The Arbitral Tribunal shall interpret the award at the request of a party to the dispute or of any intervening Party.

4. The award shall have no binding force except in respect of that particular case.

5. Unless the Arbitral Tribunal decides otherwise, the expenses of the Arbitral Tribunal, including the remuneration of the Arbitrators, shall be borne by the parties to the dispute in equal shares.

Article 12

All decisions of the Arbitral Tribunal, including those referred to in Articles 5, 6 and 11, shall be made by a majority of the Arbitrators who may not abstain from voting.

Article 13

1. This Schedule may be amended or modified by a measure adopted in accordance with Article IX(1) of the Antarctic Treaty. Unless the measure specifies otherwise, the amendment or modification shall be deemed to have been approved, and shall become effective, one year after the close of the Antarctic Treaty Consultative Meeting at which it was adopted, unless one or more of the Antarctic Treaty Consultative Parties notifies the Depositary, within that time period, that it wishes an extension of that period or that it is unable to approve the measure.

2. Any amendment or modification of this Schedule which becomes effective in accordance with paragraph 1 above shall thereafter become effective as to any other Party when notice of approval by it has been received by the Depositary.

ANNEX I TO THE PROTOCOL ON ENVIRONMENTAL PROTECTION TO THE ANTARCTIC TREATY

ENVIRONMENTAL IMPACT ASSESSMENT

Article 1 Preliminary Stage

1. The environmental impacts of proposed activities referred to in Article 8 of the Protocol shall, before their commencement, be considered in accordance with appropriate national procedures.

2. If an activity is determined as having less than a minor or transitory impact, the activity may proceed forthwith.

Article 2 Initial Environmental Evaluation

1. Unless it has been determined that an activity will have less than a minor or transitory impact, or unless a Comprehensive Environmental Evaluation is being prepared in accordance with Article 3, an Initial Environmental Evaluation shall be prepared. It shall contain sufficient detail to assess whether a proposed activity may have more than a minor or transitory impact and shall include:
(a) a description of the proposed activity, including its purpose, location, duration, and intensity; and
(b) consideration of alternatives to the proposed activity and any impacts that the activity may have, including consideration of cumulative impacts in the light of existing and known planned activities.

2. If an Initial Environmental Evaluation indicates that a proposed activity is likely to have no more than a minor or transitory impact, the activity may pro-

ceed, provided that appropriate procedures, which may include monitoring, are put in place to assess and verify the impact of the activity.

Article 3 Comprehensive Environmental Evaluation

1. If an Initial Environmental Evaluation indicates or if it is otherwise determined that a proposed activity is likely to have more than a minor or transitory impact, a Comprehensive Environmental Evaluation shall be prepared.

2. A Comprehensive Environmental Evaluation shall include:
(a) a description of the proposed activity including its purpose, location, duration and intensity, and possible alternatives to the activity, including the alternative of not proceeding, and the consequences of those alternatives;
(b) a description of the initial environmental reference state with which predicted changes are to be compared and a prediction of the future environmental reference state in the absence of the proposed activity;
(c) a description of the methods and data used to forecast the impacts of the proposed activity;
(d) estimation of the nature, extent, duration and intensity of the likely direct impacts of the proposed activity;
(e) consideration of possible indirect or second order impacts of the proposed activity;
(f) consideration of cumulative impacts of the proposed activity in the light of existing activities and other known planned activities;
(g) identification of measures, including monitoring programmes, that could be taken to minimise or mitigate impacts of the proposed activity and to detect unforeseen impacts and that could provide early warning of any adverse effects of the activity as well as to deal promptly and effectively with accidents;
(h) identification of unavoidable impacts of the proposed activity;
(i) consideration of the effects of the proposed activity on the conduct of scientific research and on other existing uses and values;
(j) an identification of gaps in knowledge and uncertainties encountered in compiling the information required under this paragraph;
(k) a non-technical summary of the information provided under this paragraph;
(l) the name and address of the person or organisation which prepared the Comprehensive Environmental Evaluation and the address to which comments thereon should be directed.

3. The draft Comprehensive Environmental Evaluation shall be made publicly available and shall be circulated to all Parties, which shall also make it publicly available, for comment. A period of 90 days shall be allowed for the receipt of comments.

4. The draft Comprehensive Environmental Evaluation shall be forwarded to the Committee at the same time as it is circulated to the Parties, and at least 120 days before the next Antarctic Treaty Consultative Meeting, for consideration as appropriate.

5. No final decision shall be taken to proceed with the proposed activity in the Antarctic Treaty area unless there has been an opportunity for consideration of the draft Comprehensive Environmental Evaluation by the Antarctic Treaty Consultative Meeting on the advice of the Committee, provided that no decision to proceed with a proposed activity shall be delayed through the operation of this paragraph for longer than 15 months from the date of circulation of the draft Comprehensive Environmental Evaluation.

6. A final Comprehensive Environmental Evaluation shall address and shall include or summarise comments received on the draft Comprehensive Environmental Evaluation. The final Comprehensive Environmental Evaluation, notice of any decisions relating thereto, and any evaluation of the significance of the predicted impacts in relation to the advantages of the proposed activity, shall be circulated to all Parties, which shall also make them publicly available, at least 60 days before the commencement of the proposed activity in the Antarctic Treaty area.

Article 4 Decisions to be based on Comprehensive Environmental Evaluations

Any decision on whether a proposed activity, to which Article 3 applies, should proceed, and, if so, whether in its original or in a modified form, shall be based on the Comprehensive Environmental Evaluation as well as other relevant considerations.

Article 5 Monitoring

1. Procedures shall be put in place, including appropriate monitoring of key environmental indicators, to assess and verify the impact of any activity that proceeds following the completion of a Comprehensive Environmental Evaluation.

2. The procedures referred to in paragraph 1 above and in Article 2(2) shall be designed to provide a regular and verifiable record of the impacts of the activity in order, *inter alia*, to:
(a) enable assessments to be made of the extent to which such impacts are consistent with this Protocol; and
(b) provide information useful for minimising or mitigating impacts, and, where appropriate, information on the need for suspension, cancellation or modification of the activity.

Article 6 Circulation of Information

1. The following information shall be circulated to the Parties, forwarded to the Committee and made publicly available:

(a) a description of the procedures referred to in Article 1;
(b) an annual list of an Initial Environmental Evaluations prepared in accordance with Article 2 and any decisions taken in consequence thereof;
(c) significant information obtained, and any action taken in consequence thereof, from procedures put in place in accordance with Articles 2(2) and 5; and
(d) information referred to in Article 3(6).

2. Any Initial Environmental Evaluation prepared in accordance with Article 2 shall be made available on request.

Article 7 Cases of Emergency

1. This Annex shall not apply in cases of emergency relating to the safety of human life or of ships, aircraft, or equipment and facilities of high value, or the protection of the environment, which require an activity to be undertaken without completion of the procedures set out in this Annex.

2. Notice of activities undertaken in cases of emergency, which would otherwise have required preparation of a Comprehensive Environmental Evaluation, shall be circulated immediately to the Parties and forwarded to the Committee and a full explanation of the activities carried out shall be provided within 90 days of those activities.

Article 8 Amendment or Modification

1. This Annex may be amended or modified by a measure adopted in accordance with Article IX(1) of the Antarctic Treaty. Unless the measure specifies otherwise, the amendment or modification shall be deemed to have been approved, and shall become effective, one year after the close of the Antarctic Treaty Consultative Meeting at which it was adopted, unless one or more of the Antarctic Treaty Consultative Parties notifies the Depositary, within that time period, that it wishes an extension of that period or that it is unable to approve the measure.

2. Any amendment or modification of this Annex which becomes effective in accordance with paragraph 1 above shall thereafter become effective as to any other Party when notice of approval by it has been received by the Depositary.

ANNEX II TO THE PROTOCOL ON ENVIRONMENTAL PROTECTION TO THE ANTARCTIC TREATY

CONSERVATION OF ANTARCTIC FAUNA AND FLORA

Article 1 Definitions

For the purposes of this Annex:

(a) 'native mammal' means any member of any species belonging to the Class Mammalia, indigenous to the Antarctic Treaty area or occurring there seasonally through natural migrations;
(b) 'native bird' means any member, at any stage of its life cycle (including eggs), of any species of the Class Aves indigenous to the Antarctic Treaty area or occurring there seasonally through natural migrations;
(c) 'native plant' means any terrestrial or freshwater vegetation, including bryophytes, lichens, fungi and algae, at any state of its life cycle (including seeds, and other propagules), indigenous to the Antarctic Treaty area;
(d) 'native invertebrate' means any terrestrial or freshwater invertebrate, at any stage of its life cycle, indigenous to the Antarctic Treaty area;
(e) 'appropriate authority' means any person or agency authorised by a Party to issue permits under this Annex;
(f) 'permit' means a formal permission in writing issued by an appropriate authority;
(g) 'take' or 'taking' means to kill, injure, capture, handle or molest, a native mammal or bird, or to remove or damage such quantities of native plants that their local distribution or abundance would be significantly affected;
(h) 'harmful interference' means:
    (i) flying or landing helicopters or other aircraft in a manner that disturbs concentrations of birds and seals;
    (ii) using vehicles or vessels, including hovercraft and small boats, in a manner that disturbs concentrations of bird and seals;
    (iii) using explosives or firearms in a manner that disturbs concentrations of birds and seals;
    (iv) wilfully disturbing breeding or moulting birds or concentrations of birds and seals by persons on foot;
    (v) significantly damaging concentrations of native terrestrial plants by landing aircraft, driving vehicles, or walking on them, or by other means;
    (vi) any activity that results in the significant adverse modification of habitats of any species or population of native mammal, bird, plant or invertebrate.
(i) 'International Convention for the Regulation of Whaling' means the Convention done at Washington on 2 December 1946.

Article 2 Cases of Emergency

1. The provisions of this Annex shall not apply in cases of emergency relating to the safety of human life or of ships, aircraft, or equipment and facilities of high value, or protection of the environment.

2. Notice of activities undertaken in cases of emergency shall be circulated immediately to all Parties and to the Committee.

Article 3 Protection of Native Fauna and Flora

1. Taking or harmful interference shall be prohibited except in accordance with a permit.

2. Such permits shall specify the authorised activity, including when, where and by whom it is to be conducted and shall be issued only in the following circumstances:
(a) to provide specimens for scientific study or scientific information;
(b) to provide specimens for museums, herbaria, zoological and botanical gardens, or other educational or cultural institutions or uses; and
(c) to provide for unavoidable consequences of scientific activities not otherwise authorised under sub-paragraphs (a) or (b) above, or of the construction and operation of scientific support facilities.

3. The issue of such permits shall be limited so as to ensure that:
(a) no more native mammals, birds, or plants are taken than are strictly necessary to meet the purposes set forth in paragraph 2 above;
(b) only small numbers of native mammals or birds are killed and in no case more native mammals or birds are killed from local populations than can, in combination with other permitted takings, normally be replaced by natural reproduction in the following season; and
(c) the diversity of species, as well as the habitats essential to their existence, and the balance of the ecological systems existing within the Antarctic Treaty area are maintained.

4. Any species of native mammals, birds and plants listed in Appendix A to this Annex shall be designated 'Specially Protected Species' and shall be accorded special protection by the Parties.

5. A permit shall not be issued to take a Specially Protected Species unless the taking:
(a) is for a compelling scientific purpose;
(b) will not jeopardise the survival or recovery of that species or local population; and
(c) uses non-lethal techniques where appropriate.

6. All taking of native mammals and birds shall be done in the manner that involves the least degree of pain and suffering practicable.

Article 4 Introduction of non-native species, parasites and diseases

1. No species of animal or plant not native to the Antarctic Treaty area shall be introduced onto land or ice shelves, or into water in the Antarctic Treaty area except in accordance with a permit.

2. Dogs shall not be introduced onto land or ice shelves and dogs currently in those areas shall be removed by April 1, 1994.

3. Permits under paragraph 1 above shall be issued to allow the importation only of the animals and plants listed in Appendix B to this Annex and shall specify the species, numbers and, if appropriate, age and sex and precautions to be taken to prevent escape or contact with native fauna and flora.

4. Any plant or animal for which a permit has been issued in accordance with paragraphs 1 and 3 above, shall, prior to expiration of the permit, be removed from the Antarctic Treaty area or be disposed of by incineration or equally effective means that eliminates risk to native fauna or flora. The permit shall specify this obligation. Any other plant or animal introduced into the Antarctic Treaty area not native to that area, including any progeny, shall be removed or disposed of, by incineration or by equally effective means, so as to be rendered sterile, unless it is determined that they pose no risk to native flora or fauna.

5. Nothing in this Article shall apply to the importation of food into the Antarctic Treaty area provided that no live animals are imported for this purpose and all plants and animal parts and products are kept under carefully controlled conditions and disposed of in accordance with Annex III to the Protocol and Appendix C to this Annex.

6. Each Party shall require that precautions, including those listed in Appendix C to this Annex, be taken to prevent the introduction of microorganisms (e.g., viruses, bacteria, parasites, yeasts, fungi) not present in the native fauna and flora.

Article 5 Information

Each Party shall prepare and make available information setting forth, in particular, prohibited activities and providing lists of Specially Protected Species and relevant Protected Areas to all those persons present in or intending to enter the Antarctic Treaty area with a view to ensuring that such persons understand and observe the provisions of this Annex.

Article 6 Exchange of Information

1. The Parties shall make arrangements for:
(a) collecting and exchanging records (including records of permits) and statistics concerning the numbers of quantities of each species of native mammal, bird or plant taken annually in the Antarctic Treaty area;

(b) obtaining and exchanging information as to the status of native mammals, birds, plants, and invertebrates in the Antarctic Treaty area, and the extent to which any species or population needs protection;
(c) establishing a common form in which this information shall be submitted by Parties in accordance with paragraph 2 below.

2. Each Party shall inform the other Parties as well as the Committee before the end of November each year of any step taken pursuant to paragraph 1 above and of the number and nature of permits issued under this Annex in the preceding period of 1st July to 30th June.

Article 7 Relationship with Other Agreements outside the Antarctic Treaty system

Nothing in this Annex shall derogate from the rights and obligations of Parties under the International Convention for the Regulation of Whaling.

Article 8 Review

The Parties shall keep under continuing review measures for the conservation of Antarctic fauna and flora, taking into account any recommendations from the Committee.

Article 9 Amendment or Modification

1. This Annex may be amended or modified by a measure adopted in accordance with Article IX(1) of the Antarctic Treaty. Unless the measure specifies otherwise, the amendment or modification shall be deemed to have been approved, and shall become effective, one year after the close of the Antarctic Treaty Consultative Meeting at which it was adopted, unless one or more of the Antarctic Treaty Consultative Parties notifies the Depositary, within that time period, that it wishes an extension of that period of time or that it is unable to approve the measure.

2. Any amendment or modification of this Annex which becomes effective in accordance with paragraph 1 above shall thereafter become effective as to any other Party when notice of approval by it has been received by the Depositary.

APPENDICES TO THE ANNEX

Appendix A: Specially Protected Species

All species of the genus *Arctocephalus*, Fur Seals. *Ommatophoca rossii*, Ross Seal.

Appendix B: Importation of Animals and Plants

The following animals and plants may be imported into the Antarctic Treaty area in accordance with permits issued under Article 4 of this Annex:

(a) domestic plants; and
(b) laboratory animals and plants including viruses, bacteria, yeasts and fungi.

Appendix C: Precautions to prevent introduction of micro-organisms

1. Poultry. No live poultry or other living birds shall be brought into the Antarctic Treaty area. Before dressed poultry is packaged for shipment to the Antarctic Treaty area, it shall be inspected for evidence of disease, such as Newcastle's Disease, tuberculosis, and yeast infection. Any poultry or parts not consumed shall be removed from the Antarctic Treaty area or disposed of by incineration or equivalent means that eliminates risks to native flora and fauna.

2. The importation of non-sterile soil shall be avoided to the maximum extent practicable.

ANNEX III TO THE PROTOCOL ON ENVIRONMENTAL PROTECTION TO THE ANTARCTIC TREATY

WASTE DISPOSAL AND WASTE MANAGEMENT

Article 1 General Obligations

1. This Annex shall apply to activities undertaken in the Antarctic Treaty area pursuant to scientific research programmes, tourism and all other governmental and non-governmental activities in the Antarctic Treaty area for which advance notice is required under Article VII (5) of the Antarctic Treaty, including associated logistic support activities.

2. The amount of wastes produced or disposed of in the Antarctic Treaty area shall be reduced as far as practicable so as to minimise interference with the natural values of Antarctica, with scientific research and with other uses of Antarctica which are consistent with the Antarctic Treaty.

3. Waste storage, disposal and removal from the Antarctic Treaty area, as well as recycling and source reduction, shall be essential considerations in the planning and conduct of activities in the Antarctic Treaty area.

4. Wastes removed from the Antarctic Treaty area shall, to the maximum extent practicable, be returned to the country from which the activities gen-

erating the waste were organised or to any other country in which arrangements have made for the disposal of such wastes in accordance with relevant international agreements.

5. Past and present waste disposal sites on land and abandoned work sites of Antarctic activities shall be cleaned up by the generator of such wastes and the user of such sites. This obligation shall not be interpreted as requiring:
(a) the removal of any structure designated as a historic site or monument; or
(b) the removal of any structure or waste material in circumstances where the removal by an practical option would result in greater adverse environmental impact than leaving the structure or waste material in its existing location.

Article 2 Waste Disposal by Removal from the Antarctic Treaty Area

1. The following wastes, if generated after entry into force of this Annex, shall be removed from the Antarctic Treaty area by the generator of such wastes:
(a) radio-active materials;
(b) electrical batteries;
(c) fuel, both liquid and solid;
(d) wastes containing harmful levels of heavy metals or acutely toxic or harmful persistent compounds;
(e) poly-vinyl chloride (PVC), polyurethane foam, polystyrene foam, rubber and lubricating oils, treated timbers and other products which contain additives that could produce harmful emissions if incinerated;
(f) all other plastic wastes, except low density polyethylene containers (such as bags for storing wastes), provided that such containers shall be incinerated in accordance with Article 3(1);
(g) fuel drums; and
(h) other solid, non-combustible wastes;

provided that the obligation to remove drums and solid non-combustible wastes contained in subparagraphs (g) and (h) above shall not apply in circumstances where the removal of such wastes by any practical option would result in greater adverse environmental impact than leaving them in their existing locations.

2. Liquid wastes which are not covered by paragraph 1 above and sewage and domestic liquid wastes, shall, to the maximum extent practicable, be removed from the Antarctic Treaty area by the generator of such wastes.

3. The following wastes shall be removed from the Antarctic Treaty area by the generator of such wastes, unless incinerated, autoclaved or otherwise treated to be made sterile:
(a) residues of carcasses of imported animals;
(b) laboratory culture of micro-organisms and plant pathogens; and
(c) introduced avian products.

Article 3 Waste Disposal by Incineration

1. Subject to paragraph 2 below, combustible wastes, other than those referred to in Article 2(1), which are not removed from the Antarctic Treaty area shall be burnt in incinerators which to the maximum extent practicable reduce harmful emissions. Any emission standards and equipment guidelines which may be recommended by, inter alia, the Committee and the Scientific Committee on Antarctic Research shall be taken into account. The solid residue of such incineration shall be removed from the Antarctic Treaty area.

2. All open burning of wastes shall be phased out as soon as practicable, but no later than the end of the 1998/1999 season. Pending the completion of such phase-out, when it is necessary to dispose of wastes by open burning, allowance shall be made for the wind direction and speed and the type of wastes to be burnt to limit particulate deposition and to avoid such deposition over areas of special biological, scientific, historic, aesthetic or wilderness significance including, in particular, areas accorded protection under the Antarctic Treaty.

Article 4 Other Waste Disposal on Land

1. Wastes not removed or disposed of in accordance with Articles 2 and 3 shall not be disposed of onto ice-free areas or into fresh water systems.

2. Sewage, domestic liquid wastes and other liquid wastes not removed from the Antarctic Treaty area in accordance with Article 2, shall, to the maximum extent practicable, not be disposed of onto sea ice, ice shelves or the grounded ice-sheet, provided that such wastes which are generated by stations located inland on ice shelves or on the grounded ice-sheet may be disposed of in deep ice pits where such disposal is the only practicable option. Such pits shall not be located on known ice-flow lines which terminate at ice-free areas or in areas of high ablation.

3. Wastes generated at field camps shall, to the maximum extent practicable, be removed by the generator of such wastes to supporting stations or ships for disposal in accordance with this Annex.

Article 5 Disposal of Waste in the Sea

1. Sewage and domestic liquid wastes may be discharged directly into the sea, taking into account the assimilative capacity of the receiving marine environment and provided that:

(a) such discharge is located, wherever practicable, where conditions exist for initial diluation and rapid dispersal; and

(b) large quantities of such wastes (genreated in a station where the average weekly occupancy over the austral summer is approximately 30 individuals or more) shallbe treated at least by maceration.

2. The by-product of sewage treatment by the Rotary Biological Contacter process or similar processes may be disposed of into the sea provided that such disposal does not adversely affect the local environment, and provided also that any such disposal at sea shall be in accordance with Annex IV to the Protocol.

Article 6 Storage of Waste

All wastes to be removed from the Antarctic Treaty area, or otherwise disposed of, shall be stored in such a way as to prevent their dispersal into the environment.

Article 7 Prohibited Products

No polychlorinated biphenyls (PCBs), non-sterile soil, polystyrene beads, chips or similar forms of packaging, or pesticides (other than those required for scientific, medical or hygiene purposes) shall be introduced onto land or ice shelves or into water in the Antarctic Treaty area.

Article 8 Waste Management Planning

1. Each Party which itself conducts activities in the Antarctic Treaty area shall, in respect of those activities, establish a waste disposal classification system as a basis for recording wastes and to facilitate studies aimed at evaluating the environmental impacts of scientific activity and associated logistic support. To that end, wastes produced shall be classified as:
(a) sewage and domestic liquid wastes (Group 1);
(b) other liquid wastes and chemicals, including fuels and lubricants (Group 2);
(c) solids to be combusted (Group 3);
(d) other solid wastes (Group 4); and
(e) radioactive materials (Group 5).

2. In order to reduce further the impact of waste on the Antarctic environment, each such Party shall prepare and annually review and update its waste managment plans (including waste reduction, storage and disposal), specifying for each fixed site, or field camps generally, and for each ship (other than small boats that are part of the operation of fixed sites or of ships and taking into account existing management plans for ships):
(a) programmes for cleaning up existing waste disposal sites and abandoned work sites;
(b) current and planned waste management arrangements, including final disposal;
(c) current and planned arrangements for analysing the environmental effects of waste and waste management; and
(d) other efforts to minimise any environmental effects of wastes and waste management.

3. Each such Partry shall, as far as is practicable, also prepare an inventory of past activities (such as traverses, fuel depots, field bases, crashed aircraft) before the information is lost, so that such locations can be taken into account in planning future sicentific programmes (such as snow chemistry, pollutants in lichens or ice core drilling).

Article 9 Circulation and Review of Waste Management Plans

1. The waste management plans prepared in accordance with Article 8, reports on their implementation, and the inventories referred to in Article 8(3), shall be included in the annual exchanges of information in accordance with Articles III and VII of the Antarctic Treaty and related Recommendations under Article IX of the Antarctic Treaty.

2. Each Party shall send copies of its waste managment plans, and reports on their implementation and review, to the Committee.

3. The Committee may review waste management plans and reports thereon and may offer comments, including suggestions for minimising impacts and modifications and improvement to the plans, for the consideration of the Parties.

4. The Parties may exchange information and provide advice on, inter alia, available low waste technologies, reconversion of existing installations, special requirements for effluents, and appropriate disposal and discharge methods.

Article 10 Management Practices

Each Party shall:

(a) designate a waste management official to develop and monitor waste management plans; in the field, this responsibility shall be delegated to an appropriate person at each site;
(b) ensure that members of its expedition receive training designed to limit the impact of its operations on the Antarctic environment and to inform them of requirements of this Annex; and
(c) discourage the use of poly-vinyl chloride (PVC) products and ensure that its expeditions to the Antarctic Treaty area are advised of any PVC products they may introduce in the Antarctic Treaty area in order that they may be removed subsequently in accordance with this Annex.

Article 11 Review

This Annex shall be subject to regular review in order to ensure that it is updated to reflect improvement in waste disposal technology and procedures and to ensure thereby maximum protection of the Antarctic environment.

Article 12 Cases of Emergency

1. This Annex shall not apply in cases of emergency relating to the safety of human life or of ships, aircraft or other equipment and facilities of high value.

2. Notice of activities undertaken in cases of emergency shall be circulated immediately to all Parties and to the Committee.

Article 13 Amendment or Modification

1. This Annex may be amended or modified by a measure adopted in accordance with Article IX(1) of the Antarctic Treaty. Unless the measure specifies otherwise, the amendment or modification shall be deemed to have been approved, and shall become effective, one year after the close of the Antarctic Treaty Consultative Meeting at which it was adopted, unless one or more of the Antarctic Treaty Consultative Parties notifies the Depositary, within that time period, that it wishes an extension of that period of time or that it is unable to approve the amendment.

2. Any amendment or modification of this Annex which becomes effective in accordance with paragraph 1 above shall thereafter become effective as to any other Party when notice of approval by it has been received by the Depositary.

ANNEX IV TO THE PROTOCOL ON ENVIRONMENTAL PROTECTION TO THE ANTARCTIC TREATY

PREVENTION OF MARINE POLLUTION

Article 1 Definitions

For the purposes of this Annex:

(a) 'discharge' means any release howsoever caused from a ship and includes any escape, disposal, spilling, leaking, pumping, emitting or emptying;
(b) 'garbage' means all kinds of victual, domestic and operational waste excluding fresh fish and parts thereof, generated during the normal operation of the ship except those substances which are covered by Articles 3 and 4;
(c) 'MARPOL 73/78' means the International Convention for the Prevention of Pollution from Ships, 1973 as amended by the Protocol of 1978 relating thereto and by any other amendment in force thereafter;
(d) 'Noxious liquid substance' means any noxious liquid substance as defined in Annex II of MARPOL 73/78;
(e) 'oil' means petroleum in any form including crude oil, fuel oil, sludge, oil refuse and refined oil products (other than petrochemicals which are subject to the provisions of Article 4);

(f) 'oily mixture' means a mixture with any oil content;
(g) 'ship' means a vessel of any type whatsoever operating in the marine environment and includes hydrofoil boats, air-cushion vehicles, submersibles, floating craft and fixed or floating platforms.

Article 2 Application

This Annex applies, with respect to each Party, to ships entitled to fly its flag and with respect to any other ship engaged in or supporting its Antarctic operations, while operating in the Antarctic Treaty area.

Article 3 Discharge of Oil

1. Any discharge into the sea of oil or oily mixture shall be prohibited, except in cases permitted under Annex I of MARPOL 73/78. While operating in the Antarctic Treaty area, ships shall retain on board all sludge, dirty ballast, tank washing water and other oily residues and mixtures which may not be discharged into the sea. Ships shall discharge these residues only outside the Antarctic Treaty area, at reception facilities or as otherwise permitted under Annex I of MARPOL.

2. This Article shall not apply to:
(a) the discharge into the sea of oil or oily mixture resulting from damage to a ship or its equipment:
   (i) provided that all reasonable precautions have been taken after the occurrence of the damage or discovery of the discharge for the purpose of preventing or minimising the discharge; and
   (ii) except if the owner or the Master acted either with intent to cause damage, or recklessly and with the knowledge that damage would probably result; or
(b) the discharge into the sea of substances containing oil which are being used for the purpose of combating specific pollution incidents in order to minimise the damage from pollution.

Article 4 Discharge of Noxious Liquid Substances

The discharge into the sea of any noxious liquid substance, and any other chemical or other substances, in quantities or concentrations that are harmful to the marine environment, shall be prohibited.

Article 5 Disposal of Garbage

1. The disposal into the sea of all plastics, including but not limited to synthetic ropes, synthetic fishing nets, and plastic garbage bags, shall be prohibited.

2. The disposal into the sea of all other garbage, including paper products, rags, glass, metal, bottles, crockery, incineration ash, dunnage, lining and packing materials, shall be prohibited.

3. The disposal into the sea of food wastes may be permitted when they have passed through a comminuter or grinder, provided that such disposal shall, except in cases permitted under Annex V of MARPOL 73/78, be made as far as practicable from land and ice shelves but in any case not less than 12 nautical miles from the nearest land or ice shelf. Such comminuted or ground food wastes shall be capable of passing through a screen with openings no greater than 25 millimeters.

4. When a substance or material covered by this article is mixed with other such substance or material for discharge or disposal, having different disposal or discharge requirements, the most stringent disposal or discharge requirements shall apply.

5. The provisions of paragraphs 1 and 2 above shall not apply to:
(a) the escape of garbage resulting from damage to a ship or its equipment provided all reasonable precautions have been taken, before and after the occurrence of the damage, for the purpose of preventing or minimising the escape; or
(b) the accidential loss of synthetic fishing nets, provided all reasonable precautions have been taken to prevent such loss.

6. The Parties shall, where appropriate, require the use of garbage record books.

Article 6 Discharge of Sewage

1. Except where it would unduly impair Antarctic operations:
(a) each Party shall eliminate all discharge into the sea of untreated sewage ('sewage' being defined in Annex IV of MARPOL 73/78) within 12 nautical miles of land or ice shelves;
(b) beyond such distance, sewage stored in a holding tank shall not be discharged instantaneously but shall be discharged at a moderate rate and, where practicable, while the ship is en route at a speed of no less than 4 knots.

This paragraph does not apply to ships certified to carry not more than 10 persons.

2. The Parties shall, where appropriate, require the use of sewage record books.

Article 7 Cases of Emergency

1. Articles 3, 4, 5, and 6 of this Annex shall not apply in cases of emergency relating to the safety of a ship and those on board or saving life at sea.

2. Notice of activities undertaken in cases of emergency shall be circulated immediately to all Parties and to the Committee.

Article 8 Effect on Dependent and Associated Ecosystems

In implementing the provisions of this Annex, due consideration shall be given to the need to avoid detrimental effects on dependent and associated ecosystems, outside the Antarctic Treaty area.

Article 9 Ship Retention Capacity and Reception Facilities

1. Each Party shall undertake to ensure that all ships entitled to fly its flag and any other ship engaged in or supporting its Antarctic operations, before entering the Antarctic Treaty area, are fitted with a tank or tanks of sufficient capacity on board for the retention of all sludge, dirty ballast, tank washing water, other oily residues and mixtures, and have sufficient capacity on board for the retention of garbage, while operating in the Antarctic Treaty area and have concluded arrangements to discharge such oily residues and garbage at a reception facility after leaving that area. Ships shall also have sufficient capacity on board for the retention of noxious liquid substances.

2. Each Party at whose ports ships depart en route to or arrive from the Antarctic Treaty area undertakes to ensure that as soon as practicable adequate facilities are provided for the reception of all sludge, dirty ballast, tank washing water, other oily residues and mixtures, and garbage from ships, without causing undue delay, and according to the needs of the ships using them.

3. Parties operating ships which depart to or arrive from the Antarctic Treaty area at ports of other Parties shall consult with those Parties with a view to ensuring that the establishment of port reception facilities does not place an inequitable burden on Parties adjacent to the Antarctic Treaty area.

Article 10 Design, Construction, Manning and Equipment of Sites

In the design, construction, manning and equipment of ships engaged in or supporting Antarctic operations, each Party shall take into account the objectives of this Annex.

Article 11 Sovereign Immunity

1. This Annex shall not apply to any warship, naval auxiliary or other ship owned or operated by a State and used, for the time being, only on government non-commercial service. However, each Party shall ensure by the adoption of appropriate measures not impairing the operations or operational capabilities of such shops owned or operated by it, that such ships act in a manner consistent, so far as is reasonable and practicable, with this Annex.

2. In applying paragraph 1 above, each Party shall take into account the importance of protecting the Antarctic environment.

3. Each Party shall inform the other Parties of how it implements this provision.

4. The dispute settlement procedure set out in Articles 18 to 20 of the Protocol shall not apply to this Article.

Article 12 Preventive Measures and Emergency Preparedness and Response

1. In order to respond more effectively to marine pollution emergencies or the threat thereof in the Antarctic Treaty area the Parties, in accordance with Article 15 of the Protocol, shall develop contingency plans for marine pollution response in the Antarctic Treaty area, including contingency plans for ships (other than small boats that are part of the operations of fixed sites or of ships) operating in the Antarctic Treaty area, particularly ships carrying oil as cargo and for oil spills originating from coastal installations, which enter into the marine environment. To this end they shall:
(a) co-operate in the formulation and implementation of such plans; and
(b) draw on the advice of the Committee, the International Maritime Organisation and other international organisations.

2. The Parties shall also establish procedures for co-operative response to pollution emergencies and shall take appropriate response actions in accordance with such procedures.

Article 13 Review

The Parties shall keep under continuous review the provisions of this Annex and other measures to prevent, reduce and respond to pollution of the Antarctic marine environment, including any amendments and new regulations adopted under MARPOL 73/78, with a view to achieving the objectives of this Annex.

Article 14 Relationship with MARPOL 73/78

With respect to those Parties which are also Parties to MARPOL 73/78, nothing in this Annex shall derogate from the specific rights and obligations thereunder.

Article 15 Amendment or Modification

1. This Annex may be amended or modified by a measure adopted in accordance with Article IX(1) of the Antarctic Treaty. Unless the measure specifies otherwise, the amendment or modification shall be deemed to have been approved, and shall become effective, one year after the close of the Antarctic Treaty Consultative Meeting at which it was adopted, unless one or more of the Antarctic Treaty Consultative Parties notifies the Depositary,

within that time period, that it wishes an extension of that period or that it is unable to approve the measure.

2. Any amendment or modification of this Annex which becomes effective in accordance with paragraph 1 above shall thereafter become effective as to any other Party when notice of approval by it has been received by the Depositary.

APPENDIX 4 ANNEX V TO THE PROTOCOL ON ENVIRONMENTAL PROTECTION TO THE ANTARCTIC TREATY

AREA PROTECTION AND MANAGEMENT

(Annex to recommendation XVI-10)

Article 1 **Definitions**

For the purposes of this Annex:

(a) 'appropriate authority' means any person or agency authorised by a Party to issue permits under this Annex;
(b) 'permit' means a formal permission in writing issued by an appropriate authority;
(c) 'Management Plan' means a plan to manage the activities and protect the special value or values in an Antarctic Specially Protected Area or an Antarctic Specially Managed Area.

Article 2 **Objectives**

For the purposes set out in this Annex, any area, including any marine area, may be designated as an Antarctic Specially Protected Area or an Antarctic Specially Managed Area. Activities in those areas shall be prohibited, restricted or managed in accordance with Management Plans adopted under the provisions of this Annex.

Article 3 **Antarctic Specially Protected Areas**

1. Any area, including any marine area, may be designated as an Antarctic Specially Protected Area to protect outstanding environmental, scientific, historic, aesthetic or wilderness values, any combination of those values, or ongoing or planned scientific research.

2. Parties shall seek to identify, within a systematic environmental-geographical framework, and to include in the series of Antarctic Specially Protected Areas:

(a) areas kept inviolate from human interference so that future comparisons may be possible with localities that have been affected by human activities;
(b) representative examples of major terrestrial, including glacial and aquatic, ecosystems and marine ecosystems;
(c) areas with important or unusual assemblages of species, including major colonies or breeding native birds or mammals;
(d) the type locality or only known habitat of any species;
(e) areas of particular interest to ongoing or planned scientific research;
(f) examples of outstanding geological, glaciological or geomorphological features;
(g) areas of outstanding aesthetic and wilderness value;
(h) sites or monuments of recognised historic value; and
(i) such other areas as may be appropriate to protect the values set out in paragraph 1 above.

3. Specially Protected Areas and Sites of Special Scientific Interest designated as such by past Antarctic Treaty Consultative Meetings are hereby designated as Antarctic Specially Protected Area and shall be renamed and renumbered accordingly.

4. Entry into an Antarctic Specially Protected Area shall be prohibited except in accordance with a permit issued under Article 7.

Article 4 Antarctic Specially Managed Areas

1. Any area, including any marine area, where activities are being conducted or may in the future be conducted, may be designated as an Antarctic Specially Managed Area to assist in the planning and co-ordination of activities, avoid possible conflicts, improve co-operation between Parties or minimise environmental impacts.

2. Antarctic Specially Managed Areas may include:
(a) areas where activities pose risks of mutual interference or cumulative environmental impacts; and
(b) sites or monuments of recognised historic value.

3. Entry into an Antarctic Specially Managed Area shall not require a permit.

4. Notwithstanding paragraph 3 above, an Antarctic Specially Managed Area may contain one or more Antarctic Specially Protected Areas, entry into which shall be prohibited except in accordance with a permit issued under Article 7.

Article 5 Management Plans

1. Any Party, the Committee, the Scientific Committee on Antarctic Research or the Commission for the Conservation of Antarctic Marine Living Resources may propose an area for designation as an Antarctic Specially Pro-

tected Area or an Antarctic Specially Managed Area by submitting a proposed Management Plan to the Antarctic Treaty Consultative Meeting.

2. The area proposed for designation shall be of sufficient size to protect the values for which the special protection or management is required.

3. Proposed Management Plans shall include, as appropriate:
(a) a description of the value or values for which special protection or management is required;
(b) a statement of the aims and objectives of the Management Plan for the protection or management of those values;
(c) management activities which are to be undertaken to protect the values for which special protection or management is required;
(d) a period of designation, if any;
(e) a description of the area, including:
    (i) the geographical co-ordinates, boundary markers and natural features that delineate the area;
    (ii) access to the area by land, sea or air including marine approaches and anchorages, pedestrian and vehicular routes within the area, and aircraft routes and landing areas;
    (iii) the location of structures, including scientific stations, research or refuge facilities, both within the area and near to it; and
    (iv) the location in or near the area of other Antarctic Specially Protected Areas or Antarctic Specially Managed Areas designated under this Annex, or other protected areas designated in accordance with measures adopted under other components of the Antarctic Treaty System.
(f) the identification of zones within the area, in which activities are to be prohibited, restricted or managed for the purpose of achieving the aims and objectives referred to in subparagraph (b) above;
(g) maps and photographs that show clearly the boundary of the area in relation to surrounding features and key features within the area;
(h) supporting documentation;
(i) in respect of an area proposed for designation as an Antarctic Specially Protected Area, a clear description of the conditions under which permits may be granted by the appropriate authority regarding:
    (i) access to and movement within or over the area;
    (ii) activities which are or may be conducted within the area, including restrictions on time and place;
    (iii) the installation, modification, or removal of structures;
    (iv) the location of field camps;
    (v) restrictions on materials and organisms which may be brought into the area;
    (vi) the taking of or harmful interference with native flora and fauna;
    (vii) the collection or removal of anything not brought into the area by the permit-holder;
    (viii) the disposal of waste;
    (ix) measures that may be necessary to ensure that the aims and objectives of the Management Plan can continue to be met; and

   (x) requirements for reports to be made to the appropriate authority regarding visits to the area;

(j) in respect of an area proposed for designation as an Antarctic Specially Managed Area, a code of conduct regarding:
   (i) access to and movement within or over the area;
   (ii) activities which are or may be conducted within the area, including restrictions on time and place;
   (iii) the installation, modification, or removal of structures;
   (iv) the location of field camps;
   (v) the taking of or harmful interference with native flora and fauna;
   (vi) the collection or removal of anything not brought into the area by the visitor;
   (vii) the disposal of waste; and
   (viii) any requirements for reporting to be made to the appropriate authority regarding visits to the area; and

(k) provisions relating to the circumstances in whic Parties should seek to exchange information in advance of activities which they propose to conduct.

Article 6 Designation Procedures

1. Proposed Management Plans shall be forwarded to the Committee, the Scientific Committee on Antarctic Research and, as appropriate, to the Commission for the Conservation of Antarctic Marine Living Resources. In formulating its advice to the Antarctic Treaty Consultative Meeting, the Committee shall take into account any comments provided by the Scientific Committee on Antarctic Research and, as appropriate, by the Commission for the Conservation of Antarctic Marine Living Resources. Thereafter Management Plans may be approved by the Antarctic Treaty Consultative Parties by a measure adopted at an Antarctic Treaty Consultative Meeting in accordance with Article IX(1) of the Antarctic Treaty. Unless the measure specifies otherwise, the Plan shall be deemed to have been approved 90 days after the close of the Antarctic Treaty Consultative Meeting at which it was adopted, unless one or more of the Consultative Parties notifies the Depositary, within that time period, that it wishes an extension of that period or is unable to approve the measure.

2. Having regard to the provisions of Articles 4 and 5 of the Protocol, no marine area shall be designated as an Antarctic Specially Protected Area or an Antarctic Specially Managed Area without the prior approval of the Commission for the Conservation of Antarctic Marine Living Resources.

3. Designation of an Antarctic Specially Protected Area or an Antarctic Specially Managed Area shall be for an indefinite period unless the Management Plan provides otherwise. A review of a Management Plan shall be initiated at least every five years. The Plan shall be updated as necessary.

4. Management Plans may be amended or revoked in accordance with paragraph 1 above.

5. Upon approval Management Plans shall be circulated promptly by the Depositary to all Parties. The Depositary shall maintain a record of all currently approved Management Plans.

Article 7 Permits

1. Each Party shall appoint an appropriate authority to issue permits to enter and engage in activities within an Antarctic Specially Protected Area in accordance with the requirements of the Management Plan relating to that Area. The permit shall be accompanied by the relevant sections of the Management Plan and shall specify the extent and location of the Area, the authorised activities and when, where and by whom the activities are authorised and any other conditions imposed by the Management Plan.

2. In the case of a Specially Protected Area designated as such by past Antarctic Treaty Consultative Meetings which does not have a Management Plan, the appropriate authority may issue a permit for a compelling scientific purpose which cannot be served elsewhere and which will not jeopardise the natural ecological system in that Area.

3. Each Party shall require a permit-holder to carry a copy of the permit while in the Antarctic Specially Protected Area concerned.

Article 8 Historic Sites and Monuments

1. Sites or monuments of recognised historic value which have been designated as Antarctic Specially Protected Areas or Antarctic Specially Managed Areas, or which are located within such Areas, shall be listed as Historic Sites and Monuments.

2. Any Party may propose a site or monument or recognised historic value which has not been designated as an Antartic Specially Protected Area or an Antarctic Specially Managed Area, or which is not located within such an Area, for listing as a Historic Site or Monument. The proposal for listing may be approved by the Antarctic Treaty Consultative Parties by a measure adopted at an Antarctic Treaty Consultative Meeting in accordance with Article IX(1) of the Antarctic Treaty. Unless the measure specifies otherwise, the proposal shall be deemed to have been approved 90 days after the close of the Antarctic Treaty Consultative Meeting at which it was adopted, unless one or more of the Consultative Parties notifies the Depositary, within that time period, that it wishes an extension of that period or is unable to approve the measure.

3. Existing Historic Sites and Monuments which have been listed as such by previous Antarctic Treaty Consultative Meetings shall be included in the list of Historic Sites and Monuments under this Article.

4. Listed Historic Sites and Monuments shall not be damaged, removed or destroyed.

5. The list of Historic Sites and Monuments may be amended in accordance with paragraph 2 above. The Depositary shall maintain a list of current Historic Sites and Monuments.

Article 9 Information and Publicity

1. With a view to ensuring that all persons visiting or proposing to visit Antarctica understand and observe the provisions of this Annex, each Party shall make available information setting forth, in particular:
(a) the location of Antarctic Specially Protected Areas and Antarctic Specially Managed Areas;
(b) listing and maps of those Areas;
(c) the Management Plans, including listings of prohibitions relevant to each Area;
(d) the location of Historic Sites and Monuments and any relevant prohibition or restriction.

2. Each Party shall ensure that the location and, if possible, the limits of Antarctic Specially Protected Areas, Antarctic Specially Managed Areas and Historic Sites and Monuments are shown on its topographic maps, hydrographic charts and in other relevant publications.

3. Parties shall co-operate to ensure that, where appropriate, the boundaries of Antarctic Specially Protected Areas, Antarctic Specially Managed Areas and Historic Sites and Monuments are suitably marked on the site.

Article 10 Exchange of Information

1. The Parties shall make arrangements for:
(a) collecting and exchanging records, including records of permits and reports of visits, including inspection visits, to Antarctic Specially Protected Areas and reports of inspection visits to Antarctic Specially Managed Areas;
(b) obtaining and exchanging information on any significant change or damage to any Antarctic Specially Managed Area, Antarctic Specially Protected Area or Historic Site or Monument; and
(c) establishing common forms in which records and information shall be submitted by Parties in accordance with paragraph 2 below.

2. Each Party shall inform the other Parties and the Committee before the end of November each year of the number and nature of permits issued under this Annex in the preceding period of 1st July to 30th June.

3. Each Party conducting, funding or authorising research or other activities in Antarctic Specially Protected Areas or Antarctic Specially Managed Areas shall maintain a record of such activities and in the annual exchange of information in accordance with the Treaty shall provide summary descriptions of the activities conducted by persons subject to its jurisdiction in such areas in the preceding year.

4. Each Party shall inform the other Parties and the Committee before the end of November each year of measures it has taken to implement this Annex, including any site inspections and any steps it has taken to address instances of activities in contravention of the provisions of the approved Management Plan for an Antarctic Specially Protected Area of Antarctic Specially Managed Area.

Article 11 Cases of Emergency

1. The restrictions laid down and authorised by this Annex shall not apply in cases of emergency involving safety of human life or of ships, aircraft, or equipment and facilities of high value or the protection of the environment.

2. Notice of activities undertaken in cases of emergency shall be circulated immediately to all Parties and to the Committee.

Article 12 Amendment or Modification

1. This Annex may be amended or modified by a measure adopted in accordance with Article IX(1) of the Antarctic Treaty. Unless the measure specifies otherwise, the amendment or modification shall be deemed to have been approved, and shall become effective, one year after the close of the Antarctic Treaty Consultative Meeting at which it was adopted, unless one or more of the Antarctic Treaty Consultative Parties notifies the Depositary, within that time period, that it wishes an extension of that period or that it is unable to approve the measure.

2. Any amendment or modification of this Annex which becomes effective in accordance with paragraph 1 above shall thereafter become effective as to any other Party when notice of approval by it has been received by the Depositary.

Notes

1 INTERNATIONAL COOPERATION AND CHANGE

1. For more comprehensive analysis of regimes and regime models, see Krasner (1983a), Haas (1980; 1983), Haggard and Simmons (1987), Rosenau (1986), Young (1983) and Zacher (1990). For a critique of regime theory see Strange (1983) and Grieco (1988).
2. For more on hegemonic stability theory see Grunberg (1990), Keohane (1984) and Strange (1977).
3. This emphasis on negotiated regimes fits within a cognitive model of regimes. A number of approaches to the study of regimes have been identified in the literature. Young (1989b, pp. 350–2) identifies utilitarian models, based on game theory, and power focused models which have been most in evidence in hegemonic stability theory. Haggard and Simmons (1987), in their recent analysis of regime theory, identify four not necessarily mutually exclusive theoretical approaches to regimes – structural, game-theoretic, functional and cognitive. Young (1990a, pp. 340–2) has extended his earlier typology to identify five streams of analysis – structural, power-based, interest-based, cognitive and contextual.
4. Young's model of institutional bargaining was developed to deal with those inadequacies he perceived in earlier models of regime formation, particularly those concerned with hegemonic imposition or game theoretic prisoners' dilemma models. Hampson (1989–90, p. 55) calls Young's model a 'useful antidote' to realist and hegemony-based accounts of the requirements for international cooperation.
5. For example, a regime which sets out the general rules with respect to retiring debt to lesser developed countries might not include such countries among its negotiating participants, or it may continue to reflect the interests of the lender countries and institutions, with which the recipients have little choice but to concur.
6. Realist scholars argue that regimes cannot be differentiated from the power relationships pertaining among states participating in the regime, and that compliance with norms and rules is conditioned only by a state's short term interests. Regimes, in this view, are epiphenomenal.
7. For a more detailed discussion of reciprocity in international relations, see Keohane (1986a).
8. The most recent and extensive work on epistemic communities and the importance of epistemic consensus investigates a wider application of the term (see Haas, 1992).
9. Hegemonic stability theory suggests that change in a regime, or in the international system generally, is dependent upon the values and structural capabilities of the hegemon or occurs when the power of the hegemon declines and it is no longer able to maintain the regime.

Game theoretic approaches to regime, on the other hand, tend to concentrate on the sets of preferences at any one time and therefore, as Jervis observes (1988, pp. 326–7), set the question of change to one side.

10. In most countries, environmental ministries have been established only in the years since the 1972 UN Stockholm Conference on the Human Environment.

2 REGIME BUILDING: THE ANTARCTIC TREATY SYSTEM

1. The British assert that Edward Bransfield saw the Antarctic Peninsula late in January 1820, a claim that is somewhat undermined by the subsequent loss of Bransfield's report to the Admiralty. The Russian contender was Thaddeus von Bellingshausen who crossed the Antarctic Circle in January 1820. The Americans have fielded two candidates: sealer Nathaniel Palmer who sailed close to the Peninsula in November 1820 and Captain John Davis, who went ashore on the Antarctic Peninsula on 7 February 1821 – the first recorded landing. For a more detailed examination of early Antarctic history see Anderson (1981), Shapley (1985, pp. 1–18), Beck (1986, pp. 21–45), Quigg (1983, pp. 5–38) and Headland (1989).
2. The US Wilkes expedition followed and mapped the coast of East Antarctica for 2400 kms, and therefore had some political significance for the inchoate US claim. The French explorer Dumont D'Urville discovered but did not set foot on the coast he named Terre Adélie, which was later claimed by France. Captain James Clark Ross, under a British flag, discovered Victoria Land, the Ross Sea and the Ross Ice Shelf (now claimed by New Zealand).
3. The 1897 Belgian expedition led by Baron de Gerlache was the first to winter over in Antarctic waters when his ship, the Belgica, stuck fast in pack ice. Borchgrevink, a Norwegian-born resident of Australia (financed by private British funds) headed south in 1898. Using sledge dogs for the first time in the Antarctic, he reached 78° south, further south than anyone had ever been before. Three major expeditions sailed in 1901, from Germany (von Drygalski), Sweden (Nordenskjöld) and Britain (Scott). Other expeditions during this period included Charcot (1908–10), Shackleton (in 1908 and 1914–15), Filchner (1911–13), the first Australasian Antarctic expedition under Mawson (1911–14), and a Japanese expedition in 1911–12.
4. In 1907, Shackleton came within 97 miles of the South Pole. The British flag was raised at the South Magnetic Pole on 16 January 1909 by (Australian) Professor Edgeworth David's three man party. Roald Amundsen planted the Norwegian flag at the South Pole on 14 December 1911, a month ahead of his rival, Englishman Robert Falcon Scott whose party perished on the return journey.
5. Those thirteen were Britain, Norway, the US, Russia, France, Japan, Belgium, Australia, Sweden, Germany, New Zealand, the Netherlands and Argentina.

6. The original 1908 claim mistakenly included parts of the Chilean and Argentine mainland (Shapley, 1985, p. 69).
7. The 1929–31 British, Australian, New Zealand Antarctic Research Expedition (BANZARE), led by Australian Douglas Mawson, was intended to prepare the way for this Australian claim.
8. France did not lay claim to other areas of coastline discovered by French explorers. French expeditions of 1903–05 and 1908–10 had been to the more accessible Palmer peninsula area (Hayton, 1960b, p. 375).
9. In 1938–39 Hermann Goerring despatched an aerial mapping expedition to the Antarctic. German planes mapped 350 000 square miles of the continent and dropped swastika claim markers (Mitchell, 1981, p. 69) even though German rights to the Antarctic had been renounced in Article 118 of the Versailles Treaty after World War I (Hunter Miller, 1927, p. 508).
10. Norway had already claimed Bouvet Island in 1928 and Peter I Island, west of the Peninsula, in 1931. Norway's claim to the mainland is only to the coast and hinterland because the 'sector principle' is not advantageous to Norway in the Arctic and so Norway has wished to avoid giving it credibility by using it in the Antarctic (Quigg, 1983, p. 112).
11. See Hayton (1956) for a good contemporary discussion of the Argentinian and Chilean claims. For a discussion on the involvement and attitudes of other South American states, see Clark (1988) and Child (1988). Under the Papal Bull of 1493 (later confirmed by the Spanish-Portuguese Treaty of Tordesillas in 1494) the world was divided between Spain and Portugal – Spain was given the New World, including undiscovered lands, which included the then unknown Antarctic.
12. The boundaries of the claim are apparently set by a Presidential decree of September 1946 which prohibits publication of Argentinian maps without all of the claimed territories as set forth in that decree (see Hayton, 1960b, p. 379).
13. On the development of US policy in this era see Shapley (1985), Quigg (1983, pp. 126–41), Hayton (1960b, pp. 382–86).
14. Records of claims were classified. None of the claims was formalised, but contributed to the development by the US government of its doctrine of 'constructive occupation' – discovery followed by subsequent exploration by air or land, coupled with a formal claim to possession as well as other acts, such as administrative acts, short of actual and permanent settlement (Shapley, 1985, pp. 46–7; 67).
15. For a more detailed analysis of the interests of the various Antarctic states and the claims see Shapley (1985, pp. 65–82), Hayton (1960b), Quigg (1983, pp. 110–41). On the international law debates about the claims and the exercise of sovereignty see, *inter alia*, Greig (1978), Conforti (1986), Johnson (1976), Bernhardt (1975) and Honnold (1978).
16. Agreement was difficult because any settlement of boundaries would have implications for the dispute between them over the Beagle Channel.
17. A British suggestion in 1947 that the competing claims be submitted to the International Court of Justice for arbitration and settlement was

rejected by Argentina and Chile. Britain repeated this proposal in 1951, 1953 and 1954 and in 1955 made a unilateral application for arbitration which was subsequently dropped.

18. The US Navy conducted coastline mapping operations and photographed an estimated 60 per cent of the Antarctic coastline (one-quarter of which had not been previously seen). Secretary of State Acheson also urged the expedition to take steps which might assist a future US claim. 68 claim markers were dropped but no public announcement about these actions was made.
19. Throughout this period the US had not rejected the possibility of making a claim of its own. President Eisenhower had approved a 1954 National Security Proposal that the US make a formal claim although in 1956 he revised Antarctic policy to delete references to permanent stations and ongoing mapping programs (Shapley, 1985, pp. 60–1; Beck, 1986, p. 41). In the 1956–57 season, US nationals deposited 31 claims in various parts of the Antarctic.
20. There were no diplomatic relations between the Soviet Union and Chile at the time.
21. Britain, New Zealand, France, Argentina, the Netherlands, the US, Chile, Norway, South Africa and Sweden (see Headland 1989, pp. 310–36).
22. For a summary of the interests of the IGY states, albeit from a British perspective, see Bertram (1958). Not all of the Antarctic stations were new ones – in 1955 there had been 20 stations on the Antarctic continent, operated by Argentina, Chile, the UK and Australia. 19 of these were peninsula stations; the other was Australia's Mawson station.
23. In both the 1956–57 and 57–58 seasons, Soviet meteorologists wintered at Little America V station, and US meteorologists wintered at the Soviet station Mirny (Department of State, 1964, p. 5).
24. For example, the US, which had already established a base at McMurdo in 1955, established the Amundsen-Scott base at the South Pole (where no-one had stood since Scott in 1912) effectively giving it a geographic (and potentially political) foothold in all sectors. Shapley (1985, pp. 85–6) gives a nice account of how decisions were made about the siting of some of the bases in order to avoid confrontation. According to Shapley (1988, p. 313) the US had not planned a South Pole station. However at a 1955 IGY planning meeting, the Soviet delegate announced that the USSR intended to put a base there. The French chairperson (Shapley suggests with a quick wink to the US delegate) countered this by saying that the US already had plans for siting a station there, even though they had not until that moment. The Soviets were therefore 'diverted' to the Pole of Inaccessibility. Quigg (1983, p. 48) tells the story differently, suggesting that the US had expressed a prior interest in the South Pole site.
25. In the early 1950s, Britain, Chile and Argentina had resumed their heated diplomatic correspondence over their conflicting claims (see Hanessian, 1960, p. 447). In 1952, British and Argentine expeditions exchanged gunfire (although without casualties) at Hope Bay. See Beck (1986, pp. 32–6) also for a description of the various actions

taken by the three competing claimants to assert their sovereignty over the area. In 1953 tension rose after the British tore down an Argentine hut on Deception Island and arrested two Argentine personnel who were expelled as illegal immigrants.

26. In 1956, New Zealand Prime Minister Nash had revived the UN trusteeship proposal but again it was not well received. Various proposals were also made for some form of expanding condominium based initially on the US, New Zealand and Australia (Quigg, 1983, p. 140). In 1956, and again in 1957 and 1958, India, concerned at the possibility of conflict in Antarctica, proposed that the question of the Antarctic be placed on the UN General Assembly agenda, but did not press the matter in the face of Argentinian and Chilean opposition and lack of support from Britain. However this served to indicate that there was a potential for interest from non-Antarctic states.
27. Documents from the meetings are still classified, as are those of the formal treaty negotiations which followed. However, researchers have recently been able to work from the personal papers of Admiral Dufek, who was the US Antarctic Projects Officer at the time, and who was kept informed of progress of the meetings through a series of detailed memoranda. Shapley's analysis (1985, pp. 91–100) is based on those unpublished notes from what she terms the 'shirtsleeves meetings'.
28. Even the US proposal for the compromise on sovereignty was based on proposals put forward by Chile in the 1940s.
29. He suggests that in relative gains bargaining, the main interest of any party is to maximise its gain by building up bargaining power (Kremenyuk, 1991, p. 30). It may be that building up bargaining power is what the USSR tried, unsuccessfully, to do in the informal negotiations prior to the Washington Conference by refusing to discuss anything of substance.
30. See Shapley, 1985, p. 111; Beck, 1986, p. 63, 73, 116, 170; Quigg, 1983, p. 146.
31. For analyses of the Treaty see, *inter alia*, Hanessian (1960), Hayton (1960a), Beeby (1972), Triggs (1985), Bush (1982, pp. 38–111). For an analysis which focuses particularly on regime components, see Peterson (1988, pp. 41–9).
32. Several references are made to the principles and objectives of the Treaty in its articles, but these are not specifically elaborated, thus leaving their definition to the Treaty parties.
33. Article IV does not *prevent* acts or activities which might be considered the prerogative of a sovereign state – it simply says that they cannot be used *in support of* the assertion of claims. The question which has exercised the minds of international lawyers is the extent to which international law can be frozen. A strict interpretation of IV(2) would be that, should the Treaty collapse, the situation as at 1961 in respect of claims would pertain. This would also mean that the practice of free access could not be argued to support the application of a law of common spaces to the Antarctic. Bush (1982, pp. 60–1) lists a 'string of questions' raised by article IV(2).

34. Disputes between parties over the exercise of jurisdiction are to be addressed by consultation among them with a view to reaching a mutually acceptable solution.
35. This provision was strongly backed by the British, Soviet and French delegations (Hanessian, 1960, p. 468). Beeby (1972, p. 12) indicates that the final wording owed much to the Soviet drafting effort.
36. Neither of the superpowers were keen on this rule being incorporated into the Treaty (Hanessian, 1960, p. 467). The prohibition is subject to the provision that if all consultative parties are signatories to any other international agreement which addresses those issues, the rules of that agreement shall apply. In any event, this has not occurred.
37. Several states expressed their concern at this limitation on open accession (see Bush, 1982, pp. 42–3). Apparently one state held out on this issue, and consensus could not be reached on anything more. Both North and South Korea acceded to the Treaty under the invitation provision.
38. The original signatories will always be able to participate in decision-making even if they do not continue to undertake scientific research. For example, Belgium, Japan and Norway closed their IGY stations but were not required to relinquish consultative status. Japan returned to the Antarctic in 1965 and Norway in the 1968–69 season. Until 1983, those who could not 'participate' in decision-making were not able even to attend consultative meetings as observers.
39. This has led some observers to conclude, erroneously, that the Treaty would expire after 30 years. It is, in fact, of unlimited duration.
40. If a review conference amendment has not become effective within two years, *any* party (consultative or non-consultative) may withdraw from the Treaty, such withdrawal to take effect a further two years from notification of such intention.
41. In this formulation, ice-shelves are not part of the high seas, even though they are not grounded on land. This was particularly important for New Zealand whose claimed territory included the Ross Ice Shelf.
42. See Bush (1982, pp. 66–70) on the high seas provisions of article VI.
43. The inspection provision had initially been linked to the demilitarisation provision, but the provisions were delinked in the face of opposition by the superpowers.
44. Shapley (1985, p. 97) suggests that a provision was drafted but its content is not known.
45. In accordance with the final Act of the Washington Conference, the participating governments met in Washington on 27 January 1960 and continued to meet to consult and make recommendations to their governments on interim arrangements regarding matters dealt with in the Treaty (Department of State, 1964, p. 11). In all, 37 meetings were held before entry into force, at which delegates produced Rules of Procedure and a draft agenda for the first consultative meeting.
46. For a useful and concise recent study of the interests of individual states in the Antarctic Treaty system, see Rowland (1988, pp. 15–25).
47. Some countries, especially Chile and Argentina, have included delegates from their defence ministries.

48. Advice is sought from SCAR through national Antarctic committees and is, in turn, officially transmitted to consultative meetings through national delegations.
49. This question of the relationship with SCAR was dealt with at the first consultative meeting and indeed in the preparatory meetings prior to that. Governments were recommended, individually, to encourage SCAR to continue its advisory work on scientific cooperation. The short recommendation (I-IV, ATS, 1989, p. 1226) then recognises that the work of SCAR constitutes 'a most valuable contribution to international scientific cooperation in Antarctica' and that SCAR should be encouraged to continue in this advisory work (see also Bush, 1982, pp. 25–7, 110–11; Myhre, 1986, pp. 77–85). Note also that this process differs from that by which a SCAR observer (as opposed to expert) presents a general report to the Meeting on SCAR activities.
50. The Treaty has not totally depoliticised the Antarctic. In 1973 the President of Argentina took his entire cabinet to the Vicecommodoro Marambio station, declaring it to be the temporary capital of Argentina (Mitchell, 1981, p. 70). A baby was born to an Argentine woman in Antarctica in January 1978, and marriage ceremonies have been conducted there. In January 1977 President Pinochet of Chile went to the Antarctic, taking with him a large chest of soil collected from various parts of Chile, as a symbol of Chilean unity. The Argentine press expressed surprise at his provocative behaviour in Argentinian territory: in Pinochet's opinion he had never left Chile (Neilson, 1977). At the second consultative meeting the British suggested that a Committee of Experts be convened to examine the position relating to jurisdiction in Antarctica and to formulate appropriate recommendations for further consideration (ATCM, 1962c). Although the item was listed in the draft agenda (ATCM, 1962b) it was not inscribed on the formal agenda and the final report of the meeting makes no reference to it having been discussed at all.
51. In 1977 Poland indicated that it believed it had fulfilled the scientific activities criteria required for consultative party status. In the absence of any rules in the Treaty on how to address this situation, the original signatories decided that it should not be automatic, but rather 'awarded' by the existing consultative parties. Their decision on this matter was, however, intended to take effect immediately rather than requiring the subsequent agreement of all governments in accordance with article IX of the Treaty. As Bush observes (1982, p. 337) this is representative of a more flexible decision-making approach.
52. The reports of meetings of experts are submitted to consultative meetings for final consideration.
53. The World Meteorological Organisation has been an exception to this (see Bush, 1982, pp. 55–6; 122).
54. Their suggestion that the Australian government be entrusted, at least for the time being, with essential administrative services was not adopted. Australia had apparently been keen to have a secretariat established in Canberra and this was a bone of some contention at early consultative meetings.

55. The consultative parties have adopted a numbering system for recommendations which identifies the meeting at which a recommendation was adopted (the first part) and the individual recommendation's number in the sequence for that particular meeting (the second part). For example, I-XIV is the fourteenth recommendation adopted at the first consultative meeting.
56. A seemingly less contentious New Zealand suggestion (ATCM, 1962d) that intersessional meetings be held at least quarterly at the Washington mission of the host government of the forthcoming consultative meeting (later revised to refer only to the diplomatic mission of the host government in the capital of any one of the Contracting powers) was unacceptable. So was a South African proposal (ATCM, 1962e) for the establishment of a working group, to carry out a variety of roles, in a capital where all Treaty parties were represented. Oddly, this document concludes that 'the above measures are designed to eliminate the possibility of even the embryo of a permanent secretariat' whereas it seems that it would do precisely that.
57. See, for example, Mansukoski (1987, p. 15) on Finland, Joyner (1990, p. 53) on India, Lee (1990, p. 581) on China.
58. This would make the total number of signatories 42 including 27 consultative parties – except that the reunification of Germany changes the numbers to 41 and 26 respectively, as both the FRG and GDR were consultative parties. The consultative party status of the former USSR has been taken up by the Russian Federation. The Ukraine has acceded to the Treaty as an independent state.
59. The rules on substantial scientific activity were relaxed to accommodate some of the newer Antarctic Treaty parties as part of the process of ensuring their participation in the system.
60. NCP delegations may attend all formal working group and committee meetings and speak, unless a consultative party delegation requests otherwise, but may not participate in the decision.
61. At the preparatory meeting for the 14th ATCM there was an attempt to include UNEP as one of the bodies to be invited as an observer but this was blocked by nations opposed to the inclusion of UN bodies. Experts do not participate in decision-making and may participate only in those discussions for which they have been specifically invited. Organisations invited to that 1987 meeting included the WMO, IUCN and SCAR.

3 ENVIRONMENTAL PROTECTION: POLITICS AND OUTCOMES

1. A footprint in an Antarctic moss bed, for example, may last for decades.
2. Ice-cores and samples of recent snowfalls reveal traces of pollutants and heavy metals whose origins lie outside the Antarctic, but which are dispersed there through atmospheric circulation (Drewry, 1988, p. 10).

Traces of radioactive fallout from atmospheric tests have been found in the Antarctic, and DDT has been detected in penguins.

3. Some Adelie penguin rookeries, for example, cover several square kilometres and the breeding season population may number several hundred thousand birds (Lucas and Dingwall, 1987, p. 222).
4. In spite of requirements that data on personnel in the Antarctic be exchanged between treaty parties, information on the total numbers in any one season is difficult to come by. A US Congressional Research Service report in 1987 was unable to compile complete data. Their total for the summer population was almost 2700 but figures from several countries were not available (Browne, 1987). Most observers suggest the summer population could be anything between 4000 and 6000 people.
5. This comment is taken from a confidential report. The British position may have been helped by the fact that SCAR was headquartered in England at the Scott Polar Research Institute (SPRI), and senior Foreign Office Antarctic diplomats have always had close ties with both SPRI and the British Antarctic Survey in Cambridge.
6. In 1986 the SCAR Executive acted upon a proposal to establish the multidisciplinary Group of Specialists on Antarctic Environmental Affairs and Conservation, to replace the arrangement whereby environmental matters were considered by a sub-committee of the Working Group on Biology. It was also suggested that SCAR should draft its own principles on the protection for the Antarctic environment and that application for SCAR membership should include a commitment to comply with those rules (Kimball, 1987c, p. 25).
7. Kimball (1988d) has undertaken a detailed survey of Antarctic NGOs and the development of their interest in Antarctic issues.
8. ASOC now points to a membership of over 200 organisations in 35 countries. It operates through secretariats in Australia, New Zealand and the US.
9. Members of ASOC include national organisations with a specific Antarctic concern, as well as environmental and conservation organisations with a wider policy interest. Some of the better known members are nature conservancy groups of long standing (such as the Sierra Club or the Audubon Society in the US) whereas others are representative of the second wave of environmental organisations such as Friends of the Earth (founded in 1969), Greenpeace (1971) and the World Wide Fund for Nature (WWF).
10. At the fifteenth consultative meeting in 1989, for example, NGO representatives were invited to attend every function hosted by the consultative delegations.
11. They were joined by Denmark, an NCP, from the 1983 meeting when NCPs were able to attend for the first time.
12. For example, Greenpeace has an office in Argentina but not in Chile. The first workshop of Latin American Antarctic NGOs was held in 1990 with the aim of setting up a Latin American NGO transnational network which would join in the efforts of the international NGOs (see Yañez and Zañartu, 1990).

13. Greenpeace has observer status at a number of international organisations and conventions. ASOC was granted observer status at the IWC in the early 1980s.
14. See Kimball (1988d) for further details. In 1984 the Commission decided that, as Greenpeace was a member of ASOC, it would consider the ASOC application only. A decision was postponed in subsequent years as the Commission sought further information from ASOC and then tried to obtain a consensus among its members for granting such status.
15. The institutions of CCAMLR first met in 1982. IUCN was not the only organisation with observer status. Others included the FAO, the Intergovernmental Oceanographic Commission, the IWC and SCAR. In 1986, the IUCN observer at the Commission meetings noted that for the first time he was able to be present at a working group meeting without being challenged (Kimball, 1988d, p. 46).
16. NGO representatives were often more experienced in the ways of the consultative process than the NCPs.
17. Kimball (1988d, p. 50n) gives examples of technical studies prepared by NGOs and workshops sponsored by them.
18. This collaboration resulted in the establishment of a joint task force to develop a long-term plan for Antarctic conservation and environmental management which would give effect to an 'integrated and comprehensive approach' (Kimball, 1986, p. 2).
19. In 1986 Greenpeace established 'Antarctic embassies' in twelve countries, and appointed naturalist David Bellamy as the British 'ambassador' to Greenpeace's Antarctic World Park (see *The Times*, 1 February 1986).
20. Their scientific program included studies into the effects of low light levels on human physiology, studies of coastal fish and meltwater lakes, as well as tests on soil and water samples taken from consultative party bases.
21. The description and analysis here draws on those consultative meeting documents which are now publicly available, primary material from non-governmental organisations, the final reports of consultative meetings, and secondary sources.
22. This also reinforces the general norm of the Treaty system that Antarctic issues should be kept firmly within the purview of the Treaty system.
23. These principles and norms were repeated in a declaration on environmental protection adopted at ATCM IX in 1977 (recommendation IX-5).
24. Roberts (1966, p. 2) (then senior Antarctic diplomat with the British Foreign office) suggests that it was included in the Treaty because of pressure from biologists on the need for conservation measures and because it was a relatively uncontroversial subject.
25. This became a standard formula in respect of conservation and other recommendations that still had to be ratified by Governments in accordance with legislative procedures and which might take some time. Of course it also raises the problem that the recommendation establishing interim guidelines had itself to be ratified.

26. It was suggested that the Treaty Area be declared an 'International Wildlife Reserve'. Argentina and Chile wanted the word 'international' removed; the Soviet representative wanted it kept in. The US representative said his government would accept a phrase such as Special Conservation Area, Protected Natural Haven or Nature Sanctuary (see ATCM, 1964a).
27. See Boczek (1986) for a detailed analysis of SPAs.
28. A suggestion that governments should 'take appropriate action *including such legislative and regulatory measures as may be required* to carry out these Agreed Measures' proved to be contentious. The Chilean representative proposed that the phrase in italics be removed: the Soviet and the US representatives supported its retention. In the final event the Chileans prevailed (see ATCM, 1964a).
29. The bases were the Soviet Bellingshausen and the Chilean Teniente Marsh.
30. This redesignation seems to have been in response to SCAR's suggestions (see ATCM, 1975b).
31. SPAs were to be kept to the minimum in number and size to meet the requirements, in keeping with a British suggestion that some of the SPAs were too big. SCAR's review suggested that there be no change to the criteria for selection of SPAs (ATCM, 1975b).
32. These included the three SPAs terminated by recommendation VII-2.
33. At least one of these sites (Deception Island) had first been proposed by Chile for inclusion as a SPA under the Agreed Measures in 1970.
34. The rationale and criteria for designation of marine SSSIs and the information required in a management plan was set out in recommendation XIV-6 to provide guidance for SCAR in making proposals to the consultative parties, even though the guidelines were drawn up by and had already been applied by SCAR.
35. See also Schofield (1976, p. 20) who suggests that the impact and overall scope of conservation measures and system of protected areas are limited. Boczek (1986, p. 84) notes that the designation of some SPAs has been advantageous in the protection of wildlife. He refers particularly to the giant petrel colonies in the Rookery Islands and the fulmarine petrels and southern skuas on the Budd Coast. See also Lucas and Dingwall (1987, pp. 233–37) for arguments on deficiencies in the protected area system.
36. This has been well documented by NGOs. See, for example, ASOC (1985a), Greenpeace (1987a, 1987b, 1989b).
37. Snow gauges, water gauges and measuring pipes were broken. China was still granted consultative party status in 1985.
38. This was in spite of a lack of support for a similar proposal made at the preparatory meeting.
39. At the preparatory meeting for the 14th ATCM, Australia's proposal on additional protective measures was apparently seen by some delegations as a claimant plot to exclude other activities and nations from large areas of the continent. If this is the case, given that Australia claims 42 per cent of the continent, they would have to have been sizeable protected areas!

40. Australia tabled a paper (ATCM, 1987b) to show how the suggested new category could improve the effectiveness of protection under the Antarctic Treaty, setting out a lengthy 'hypothetical' example of its possible application. Australia also supported SCAR's recommendations and stressed that the SCAR report needed 'urgent and serious attention ... in order to properly meet our responsibilities under article IX of the treaty and to respond to international concerns over the adequacy of our care for the Antarctic environment' (ATCM, 1987c).
41. Batteries, plastic packaging and plastic and rubber products were to be removed from the Treaty area.
42. Some national Antarctic operating agencies, particularly the US National Science Foundation, had been applying environmental impact procedures to their activities in the Antarctic.
43. It was at this meeting that consultative parties also adopted the general norms that no activity with an inherent tendency to modify the environment should be undertaken unless appropriate steps had been taken to exercise appropriate controls (recommendation VIII-13).
44. The US Antarctic program brings over 6 million gallons of fuel into the Antarctic each year (Merriam, 1990, pp. 19–20).
45. A report in *The Guardian* of 20 February 1989 noted that seven 40 gallon drums of untreated waste had been collected from the McMurdo outfall, and that the waters contained levels of cadmium which would be illegal in the US.
46. The final Report was adopted by consensus so it is fair to assume that all parties were in agreement with this statement. Its status with regard to the requirement for governments to take action is unclear.
47. See ATCM (1983b; 1983c).
48. The final report of the 1983 meeting contained steps found in some national EIA procedures which, it was suggested, might be useful to SCAR in considering this matter.
49. The US, Britain, New Zealand and France tabled examples of environmental impact assessments prepared for proposed Antarctic activities.
50. The British position is significant in that Nigel Bonner, one of the authors of the SCAR report, was also a member of the UK delegation and would, one assumes, have had some input into the drafting of the British paper.
51. The two new consultative parties, Uruguay and China, for example had established bases on King George Island where the problem of station crowding was most acute. The kind of work being undertaken by some of the newer parties focussed primarily on measuring and monitoring, particularly in the field of meteorology, rather than new and Antarctic-specific primary research.
52. The UNEP principles upon which these guidelines were based suggested that countries should consider the comments of government agencies, members of the public or relevant experts or interested groups, a provision not adopted by the Treaty parties.
53. Non-governmental expeditions differ from tourist activity (which is, of course, also non-governmental) in that the latter are 'for the most part

... recreational and adventurous ... having little more than notional support from the government of the country in which they are organised [and] ... being heavily dependent on sponsorship or other private contributions in cash or kind' (ATS, 1989, p. 2201).

54. For a more detailed discussion of tourist activity in the Antarctic see, *inter alia*, Mussack (1988), Boczek (1988, pp. 457–62) and Reich (1980).
55. See Boczek (1988) on the legal status of tourists and other visitors in the Antarctic.
56. This recommendation also included a section on 'Guidance for Visitors' for distribution to tourists as a guide to good conduct in the Antarctic. This guide is a short and simple list of 'do's and don'ts'.
57. Ziolkowski (1990, p. 18) reports that NGOs considered the voluntary guidelines a welcome first step but generally thought they did not go far enough.
58. This report recognised, for example, that the sewage discharge systems at both Palmer and McMurdo did not meet the Code of Conduct (NSF, 1989b, p. 27) and that solid and hazardous waste management practices at McMurdo did not conform to sound environmental practice in the Antarctic.
59. The NSF held meetings with NGOs in February 1989 to discuss the clean-up initiative.
60. In another example, 40 tonnes of waste, including discarded machinery and leaking tins of caustic soda were cleaned up at the Australian Casey base following public criticism by a British scientist.

4 LIVING 'RESOURCES' AND THE LIMITS TO CONSERVATION

1. Pelagic sealing refers to the killing or capturing of seals in the sea or on floating ice.
2. The word 'Voluntary' was added to draft recommendations proposed by the UK and the US delegations.
3. Under the guidelines the Ross Seal was to be totally protected except when taken for scientific purposes. At this meeting fur seals and Ross seals were also designated specially protected species (on land) under the Agreed Measures.
4. Both were appended to the final report of the meeting.
5. No negotiating documents are available from this meeting.
6. In spite of this, states have continued to accede to the Convention as they have become members of the Antarctic Treaty. As at 22 July 1991, there were 15 contracting parties to CCAS.
7. Permits to take the seals were issued for scientific purposes for a much smaller number than were killed. See ASOC (1988j) on the limited scientific research that was actually undertaken on the seals killed.
8. The Convention provides for a review conference to be held every five years. A review conference was not held in 1983 because, in the

absence of any sealing activity, no need was perceived. Kimball (1990, p. 11) suggests that it was held 'by correspondence'.

9. The final report of the review conference did note, however, that the use and purpose of special permits needed clarification. An NGO report of this meeting suggested that the parties feared establishing a precedent for easy amendment. There were apparently also concerns about allowing a path for EC entry into the Convention (ASOC, 1988i).
10. In 1976 the UNDP approved a much-reduced $202 500 budget for the FAO information program on the Southern Ocean. Argentina, Chile, Australia and the UK reminded the FAO not to overlook the consultative parties' role and responsibility in the area (Barnes, 1982b, p. 278n).
11. The SCAR meeting called as a result, in 1976, established the BIOMASS (Biological Investigation of Marine Antarctic Systems and Stocks) research programme in conjunction with the Scientific Committee on Oceanic Research (SCOR).
12. There are a number of useful analyses of the negotiation of CCAMLR, including Gardam (1985) and Frank (1983–4). The most comprehensive is that by Barnes (1982b) who was a non-governmental organisation representative on the US delegation.
13. Their specific concern related to their right to establish conservation measures in the 200 mile EEZ around Kerguelen and Crozet islands, sub-Antarctic islands over which French sovereignty was not disputed. The draft contained no recognition of jurisdictional rights of coastal states because of the territorial dispute on the Antarctic continent. Other countries with islands in the same geographic and juridical position were happy with the draft. Some delegates apparently wondered whether France was serious about having the convention finalised. Barnes suggests (1982b, p. 257) that French intransigence may have been a delaying tactic while a bilateral fishing agreement with the USSR on access to Kerguelen fishing grounds was completed.
14. Taiwan (also not an Antarctic Treaty signatory) was also, excluded from the final conference on political grounds, according to Barnes (1982b, p. 272) although he does not say what they were.
15. Representatives from SCAR, the FAO, the European Communities, the Intergovernmental Oceanographic Commission (IOC), International Union for Conservation of Nature and Natural Resources (IUCN), International Whaling Commission (IWC) and the Scientific Committee on Oceanic Research (SCOR – another ICSU committee) attended. The question of ASOC attendance was raised but according to Barnes (1982b, p. 259) the UK, Chile and the USSR privately told Australia that they had reservations about this. The USSR initially proposed that all working group and committee meetings be closed to observers. This would have included EEC representatives, a suggestion that the British would not countenance (Barnes 1982b, pp. 258–9).
16. Barnes also refers to the role of Professor Moiseev in convincing his Soviet colleagues (and the Japanese as well, Shapley suggests (1985, p. 151)) to accept the ecosystem approach. While Barnes doesn't identify him as a scientist, Shapley does and notes also that it was

Moiseev who first raised the question of krill and its exploitation at a SCAR symposium in 1968.

17. Indeed, the USSR proposed that the Convention should state that the claimants did *not* have the right to exercise sovereignty in coastal areas. The claimant response was to suggest that special rights for claimants be incorporated, including economic benefits from catches by other states in their EEZs (Barnes, 1982b, pp. 253–4).
18. Australia unsuccessfully proposed that this be deleted on the grounds that it could be interpreted as ascribing a degree of exclusivity to the Antarctic Treaty parties that was unreasonable and of dubious legal effect (Bush, 1982, p. 407).
19. The coordinates used in the Convention roughly approximate those used to define FAO statistical reporting areas.
20. Argentina successfully requested a redrawing of the coordinates of the Convergence, in the Drake Passage, further away from its territory. The Argentinian instrument of ratification included a statement to the effect that nothing in the Convention affected or impaired its rights of sovereignty and maritime jurisdiction in the areas covered by the Convention under its sovereign control (in Bush, 1982, p. 431).
21. The question of whether an issue is one of substance or not is also one of substance, and is therefore to be decided by consensus. Non-substantive matters are decided by simple majority. An early proposal would have established a weighted voting procedure to favour ATCPs over any other Contracting Party to the Convention (Barnes, 1982b, p. 245). Proposals were also put forward for a two-thirds majority and the first draft of the Convention incorporated this proposal for all but decisions on quotas and establishment of fishing sub-areas, with no provision for objection.
22. See Howard (1989, pp. 140–45) for details.
23. States such as Australia, the US and the UK wanted the Scientific Committee to fulfil its consultative functions and to provide scientific evidence upon which management decisions would then be based. The USSR perceived it much more as a political body, responding to national interests (see Orrego Vicuña, 1991, pp. 30–1).
24. Gulland suggests (1988, p. 232) this was because they felt that data reporting requirements had been used against them in the IWC to strengthen restrictions on whaling.
25. For example, in 1989 the convenor of the Working Group on fish Stock Assessment made an official statement that deplored the fact that the opinions of the Scientific Committee had been disputed or rejected by certain members of the Commission (Puissochet, 1991, pp. 73–4).
26. These measures were of two types only – restrictions of mesh size and the closing of certain fisheries (see Howard, 1989, pp. 131–35 for details).
27. See Basson and Beddington (1991) on the problems of implementing article II of the Convention which sets out the principles of the ecosystem approach.

28. Part of the difficulty with adopting conservation measures for krill is that not enough is known about either their biology or ecology to enable the design of workable management models (Powell, 1990, p. 66). Krill catches are presently thought low enough not to have an impact on related and dependent species.
29. See, for example, *Antarctic News*, No.2 (1984), No.3 (1986) and No.5 (1987), Mosley (1986a, p. 48) and Kimball (1987b).
30. Australia may have a situational or 'psychological' advantage as the state in which the CCAMLR institutions are located and where the Commission and Scientific Committee meetings are held. Australian decision-makers and scientists are able, at least, to be in close informal contact with the secretariat.
31. Other measures which have been adopted by the Commission have been superseded or ceased to be effective.
32. Although conservation measures are the responsibility, under the Convention, of the Commission, any ATCP may request that a Commission resolution on CEMP site protection be discussed at a regular consultative meeting (CCAMLR, 1991, pp. 6–12).
33. Non-governmental, in this case, includes bodies such as SCAR, SCOR, ICSU and the IUCN.
34. This application was not without precedent. Greenpeace and ASOC have observer status with the IWC, for example, which is responsible for whales in the Southern Ocean. Greenpeace was not granted observer status to CCAMLR on the grounds that it is a member organisation of ASOC.

5 ANTARCTIC MINERALS: PROBLEMS AND PROCESSES

1. The Gondwana analogy is not without challenge. While comparisons are made between the geology and plate techtonics of the Antarctic Peninsula and the South American Andes, this applies mainly to the southern reaches of the mountain chain which are 'relatively barren in mineral deposits when compared with the Andes farther north' (Splettstoesser, 1985, p. 43). Sediments in the Bass and Gippsland Basins of Australia which are used as analogous examples for hydrocarbon potential were formed after the breakup of Gondwana (Thomson and Swithinbank, 1985, p. 34). Crabbe (1985, p. 366) points out that the Gippsland Basin is not part of the southern Australian marginal rift and is therefore not relevant to the Antarctic margin at all. Quilty suggests (1986–87, p. 117) that such minerals as may have existed in Antarctica following the break-up of Gondwana 'would have been removed through long periods of glaciation'.
2. A mineral occurrence is the presence of a mineral, often in very small quantities. A deposit is a more substantial quantity and a reserve is a deposit whose size and extent has been measured or calculated and that is known to be of commercial value now or is expected to become so in the future (Holdgate and Tinker, 1979, p. 14). Stewart (1963)

listed 222 mineral species, sub-species and varieties, including some questionable occurrences, that had been reported in the previous 70 years. He noted that many of these had been found only in minute amounts with some having been described only after microscopic examination. See also Potter (1969, pp. 17–19).

3. In 1973, the US research ship Glomar Challenger found traces of ethane and methane in three of four holes drilled while carrying out scientific work in the Ross Sea. The Glomar Challenger was not searching for oil or gas. The Ross Sea holes were cemented to prevent gas leaks, and scientists aboard said that no conclusions could be made about possible deposits (Shapley, 1985, p. 124). These finds have been cited as evidence of hydrocarbon potential although as Elliot (1977, p. III-5) points out, methane is 'not considered an indicator of ... hydrocarbon deposits' and neither is the presence of hydrocarbon reservoirs necessarily mandated by the presence of ethane. Sediments taken from the Bransfield Strait in 1985 were considered by West German researchers to provide evidence of oil deposits at lower depths. A 630 metre core drilled from McMurdo Sound by a joint US-NZ glaciological research team contained asphaltic residue indicating that hydrocarbons had been present at some stage (possibly millions of years ago) but had later leaked out.
4. Stewart (1963) observed that 'as far as may be ascertained, no commercial deposits have been discovered'. Potter (1969, p. 29) in a study for Resources for the Future suggested that the 'chances of [the Antarctic] having commercial minerals are probably less than one percent of those of the 'related' continents, South America, Australia and Africa'. A report for the US Geological Survey concluded that 'Antarctica ... has no known economically recoverable resources of any category' (Wright and Williams, 1974, p. 2). Crabbe states (1985, p. 365) that 'despite much speculation, there are still no known petroleum or mineral resources in Antarctica'. Laws (1990, p. 9) has more recently argued that 'there are no proven commercial resources, either of hard rock minerals or of hydrocarbons. Their actual occurrence is speculative'.
5. A SCAR report (see SCAR/EAMREA, 1979, p. 14) notes that there is some doubt that these were official USGS estimates. They were not included in the USGS circular on mineral resource potential (Wright and Williams, 1974). The report of the group of experts convened at ATCM IX cautioned that these were speculative figures and 'should not be cited unless supported by much firmer evidence' (in ATS, 1989, p. 3345) and noted that it had been advised that these figures originated from an unpublished, highly provisional calculation in an internal USGS document (see also Mitchell, 1977, p. 91). They were formally attributed to USGS by Spivak (1974). Nossiter (1977) argues that these papers were written at a time when the Ford Administration was eager to find new oil resources under the threat of a future Arab boycott. He suggests that the Carter administration later downplayed these figures because of a desire, shared by the oil companies, to convince the public that oil was a scarce resource and it was therefore reluctant to advertise any possible new finds.

6. In 1979 a Gulf Oil representative was reported as suggesting that the two most likely Antarctic sites in the Ross and Weddell Seas could yield up to 50 billion barrels of oil (Bell, 1982, p. 24). Japanese estimates in 1980 were in the order of 45 billion barrels of oil and 3.3 trillion cubic metres of gas. A 1980 report prepared for the West German Federal Institute of Geosciences and Natural Resources estimated recoverable oil reserves of 4 billion tonnes (30 billion barrels) and 4 trillion cubic metres of gas (Zorn, 1984, p. 4).
7. In 1978–79 a Norwegian expedition found iceberg scouring at depths of up to 400 metres in the Weddell Sea with 'plough marks' 25 to 50 metres wide and 5 metres deep on the sea bed (Zorn, 1984, p. 6).
8. By way of contrast, the Alaska North Slope oil field is believed to contain around 8 billion barrels (Wilson, 1983, p. 78).
9. The official was John Garrett who also served on the US Department of State's Antarctic Advisory Committee.
10. See Zumberge (1982, pp. 134–44) for a concise summary of the environmental impacts of minerals exploration and exploitation in the Antarctic. Zumberge is, nevertheless, conservative in his assessment of the need to restrict minerals activity.
11. If the spread of oil could not be contained before the winter pack ice advanced, nothing could be done during the months when the continent is inaccessible and its surrounding waters covered by ice. Indeed, the technology to clean up an Antarctic ice spill may not even exist. Polluted ice could reduce the reflectivity of the ice-sheet which might have impacts on local temperatures, thus affecting biological communities.
12. The NIEO was supported and given shape by the Group of 77, which was formed at the first UN Conference on Trade and Development (UNCTAD) in 1964 and consolidated in the 1973 Non-Aligned Movement meeting in Algiers. The NIEO called specifically for a restructuring of the post-war economic order, and particularly the General Agreement on Tariffs and Trade (GATT), in order that less-developed countries would not be disadvantaged and would benefit positively from the international order. The NIEO was given concrete expression in the Declaration of the Establishment of a New International Economic Order adopted at the Sixth Special Session of the UN General Assembly in May 1974 and the Charter of Economic Rights and Duties adopted later that year at UNGA's 29th session. See Pirages (1978, pp. 53–9) for a short summary of the NIEO principles and demands.
13. During a 1971 UN debate on the Indian Ocean as a Zone of Peace, Sri Lanka (then Ceylon) suggested that areas beyond national jurisdiction, including Antarctica, were part of the international domain and should be subject to international regulation and responsibility. In the same year an annex on the natural resources of Antarctica was included in an information document prepared for the ECOSOC Committee on Natural Resources (see Orrego Vicuña, 1988, p. 473).
14. In 1977, Libya and Algeria proposed that the Antarctic be placed under the International Authority which would be established by the

Law of the Sea convention. In 1984, Pakistan and the Philippines made a similar suggestion during the General Assembly debate on the Antarctic.

15. This argument was first raised by Malta's Ambassador to the UN, Arvid Pardo, in 1967. At the 1975 General Assembly session Sri Lanka's UN representative, Ambassador Amerasinghe, who was also the President of UNCLOS, argued in support of the internationalisation of the Antarctic.
16. As well as being on the UN agenda, the question of Antarctic has also been discussed in a number of Third World fora including the Non-Aligned Movement, the Organisation of African Unity, the League of Arab States, the Organisation of the Islamic Conference, the Organisation of Eastern Caribbean States and the Caribbean Community and Common Market.
17. Resolutions also called for the consultative parties to expel the apartheid regime of South Africa from the Antarctic Treaty.
18. A number of Eastern Bloc Treaty states have voted in favour of the resolution against South African participation. Likewise, some non-Treaty parties have joined in the non-participation roll call, including Afghanistan, Albania, Colombia, Costa Rica, Ecuador, Greece, Guatemala, Israel, Laos, Nicaragua and Vietnam. The Latin American states may well have been influenced by Argentina, an active member of the Non-Aligned Movement. Colombia, Ecuador and Guatemala have since become signatories to the Antarctic Treaty. Until recently, Australia spoke on behalf of the informal New York group of Treaty parties. Initially it was intended that the spokespersonship would rotate alphabetically through the consultative parties but some, further down the alphabet, were reluctant to take on this task because (in the case of Brazil, for example) they did not want to alienate Third World colleagues, or because of a lack of expertise or experience. Australia's then Ambassador to the UN (and later Secretary of the Department of Foreign Affairs and Trade), Richard Woolcott, also accumulated a great deal of experience and ability in speaking for the Treaty Parties. The mantle, however, has now been passed to Germany.
19. This comment was made by Oscar Pinochet in his opening statement on behalf of the Chilean delegation at ATCM-VII in 1972. France suggested that 'the search for a solution to this problem must be subordinated to the solution of the problem of how to avoid rivalries the consequences of which would inevitably be borne by the powers in possession of Antarctic territory' (ATCM, 1972b). Argentina, urging caution in dealing with this issue, asked 'should we ... let ourselves be carried away by the interests of economic concern ... to the detriment of the purpose of maintaining peace and avoiding conflict' (ATCM, 1972f).
20. R. B. Thomson (of the New Zealand Department of Scientific and Industrial Research) was reported as saying that expected diamond and platinum finds would lead NZ to raise the subject of exploitation (*Christian Science Monitor*, 6–8 June 1970). New Zealand had also

received an application to prospect and develop off the Ross Sea (Colson, 1980, p. 884). The issue was dropped at the third preparatory meeting which nevertheless confirmed a consensus understanding that without prejudice to the position of members concerned on territorial issues, no unilateral action should be taken with respect to the problem of mineral prospecting in Antarctica (ATCM, 1970). The US and UK delegations raised the question again as a possible agenda item at the fifth preparatory meeting.

21. See Wallace (1988, p. 6). Chile favoured an indefinite moratorium, France suggested at least 10 to 15 years (ATCM, 1972b), a position also supported by Argentina and the USSR. The US was opposed to any prohibition on minerals activity. The draft recommendation tabled by the UK suggested that there be a moratorium until such time as any agreed measures might become effective and it was thought that world interests would not be harmed by such a moratorium.

22. Some participants in the Legal and Political Working Group argued that, as minerals exploitation was not expressly permitted, it could not proceed. Others suggested that as it was not expressly prohibited, it could proceed but that it might well be contrary to the purposes and objectives of the Treaty unless all consultative parties consented to any activity. Others suggested that such activity could be undertaken, without multilateral consent, as long as it was consistent with the objectives of the Treaty.

23. Belgium reminded delegates that this subject 'by its nature, could either destroy the spirit of the Treaty or develop and strengthen it by extending cooperation to new fields' (ATCM, 1975a, p. 52). The British raised the spectre of an unregulated scramble bringing 'large numbers of men [sic] swarming into the continent' (ATCM, 1975a, p. 61). The Argentinian delegation exhorted the meeting to 'take up resolutely and calmly the challenge implicit in these controversial issues' (ATCM, 1975a, p. 50).

24. It is not clear whether the reference to 'states and persons' applies to consultative states or whether it is designed primarily to discourage non-Treaty states. This sentiment was not included in the lengthy recommendation on minerals (VIII-14) adopted at this consultative meeting. Recommendation VIII-14 refers to the 'need for restraint' but it does not specify by whom or exactly what this means. Recommendation VIII-14 also refers to 'possible action by others' indicating that it is most likely intended to discourage states other than consultative parties.

25. The degree of importance that the consultative parties attached to this exercise may be measured by the suggestion in this recommendation that 'if possible and appropriate, Governments may wish to assist their National Antarctic Committees in this undertaking by appropriate means' (para 3(i)). SCAR functioned on a limited financial base while being required to deal with more complex and extensive requests from the consultative parties. SCAR is also asked to continue to coordinate geological and geophysical research to provide fundamental scientific data.

26. Australia (ATS, 1976a) suggested that the present Agreed Measures were not sufficient to protect the environment from the effects of resource exploitation. The USSR also expressed concern about the environmental effects of an oil spill on krill fisheries.
27. At the Paris preparatory meeting, New Zealand also proposed a possible solution to the 'ownership' problem – see Mitchell (1977, p. 101). Minerals activity would be prohibited in certain areas for environmental reasons. Applicants wishing to explore in other areas would submit a proposal to a Regulatory committee consisting of the twelve consultative parties. This proposal would contain details of proposed activities and an environmental impact statement, and should demonstrate that such activity would not substantially or materially damage the Antarctic environment. It would require the unanimous approval of all consultative parties, although operators might come to an informal arrangement (on payment of royalties for example) with the relevant claimant (or claimants) to forestall a veto. New Zealand also suggested in a discussion paper that the consultative parties should recognise the need for the proceeds of resource exploitation to be shared on a just and equitable basis with the international community (Bell, 1982, p. 26).
28. The US tabled a paper which provided an informal assessment of mineral resources in the Antarctic which helped to focus the discussion of the technical group (Colson, 1980, p. 891).
29. There are references throughout the report to the lack of knowledge on many aspects likely to be important in the event of minerals activity. For example, there was insufficient knowledge to permit a reliable estimation of the impact of oil spills on the ecosystem. The report also drew attention to the need for a sufficient environmental data base.
30. The report of this meeting, known as the Bellagio Report, was published by SCAR (Holdgate and Tinker, 1979). It noted that the unique scientific nature of Antarctica, as well as its unique political nature, represented 'major assets which might be affected by mineral development' (Holdgate and Tinker, 1979, p. 5). This meeting brought together experts from the SCAR/EAMREA group and the Group of Experts from ATCM-IX (1977) as well as other participants.
31. ATCM X also adopted a recommendation on hydrocarbon contamination of the Antarctic marine environment (recommendation X-7). Chile observed, in a paper tabled for that topic (ATCM, 1979b) that damage to the ecosystem would be greater because of the scant oil degradation in the Antarctic environment, noting also that distance makes it impossible to take rapid steps to combat the spill.

6 THE ANTARCTIC MINERALS CONVENTION (CRAMRA)

1. The fourteen were: Argentina, Australia, Belgium, Chile, France, the Federal Republic of Germany, Japan, New Zealand, Norway, Poland, South Africa, the Soviet Union, the UK and the US. They were joined

by Brazil and India from January 1984, by China and Uruguay from April 1986, and by the German Democratic Republic and Italy from January 1988.

2. Argentina and Chile were also taken to fall within the 'developing country' definition, even though the term is not defined. However their interests were dominated by their claimant status. Developing countries who were NCPs at the time included Cuba, Ecuador, Papua New Guinea and Peru. No African or Middle-Eastern countries are signatories to the Antarctic Treaty, and those Asian countries which are would no longer be considered developing (with the possible exception of North Korea).
3. At the plenary session of ATCM-XIII (1985) Peru (a NCP) observed that 'as a traditional mining country, Peru is particularly concerned that the future regime for ... Antarctic resources should take into account the interests of mineral producing developing countries' (ATCM, 1985a, p. 195). Peru also referred to the need to safeguard the regional ecology.
4. The American Petroleum Institute expressed its views forcefully: 'with the debacle of UNCLOS III fresh in our collective minds, the spectre of creating formalised institutions that would encourage joint ventures with developing countries should be anathema' (API, [n.d.]).
5. Australia and the US had the most complete participation of domestic interests. Their delegations included representatives from NGOs, environment ministries, resource and finance ministries, industry groups as well as representatives from national Antarctic operators.
6. These reports were known formally as MR/17 and subsequent revisions I to V/Corr.1. Informally they were referred to as Beeby I to VII. Beeby VII was presented as a formal draft convention.
7. They were usually granted access to delegates at coffee breaks, except during the Tokyo session in May 1984 when the conference venue was declared out of bounds to NGOs. They hosted dinners, had informal meetings with delegates and were sometimes invited to receptions held by consultative party embassies.
8. See, for example ASOC 1984a, 1984b, 1985b, 1986c, 1987, 1988a, 1988c, 1988d, 1988f, 1988g, 1988h, 1988k, 1988l.
9. Although Chris Beeby, as head of the New Zealand delegation, had chaired the first two sessions in Wellington, the Bonn meeting was chaired by the head of the West German delegation, Ernst Jung. This was the first negotiating session outside New Zealand and it is likely that the consultative parties thought at first to adopt the practice of consultative meetings in appointing chairpersons from the host country. Francioni (1986, p. 178) refers to a West German draft text of 15 May 1984. It is not clear whether this was offered as the next stage in the negotiating process, as an alternative to Beeby II, or simply as one of many documents tabled containing suggested draft provisions.
10. They would be permitted to speak, receive copies of documents and submit information papers and attend all plenary and formal committee and working group meetings but could not participate in the decision-making process. Some NCPs commented on the difficulty of

trying to figure out the underlying aspects of the negotiations (Kimball, 1985b, p. 7). It is likely that at this stage all relevant documentation was not automatically made available to NCPs (IIED, 1986, p. 4).

11. A 'Statement of Concern' signed by 32 prominent New Zealanders was sent to the Whangaroa meeting, calling for Antarctica to be administered as a 'region of wilderness, peace and scientific cooperation ... to be held in perpetuity as part of the common heritage of humankind' (*Antarctic News* # 5, 1987, p. 14).
12. Hungary, North Korea, Spain, Cuba and Austria were absent.
13. The drafting committee met again in Sydney, Australia, in September 1988 to develop the Chinese text provided for in article 67.
14. Treaty parties have acknowledged that this is not the most auspicious title for a convention which, they argue, does not presuppose minerals activity and was not negotiated for the purpose of encouraging such activity. It is most often referred to as the Minerals Convention, the Wellington Convention, or by its less revealing acronym CRAMRA (pronounced as it is spelled). At the Rio meeting in 1985, the head of the Swedish delegation (at that stage a non-consultative party) suggested the title 'The Antarctic Protection regime ... in relation to Mineral Resources' (ATCM/SCM-IV, 1985, p. 5).
15. Article 20(5) provides that the Commission may decide to establish a permanent headquarters in New Zealand.
16. By this date the number had increased to 22 with the addition of Spain and Sweden to the ranks of the consultative parties.
17. These categories of membership were effectively established in Beeby I although the relevant article went through a series of redraftings which continued to distinguish between those parties which were ATCPs at the time the Convention was adopted, and those which attained that status subsequently.
18. Under Treaty practice, NCPs have not been bound by consultative meeting recommendations unless they specifically declare themselves so bound.
19. In Beeby I (article XXIX) a separate regulatory committee was to be convened in respect of each exploration application. Beeby II adopted a US suggestion that one regulatory committee be established for each area identified for possible activities. The US initially proposed (at ATCM-XI in 1981) that the Antarctic could be divided into four quadrants, an idea subsequently dropped at the January 1983 meeting of the SCM (*Campaigners Antarctic Notes*, #5, 1984). The suggestion was that 8 of the then 14 consultative parties would sit on each panel, although none would sit on more than two. Each panel would set up terms for the operation of would-be oil producing groups under the guidelines of a minerals commission which would comprise all fourteen countries (cited in *Dawn*, 22 November 1983).
20. Non-claimant miner states opposed the concentration of claimant decision-making in the regulatory committees (*Antarctic News*, #7, 1988, p. 2). However, both claimants and pro-mining non-claimants wanted as much power as possible to reside in the regulatory committees, without significant review by the Commission, a position opposed by

the USSR, Poland, developing states and many NCPs (Antarctica Project, 1987, p. 4).

21. This was because of the area of overlapping claims. All of the three overlapping claimants (Argentina, Britain or Chile) would have demanded some say over this area.
22. Given that both states maintain that their activities in the Antarctic provide them with a basis of territorial claim, this participation in all regulatory committees could raise some interesting theoretical legal debates about the strength of their possible claims if the Treaty should ever collapse. They were referred to in Beeby III as the two members which 'at the time of entry into force of the Antarctic Treaty, maintained the largest presence in Antarctica', a description which was later dropped.
23. It would also have been difficult to include one of the superpowers and exclude the other from this guaranteed position on all regulatory committees.
24. They may be claimant or non-claimant members. It was initially included in the drafts as one or two developing countries from among the claimants, and one or two from among the non-claimants. The final number adopted was argued to reflect the balance between developed and developing countries at the time (ATCM/SCM-IV, 1988c, p. 5). The final Act recognised that if the balance was altered significantly there would be a case for an amendment to the membership provision. Developing countries, a term which is not defined, is normally taken to include the claimants Chile and Argentina. This means that if the area to be opened is outside the area of the overlapping claims, at least one of these two is nevertheless likely to be included on the regulatory committee.
25. In other words, three out of the four claimants and four out of the six non-claimants.
26. If the same area is to be open for exploration of different minerals to those already specified, the application procedure to the Commission must be repeated, although the same regulatory committee will be responsible for establishing any management scheme.
27. Article 21 includes 25 sub-paragraphs which list the functions of the Commission.
28. Developing countries opposed this provision, concerned that they would have almost no influence over the management scheme.
29. This was included in Beeby IV but was dropped in Beeby V.
30. The Law of the Sea Convention is not yet in force and so it might be disputed to what extent it creates customary international law with respect to the deep seabed and the common heritage principle. Nevertheless, with respect to the Antarctic, the dispute was not whether the deep seabed was a global commons, but where the deep seabed stopped with respect to a continent argued by much of the international community to be beyond national jurisdiction.
31. The Final Act noted that the area of regulation does not extend to the continental shelf of islands situated north of 60° south (ATCM/SCM IV, 1988c, p. 3).

32. The Final Act noted that the geographic extent of the continental shelf would be determined by reference to paragraphs 1 to 7 of article 76 of UNCLOS (ATCM/SCM IV, 1988c, p. 4).
33. Jurisdictional disputes are specifically excluded from the formal dispute settlement procedures established in the Convention. The Convention echoes the provisions of the Antarctic Treaty in providing simply for the disputing parties to consult together with a view to reaching a mutually acceptable solution on disputes over jurisdiction (article 7(3)).
34. Argentina and Chile made explicit declarations on signing the Convention on economic benefits as understood in article 35. Argentina identified the claimants as those with 'the most direct interest' (Beck, 1991a, p. 249). Chile expressed its understanding that nothing in the Convention impeded or affected agreements which the claimants might negotiate 'to give effect to the economic benefits to which they were entitled' (Beck, 1991a, p. 249). Paul Keating, then Treasurer of Australia, suggested in a letter to the Minister of Foreign Affairs that the UK and France had conceded at Wellington instead of continuing with Australia to press for a specific revenue share to be specified in the Convention. Their decision, he said, reflected the favourable outcome in respect of equity participation (that is, no formal requirement for Third World participation in consortia) and their position as major potential explorers and developers in Antarctica (P. Keating, 1988).
35. In 1984 Chile suggested that a fund be established under the Minerals Convention to facilitate developing nations' participation in Antarctic science (Kimball, 1985a, p. 10).
36. Geological research under the rules of the Antarctic Treaty, on the other hand, must be exchanged among the consultative parties although this has not always been fully observed in practice.
37. Article 34 also establishes that the Commission shall cooperate with the United Nations, its relevant specialised agencies and 'as appropriate, any international organisation which may have competence in respect of mineral resources in areas adjacent to those covered by this Convention'. This is an oblique reference to the International SeaBed Authority established under the Third UN Convention on the Law of the Sea.
38. It is unclear what role has been ascribed to observers but under the consultative process, observers (whether NCPs or other invited observers) have been able to speak at meetings although they do not participate in formal decision-making.
39. NGOs argued that their argument on property rights was vindicated after they obtained, through US Freedom of Information rules, a partially declassified 1975 National Security Council memo to the US President. With respect to economic interests in non-living resources, it suggests that US interests are to 'facilitate an increase in the global supply of resources through (i) *defining property rights* to Antarctica [sic] mineral resources; (ii) ensuring reasonable conditions of investment consistent with US interests...' (see *Antarctic News*, #9, 1989, p. 4; emphasis added) although those interests are also stated to include environmental protection.

40. This principle was included in Beeby I, dropped in Beeby II and then reinstated. This reinstatement seems to have been on the insistence of Australia which tabled a paper at the Tokyo session proposing comprehensive principles including the protection of endangered species.
41. This article provides that, if the Chair of the Commission determines that there is likely to be an objection to the identification of an area for minerals activities, he or she shall convene the members most directly interested with a view to reconciling the differences and producing a generally acceptable proposal.
42. Procedures are established in the Convention for the Commission to review the decision of the regulatory committees but in doing so they may not perform the function of the regulatory committee and may only request the Committee to reconsider its decision.

7 COMPREHENSIVE ENVIRONMENT PROTECTION: FROM CRAMRA TO MADRID

1. Within days of the Minerals Convention being adopted, a bill was introduced into the Belgian Parliament to prohibit Belgian nationals or companies from undertaking any minerals activity in the Antarctic. Belgium, which was a consultative party and had participated in the negotiations, did not sign the Convention.
2. Within three months the Cousteau Foundation petition had attracted a million signatures, a record for a petition in France. By April 1990 this total had reached 1.8 million. The Greenpeace Antarctic Declaration had been signed by 1.4 million people by May 1989, and the World Park petition circulated by members of ASOC attracted over 1 million signatures (ASOC, 1989b, p. 2).
3. Tourists on board had to be rescued by personnel from the nearby US Palmer Station. The Argentinians tried to downplay the environmental significance of the spill and their lack of technology to respond to the emergency. Equipment was brought in from the US to clean up the spill in order to protect scientific research at Palmer and wildlife in the vicinity, but the operation was difficult. There were two minor and less well-publicised Antarctic accidents in February. The British ship *Endurance* sustained some damage when it hit an iceberg near Deception Island but was able to navigate to Esperanza Bay at the tip of the Peninsula. A Peruvian research vessel, the *Humboldt*, ran aground and leaked oil in Fildes Bay on King George Island (see Redgwell, 1990a, p. 474n).
4. See, for example, ASOC (1989a).
5. For an analysis of Australia's decision to reject the Minerals Convention see Elliott (1993) and Bergin (1991).
6. This committee included officials from the Department of Prime Minister and Cabinet, Treasury, Department of the Arts, Sports, Environment, Tourism and Territories, the Department of Foreign

Affairs and Trade, and the Department of Primary Industry and Energy.

7. The Australian Mining Industry Council was in favour of signature, although it would have preferred an anti-subsidy clause. Its opposition to the Australian government stance was grounded in its belief that no area should be permanently closed off to mining.
8. He had also raised this issue in an earlier letter to the Foreign Minister (see P. Keating, 1988).
9. Rowland also notes (1988, p. 22) that the Australian submission to the UN Secretary-General in 1983 'was notable for containing passages about scenic and wilderness values, of a kind not mentioned in the statements of other countries'.
10. One report suggests that at the final ceremony of the minerals negotiations, members of the Australian delegation wore World Park badges (ASOC, 1989d, p. 4).
11. When he made his announcement calling for the minerals negotiations to be re-opened, Michel Rocard said that he had discussed this issue with Cousteau and referred to the Cousteau petition.
12. The son of another British Antarctic explorer, Shackleton, was not so supportive. Lord Shackleton, while expressing his hope that mining would never happen in the Antarctic, said that 'undermining the consensus ... when no [new] consensus is in sight would ... lead to tensions and instabilities that might well not be contained' (Shackleton, 1989).
13. Australia's Foreign Minister, Senator Evans, also stressed that 'our initiative on Antarctica is in no way a repudiation of Australia's long-standing support for the Antarctic Treaty system' (Evans, 1990, p. 5).
14. This characterisation is, as Higgott and Cooper note (1990, pp. 599–600), as much an attitudinal description as any measure of size or economic or military capability. In particular, Australia's approach to foreign policy has been characterised, by a former Secretary of the Department of Foreign Affairs and Trade, as one which emphasises 'coalitions ... to enhance our capacity to persuade and influence' (Harris 1988, p. 182). The Australian Foreign Minister has characterised his government's foreign policy as 'middle-power' diplomacy which 'exploit[s] networks of support and influence rather than hierarchical gradations of power' (Evans and Grant, 1991, p. 326).
15. The report was presented to him on 10 June 1989 and his announcement favouring the World Park proposal was made at a Planet Earth international seminar at the Elysée three days later.
16. The question of the Antarctic was only one of the matters discussed, although it took up about half the joint statement.
17. In 1981 the IUCN recommended that the ATCPs foster measures to maintain the intrinsic values of the Antarctic environment and 'ascribe to the Antarctic environment as a whole a designation which connotes worldwide in its unique character and values' (IUCN, 1981). The 16th IUCN General Assembly in 1984 restated these concerns and called for minerals activity to be prohibited 'until such time as full consideration has been given to protecting the Antarctic environment completely

and the environmental risks have been fully ascertained and safeguards developed to avoid adverse environmental effects' (IUCN, 1984a). These resolutions were affirmed as a statement of IUCN policy at the 17th Assembly in 1988.

18. The first detailed NGO proposals on a comprehensive convention were put before the Australian government in 1984 and were then distributed to the consultative parties at ATCM XIII in 1985. See, for example, ASOC (1986b, 1988b, 1989b, 1990a, 1990b), Ecofund (1986) and Greenpeace (1986).

19. As noted in chapter 5, New Zealand tentatively made this suggestion in 1975 during the early stages of debates about how to manage the minerals issue. There was little support for the idea then.

20. The British referred to a 'blinkered concern for the minor effects that [the] pursuit of knowledge is having on the Antarctic environment' which could, they implied, undermine the Treaty system itself (ATCM, 1989a).

21. This fifteen included Finland and the Republic of Korea who were NCPs when they signed and had attained consultative status just prior to ATCM-XV. West Germany, Belgium, Italy and India had not signed. Spain, which had attained consultative status after the Convention was adopted but before it was opened for signature had also not signed.

22. An open letter to delegates from HH Prince Sadruddin Aga Khan, ecologist Professor Arne Naess, David McTaggart (Chair of Greenpeace International) and Charles de Haes (Director General of WWF) called 'on the member states of the Antarctic Treaty, indeed all nations of the world, to take the first unambiguous step toward safeguarding this environment. We ask them to discard the Antarctic Minerals Convention and begin immediate negotiations to forever preserve the Antarctic environment' (Aga Khan et al., 1989, p. 1).

23. This statement addresses issues such as appropriate technologies, the dumping of wastes at sea, national waste management plans, waste disposal standards for vessels and the storage and handling of wastes.

24. The Meeting suggested that consultative parties who were signatories to the International Convention for the Prevention of Pollution from Ships (1973) and its Protocol of 1978 (MARPOL 73/78) consider formally proposing that the Treaty area be designated a special area under its provisions. This was approved by the Marine Environment Protection Committee (MEPC) of the IMO on 15 November 1990. The Meeting also agreed that the issues of dumping at sea and liability for marine damage warranted further attention.

25. Hawke included the Antarctic in his list of topics when he spoke to US President Bush in Washington in June 1989. He mentioned it when he addressed the National Press Club in Washington. He raised it at a dinner hosted by the Minister for Industry and Regional Development and the Associate Minister for Foreign Affairs in Paris on 19 June 1989, and at another dinner hosted by the Lord Mayor of London on 21 June. Senator Evans discussed the Australian/French initiative when he met US Secretary of State James Baker in October for Australian-US Ministerial talks. In September 1990 Evans and the Australian

Ambassador for the Environment, Sir Ninian Stephen, visited a number of Latin American countries (Argentina, Chile, Brazil, Uruguay and Peru) during which time the initiative was also discussed.

26. On 23 June 1989, nine major environmental organisations in the US (The Antarctica Project, the Sierra Club, Greenpeace US, the Environment Defense Fund, the Environmental Policy Institute, Friends of the Earth, National Audubon Society, the Humane Society of the US and the Oceanic Society) delivered a letter to President Bush, urging the administration to respond favourably to the Australian and French decisions (Antarctica Project et.al, 1989). A week later representatives of the Antarctica Project (Evelyn Hurwich and James Barnes, both of whom had signed the Bush letter) wrote to the director of the US Environmental Protection Agency, Bill Reilly, urging the agency to take a leadership role in developing a positive environmental agenda for Antarctica (Hurwich and Barnes, 1989, p. 2).

27. A Greenpeace publication (Patel and Mayer, 1991) including articles by a number of leading Antarctic scientists supporting comprehensive environmental protection and wilderness reserve status for the Antarctic.

28. Almost two months later, on 3 January 1990, Udall and his colleagues received a response from Janet Mullins, an Assistant Secretary in the Legislative Affairs division of the Department of State. She suggested that the majority of ATCPs supported the Minerals Convention and considered it a key element in comprehensive environmental measures (Mullins, 1990).

29. Congressman Conte (with co-sponsors Walter Jones (D-NC) and Representative Wayne Owens (D-Utah)) introduced the Antarctic Protection and Conservation Act 1990 (HR 3977) on 8 February. This bill stated that 'the Convention on the Regulation of Antarctic Mineral Resource Activities is not adequate to provide the necessary level of protection for the fragile environment of Antarctica and could actually stimulate movement toward commercial exploitation' (article 2(5)). Companion legislation was introduced into the Senate by John Kerry (S 2575 – the Antarctica Protection Act). On 4 April 1990 Congressman Vento introduced his draft Antarctica World Park and Protection Act of 1990 (HR 4514) (see US Congress, 1990b, p. E990). In May Senator Gore introduced the Antarctic Environmental Protection Act 1990 (S 2571) (see US Congress, 1990d, pp. S5658–5659) which was designed to amend the Antarctic Conservation Act of 1978. Identical legislation was introduced into the House of Representatives by Congressman Jones.

30. The resolution was submitted by the Australian Conservation Foundation, Greenpeace International and the World Wide Fund for Nature. The UK, the US and Norway indicated that, had a vote been necessary (the resolution was agreed to by consensus) they would have abstained on the grounds that they did not want to pre-empt the results of the Chile special consultative meeting (ASOC, 1990c, p. 2).

31. Jim Barnes from the Antarctica Project and Friends of the Earth, Maj de Poorter from Greenpeace International and Andrea Figari from

Greenpeace Argentina were the official ASOC representatives. They were able to attend the plenary and working group meetings, but not the Head of Delegation sessions. They were able to speak, but only after the Treaty party delegations had spoken on any particular subject.

32. Nevertheless the meeting on the liability protocol under the Minerals Convention lasted only half a day and was attended by about half the delegations only (ASOC, 1991a).
33. The article read 'Any activities relating to minerals resources other than scientific research shall be prohibited ...' (ellipses included). Initially the wording had been enclosed in square brackets, an indication of serious reservations on the part of some delegates. The ellipses (...) included in the Anderson draft indicated that further elaboration was expected.
34. For a detailed discussion of the Andersen text see Kimball (1991b). For a critique see Wallace (1991).
35. Australian discussions with the US elicited an unfavourable response to a permanent prohibition from the State Department and from the Departments for Energy and for the Interior.
36. Gerald Kaufman, British Labour Party shadow Foreign Minister argued that a 'moratorium was a compromise which leaves open the option to mine at some future date'. It was, he said, 'not acceptable to the Labour Party' (Anon., 1991a, p. 1). The Leader of the Liberal Democrats wrote a public letter to Prime Minister Major opposing the British delegation's position.
37. While Foreign Minister Genscher had announced his support for the declaration of an Antarctic World Park some time prior to this, the final decision on the minerals issue lay with the Ministry of Industry. The Industry Minister had received tens of thousands of post-cards urging that Germany support an anti-mining position.
38. Article 24(1) also provided for amendment to the articles of the Protocol (including, therefore, article 6 prohibiting mining) at any time in accordance with article XII(1)(a) and (b) of the Treaty which, in effect, required consensus to enter into force.
39. A drafting committee was scheduled to convene from 10 to 14 June to consider the Protocol and the four annexes so far agreed to on environmental impact assessment, conservation of Antarctic fauna and flora, waste disposal and management, and marine pollution. The drafting committee was unable to reach agreement on a definitive version of the marine pollution annex. An ad hoc contact group drew up an informal paper for circulation to the meeting.
40. Italy argued that this was in contravention of the Vienna Convention on the Law of Treaties by permitting a state to abandon some of its obligations while remaining a full and functioning member of the Protocol (see ASOC, 1991g).
41. Japan, the Republic of Korea and India were the three consultative parties which did not sign on that day although they have since done so. The eight non-consultative parties were Austria, Canada, Columbia, Switzerland, the Democratic Republic of Korea, Greece, Hungary and Romania.

42. As Pascoe (an Australian embassy official in Chile) noted (1990, p. 106) 'Greenpeace and the ACF had a particular influence in the Australian context. In France ... the role of non-governmental organisations [was] symbolised by the report on Antarctica of the Cousteau Foundation'.
43. The head of the Belgian delegation observed at the Viña del Mar meeting that, when the Minerals Convention was adopted, 'we had not seen the wave of environmental concern that has come upon us and we must change our attitudes and reach results as rapidly as possible' (Anon., 1990j). Waller observes (1989, p. 662) that 'the one consideration most responsible for eroding state support for the ... minerals regime [was] growing international concern for the Antarctic environment'.
44. Laws suggested that 'excessive regulation and direction could lead to unproductive use of scientific capacity and funds' (ATCM, 1991b, p. 235).
45. In this regard, he suggested that if the consultative parties failed to give reasonable weight to the views of SCAR (as opposed to those of the NGOs) and if the funding to undertake the tasks requested of it was not forthcoming, then SCAR might have to 'concentrate on primary science and withdraw from giving advice on applied or management problems' (ATCM, 1991b, p. 236).
46. He told the meeting that 'Antarctic scientists are generally concerned that governments may respond positively to public relations pressure campaigns of some environmental NGOs with quite different objectives to those of scientists' (ATCM, 1991b, p. 236). John Heap argued (1991, p. 123) that scientists were the real constituents of the Antarctic and that it was their views which should be taken into account when new rules were devised. He referred also to the 'inflated media-hyped campaign about damage to the Antarctic environment'.

8 THE MADRID PROTOCOL ... AND AFTER

1. The Protocol may, in fact, be amended at any time in accordance with the provisions of the Antarctic Treaty – this would require a consensus of all consultative parties to change any of the articles of the Protocol including that which establishes the ban on minerals activity.
2. Article 6(3) of the arbitration schedule provides for emergency provisional measures to be prescribed in a case of emergency.
3. This annex now includes a requirement that all taking of native mammals and birds must be done in a manner that involves the least degree of pain and suffering. The provisions on Specially Protected Areas were removed and were later included in the fifth annex adopted at ATCM-XVI in Bonn.
4. Article 11 stipulates that the annex 'shall not apply to any warship, naval auxiliary or other ship owned or operated by a state and used ... only on government non-commercial service'. This provision could cover many of the vessels used by governments in their Antarctic programmes.

5. The delegations which included NGO representatives were New Zealand, Australia, Denmark, Chile, France, Germany and the US.
6. The list of invited organisations were SCAR, COMNAP (the Council of Managers of National Antarctic Programs), ICAO (the International Civil Aviation Organisation), IHO (the International Hydrographic Organisation), IOC (the International Oceanographic Commission), IUCN (International Union for the Conservation of Nature), IMO (International Maritime Organisation), WMO (the World Meteorological Organisation) and WTO (the World Tourism Organisation).
7. John Heap, in his opening address, noted that '[f]or democratically expressed reasons within their own countries, the care that CRAMRA expressed for the environment was seen by some as not being innovative enough' (ATCM, 1991b, pp. 199–200).
8. The Meeting agreed to encourage the Standing Committee on Antarctic Logistics and Operations (SCALOP) a sub-committee of COMNAP, to convene a meeting of experts on the issue of marine dumping and prevention of marine pollution in close consultation with the Marine Environment Protection Committee (MEPC) of the IMO. The Meeting also devised terms of references for a forthcoming group of experts meeting on environmental monitoring.
9. The three consultative parties that did not sign the Protocol in Madrid – Japan, the Republic of Korea and India – made a commitment to early signature and ratification.
10. This was in part a response to the EIA requirements under the Protocol which require that CEEs must be submitted at least twelve months in advance of the proposed activity for consideration by a consultative meeting. With the move to annual meetings, preparatory meetings will no longer be held.
11. There has been some suggestion that Argentina might accept a secretariat if it is located in Buenos Aires or Santiago.
12. The US tabled two proposals to adopt one MPA and one SRA which had been reviewed and endorsed by SCAR. The US recognised that these categories, adopted at the fifteenth meeting, had now been replaced by categories created by the new Annex. The meeting agreed, in the light of this, that the parties would voluntarily comply with the provisions of the management plans tabled by the US until the Protocol came into effect.
13. The new category of Antarctic Specially Protected Area includes all existing SPAs, Sites of Special Scientific Interest and subsumes the category of Specially Reserved Area established at ATCM-XV. The new category of Antarctic Specially Managed Area subsumes the ATCM-XV category of Multiple-Use Planning Area. Historic Sites and Monuments may fall into either of the new categories. ASPAs may be located within ASMAs. Both categories may include marine as well as terrestrial sites.
14. NGOs suggest that only one subset of non-governmental activity requires specific rules and regulations, that of large-scale commercial tourism, especially with respect to compliance and liability. The

recently formed International Association of Antarctic Tourist Operators (IAATO) has adopted a Voluntary Code of Conduct.

15. The consultative status of the former Soviet Union was taken by the Russian Federation. The Ukraine has since acceded to the Antarctic Treaty as a state in its own right, but does not have consultative party status.
16. At time of writing, four parties have deposited instruments of ratification – Spain, France, Peru and Ecuador. Several others have begun and, indeed, almost completed the process.
17. The draft final Report referred to a '*first* meeting of experts' (ATCM, 1992, p. 8; emphasis added).

CONCLUSION

1. See Susskind and Ozawa (1992) for an analysis of how institutional bargaining can be improved to overcome some of these difficulties.

Bibliography

AFP (1989a) Text of AFP Report (interview with PM Rocard on 'Questions à Domicile', April)

—— (1989b) 'France Antarctica' [AFP Report], BLTN.3, 22 July

Aga Khan, H. H. Prince Sadruddin, Professor Arne Naess, Charles de Haes, David McTaggart (1989) 'Joint statement on the Antarctic', *ECO* LXXIV (1), 9 October, p. 1

Aldhous, Peter (1991a) 'Mining ban in the air', *Nature*, 351 (6322), 9 May, p. 88

—— (1991b) 'US blocks Antarctic accord', *Nature*, 351 (6329), 27 June, p. 680

Alexander, Lewis M. and Lynne Carter Hanson (eds) (1985) *Antarctic Policies and Marine Resources: critical choices for the 1980s* (Kingston: Center for Ocean Management, University of Rhode Island)

Alley, Roderic (1984) 'New Zealand and Antarctica', *International Journal*, 39 (4), Autumn, pp. 911–31

Almond, Harry H. (1985) 'Demilitarisation and Arms Control: Antarctica', *Case Western Reserve Journal of International Law*, 17 (2), Spring, pp. 229–84

Ambassade de France à Londres (1989a), Report on the closing speech of M. François Mitterand, President of the Republic, at the 'Planet Earth' International Seminar in Paris (13 June 1989), SR/LON/14/89

—— (1989b) 'Reply of Mme Edwige Avice, Minister Delegate attached to the Ministre d'Etat, Minister of Foreign Affairs, to a Parliamentary Question (National Assembly, 14 June 1989)', *Speeches and Statements* Sp.St/LON/79/89 (London: Service de Presse et d'Information)

American Petroleum Institute (n.d.) *Antarctic minerals regime: issue papers*, (unpublished document)

Andersen, Rolf Trolle (1991) 'Negotiating a new regime: how CRAMRA came into existence' in Arnfinn Jorgensen-Dahl and Willy Ostreng (eds), *The Antarctic Treaty system in World Politics* (Basingstoke: Macmillan)

Anderson, Peter J. (1981) 'How the South was won', *The Wilson Quarterly*, V (4), Autumn, pp. 53–68

Angel, Martin V. (1987) 'Criteria for Protected Areas and other Conservation Measures in the Antarctic region', *Environment International*, 13 (1), pp. 105–14

Anon. (1940) 'Norwegian sovereignty in the Antarctic', *American Journal of International Law*, 34, pp. 83–5

Anon. (1948) 'Discussions asked on territorial problem of Antarctica', *Department of State Bulletin*, 19 (479), 5 September, p. 301

Anon. (1984a) 'Australia and the Antarctic Treaty system', *Australian Foreign Affairs Record*, 55 (2), February, pp. 90–8

Anon. (1984b) 'The value of the Antarctic Treaty system', *Australian Foreign Affairs Record*, 55 (9), September, pp. 905–11

Anon. (1984–85) 'France's Antarctic Airstrip: a test of the Treaty', *Search*, 15 (11/12), December/January, pp. 299–300

Anon. (1985a) 'Challenge to the Antarctic Treaty within the United Nations', *Australian Foreign Affairs Record*, 56 (1), January, pp. 27–8

Anon. (1985b) 'Antarctic minerals regime', *Australian Foreign Affairs Record*, 56 (1), January, pp. 29–30

Anon. (1985c) 'The Convention on the Conservation of Antarctic Marine Living Resources', *Australian Foreign Affairs Record*, 56 (1), January, pp. 31–2

Anon. (1986) 'Antarctic Scientists clean up messy site', *New Scientist*, 110 (1504), 17 April, p. 20

Anon. (1987) 'World Park Antarctica', *Greenpeace*, 12 (3), July–September, p. 21

Anon. (1988a) 'Convention too hard: treaty park next', *ECO*, XLVII (6), January, p. 1

Anon. (1988b) 'Beeby 5: miners' delight, safeguards slight', *ECO*, XLVII (1), January, pp. 1–2

Anon. (1988c) 'Need to know', *ECO*, XLVII (5), January, p. 2

Anon. (1988d) 'No consensus on Antarctica', *UN Chronicle*, XXV (1), March, p. 77

Anon. (1988e) 'French Airstrip in Antarctica', *Greenpeace*, 13 (2), March–April

Anon. (1988f) 'Greenpeace Antarctic Expedition: Year Three', *Greenpeace*, 13 (2), March–April, p. 21

Anon. (1988g) 'Mineral Rights', *Greenpeace*, 13 (2), May–June, p. 12

Anon. (1988h) 'Consensus must remain', *ECO*, XLVIII (6), May–June, p. 1

Anon. (1988i) 'Ministers welcome negotiation conclusions', *ANARE News*, 54, June, p. 15

Anon. (1988j) 'Antarctic Hypocrisy', *Greenpeace*, 13 (4), July/August, pp. 18–19

Anon. (1988k) 'Minerals Convention adopted', *Environmental Law and Policy*, 18 (4), Fall, pp. 100–2

Anon. (1989a) 'Wilkes clean-up planned by Australia', *ANARE News*, 57, March, p. 17

Anon. (1989b) 'New Zealand Antarctic scientists say no to Antarctic mining', *Polar Record*, 25 (153), April, p. 154

Anon. (1989c) 'Another Antarctic oil spill', *Habitat*, 17 (2), April, pp. 14–15

Anon. (1989d) 'Canberra's green turn on Antarctic', *Financial Times*, 23 May

Anon. (1989e) 'Antarctic veto', *New Scientist*, 122 (1666), 27 May, p. 34

Anon. (1989f) 'Antarctic oil spill highlights need for reform', *EDF Letter*, XX (2), May, pp. 1; 5

Anon. (1989g) 'Antarctic Antics', *New Scientist*, 122 (1667), 3 June, p. 20

Anon. (1989h) 'Antarctica – a great victory', *Conservation News*, 21 (5), June, p. 1

Anon. (1989i) 'Tides of change' *ECO*, LXXIV (2), October, pp. 1–2; 4

Anon. (1989j) 'FOE Resolution', *ECO*, LXXIV (2), October, p. 4

Anon. (1989k) 'Scientists speak on CRAMRA', *ECO*, LXXIV (3), October, p. 4

Anon. (1989l) 'French plea for Antarctica park', *New Scientist*, 124 (1686), 14 October, p. 3

Anon. (1989m) 'Antarctic ship operators release joint environmental guidelines', *Antarctic News* (Press Release), November 6

Anon. (1989n) 'Antarctic Expedition 1989–90', *Greenpeace*, 14 (6), November/December

Anon. (1990a) 'Cousteau visit heralds new Antarctic momentum', *Backgrounder*, 1 (9), 9 March, pp. 4–5

Anon. (1990b) 'Antarctic update', *Calypso Log*, 17 (2) , April, pp. 4–5

Anon. (1990c) 'NSF reports focus on improving USAP environmental practices', *Antarctic Journal of the US*, XXV (2), June, pp. 1–2

Anon. (1990d) 'Protection of the Antarctic environment: meetings in Chile', *Backgrounder*, 1 (29), 16 November, pp. 7–10

Anon. (1990e) 'A unified Germany?', *ECO*, LXXVII (3), November, p. 2

Anon. (1990f) 'Letters from home', *ECO*, LXXVII (3), November, p. 1

Anon. (1990g) 'Ban on Antarctic mining opposed', *Sydney Morning Herald*, 22 November

Anon. (1990h) 'Ice treaty divides nations', *Bay of Plenty Times*, 30 November

Anon. (1990i) 'Scientists support World Park', *ECO*, LXXVII (5), December, p. 4

Anon. (1990j) 'More support on mines ban in Antarctica', *West Australian*, 23 November

Anon. (1991a) 'UK Opposition parties reject government policy', *ECO*, LXXX (1), p. 1

Anon. (1991b) 'Antarctic deadlock in Madrid?', *New Scientist*, 130 (1766), 27 April, p. 6

Anon. (1991c) 'What's the hurry in Antarctica', *The New York Times*, 6 June

Anon. (1991d) 'US on thin ice in Antarctica', *The Australian Financial Review*, 24 June, p. 14

Anon. (1991e) 'US under fire for Antarctic decision', *Canberra Times*, 24 June, p. 6

Anon. (1991f) 'US "sabotaging treaty"', *The Australian*, 24 June, p. 6

Anon. (1991g) 'US to sign protocol protecting Antarctic environment', *USIS Backgrounder*, 8 July

Anon. (1991h) 'Australia's Antarctic initiative gets go-ahead', *Environment: Australia's international agenda*, 2, September, p. 1

Anon. (1992) 'A look at COMNAP's revised practical guidelines on EIA', *ECO*, LXXXII (2), November, pp. 1–4

Antarctic and Southern Ocean Coalition (ASOC) (1984a) *Antarctic Minerals negotiations: the second draft*, 22 October (Wellington: ASOC NZ)

—— (1984b) *Antarctica brief: the minerals regime*, 26 October 1984 (Wellington: ASOC NZ)

—— (1985a) *Background Paper on the French Airfield at Pointe Géologie, Antarctica*, March 1

—— (1985b) *How might an AEPA work in a minerals regime?*, NGO Non-paper, ATCM 13/NGO/2, Brussels, 10 October

—— (1986a) *Environmental Assessment*, ASOC Paper No 1, ANT/SCM 8, Minerals Negotiations, Hobart, Australia, 16 April

—— (1986b) *The World Park option for Antarctica*, ASOC Information Paper No 4, ANT SCM/8, 21 April

—— (1986c) *Antarctic briefing paper: minerals*, 1 August

—— (1987) *Some solutions to problems with the draft Antarctic minerals convention (MR/17, Revision III)*, ASOC Information Paper No 1, ANT SCM/10, Montevideo, Uruguay

—— (1988a) *Considerations regarding liability provisions in an Antarctic minerals convention*, ASOC Information Paper, No 1988–1, 4 January 1988, (prepared by Clifton Curtis, Paul Consbruck, Max Selz) (Washington DC: The Oceanic Society)

—— (1988b) 'The World Park option for Antarctica', Press Information, 17 January 1988

—— (1988c) 'Key environmental issues for the Antarctic mineral negotiations', 26 January

—— (1988d) *Report on the Fourth Antarctic Treaty Special Consultative Meeting, Antarctic Mineral Resources,* 11th session, Wellington, 18–29 January 1988

—— (1988e) *Some solutions to problems with the draft Antarctic Minerals Convention, (MR/17, Revision V),* ASOC Information Paper 1988–2, 26 March

—— (1988f) *The Antarctic Minerals Negotiations: conflicts with the goal of environmental protection,* 17 April

—— (1988g) *Some solutions to problems with the liability provisions in the draft minerals convention (MR/17, Revision V, Corr.1),* ASOC Information Paper 1988–3, 5 May

—— (1988h) *Decision-making and balance of power (MR/17, Revision V, Corr.1),* ASOC Information Paper 1988–4, 10 May

—— (1988i) *Report of the First Review Meeting of the Convention for the Conservation of Antarctic Seals, London, September 1988*

—— (1988j) *Sealing conducted under the Convention for the Conservation of Antarctic Seals,* CCAS/NGO/1

—— (1988k) *Analysis of the Convention on the Regulation of Antarctic Mineral Resource Activities (AMR/SCM/88/78–2 June 1988),* ASOC Information Paper 1988–4 [sic], 29 October

—— (1988l) *Report of final negotiating session for an Antarctic Minerals Convention, Wellington, New Zealand, May 1988,* October

—— (1989a) *Implications of Alaskan Oil Spill for the Antarctic,* Information Paper No 1, PREP ATCM XV/ASOC INF.1

—— (1989b) *Permanent Protection for Antarctica: a conservation convention is urgently needed,* Information Paper No 2, PREP ATCM XV/INF.2, 11 May

—— (1989c) *Antarctica and the minerals convention,* Briefing Paper (to Members of Parliament), 29 May 1989, Wellington

—— (1989d) *New Zealand on Antarctic minerals: vanguard or rearguard?,* (Prepared by Cath Wallace and Barry Weeber) 20 June, Wellington

—— (1989e) *Antarctic Minerals Convention: property rights for miners thesis confirmed: conservationists release US government document,* Media Release, 17 July

—— (1989f) *Results of 15th Antarctic Treaty Consultative Meeting,* Report to ASOC Members from Jim Barnes and Lyn Goldsworthy

—— (1990a) *The Convention on Antarctic Conservation* [Draft], 16 May

—— (1990b) *The Convention on Antarctic Conservation,* ASOC Information Paper ANT/SCM/NGO INF Paper 1990–1, 7 November

—— (1990c) *Report for ASOC: Antarctic issues at IUCN General Assembly, Perth, Australia, 28 November–5 December 1990* (prepared by Cassandra Phillips)

—— (1991a) *Report on the Antarctic Treaty Special Consultative Meeting on the Antarctic Environment,* 6 February

—— (1991b) *Report of the Third Session of XI Special Antarctic Treaty Consultative Meeting, June 19–22 1991: Madrid Spain*

—— (1991c) *A critique of the Protocol to the Antarctic Treaty on Environmental Protection*, ASOC Information Paper No 1, 8 October, XVI ATCM/INFO 21

—— (1991d) *Annex on liability and compensation for Antarctic activities*, ASOC Information Paper 1991–2, 22 April, XVI ATCM/INFO 55, 9 October

—— (1991e) *The place of science on an environmentally regulated continent*, ASOC Information paper No 4, 8 October, XVI ATCM/INFO 79, 11 October

—— (1991f) *Antarctic Protected Areas*, ASOC Information Paper No 3, 10 October 1991, XVI ATCM/INFO 63

—— (1991g) *The regulation of tourism in Antarctica*, ASOC Information Paper No 6, XVI ATCM/INFO 77, 11 October

—— (1991h) *Report on the signing of the Madrid Protocol on Environmental Protection to the Antarctic Treaty*, October

Antarctic Treaty Consultative Meeting (1961a) *Final Report of the First Antarctic Treaty Consultative Meeting, 10–24 July, 1961*

—— (1961b) *Preservation and conservation of living resources*, Doc./P. 12, 13 July, submitted by Australia

—— (1962a) *Final Report of the Second Antarctic Treaty Consultative Meeting, 18–28 July, 1962*, in W. M. Bush, *Antarctica and International Law* (London: Oceana Publications)

—— (1962b) *Final draft of provisional agenda*, Doc.1, 17 July

—— (1962c) *Matters relating to the exercise of jurisdiction in Antarctica*, Doc P. 10, 17 July, presented by the UK

—— (1962d) *Other matters* (on intersessional meetings), Doc P. 13, 18 July, submitted by the New Zealand delegation

—— (1962e) *Other matters* (on administrative arrangements), Doc P. 17, 25 July, submitted by the South African delegation

—— (1964a) *Agreed Measures for the Conservation of flora and fauna in the Antarctic*, Doc/P. 1, 1 June, Working document drawn up by a working party established by delegates to the preparatory meetings for the 3rd Meeting

—— (1964b) *Pelagic sealing*, Doc/P4, 1 June, UK

—— (1970) *Annotations of provisional agenda*, ANT/2, 8 October

—— (1972a) *Final Report of the Seventh Antarctic Treaty Consultative Meeting, Wellington, 30 October–10 November 1972* (Wellington: Ministry of Foreign Affairs, 1973)

—— (1972b) *Antarctic resources – effects of mineral exploration*, ANT/2, 27 September, Note by France

—— (1972c) *Conservation of Fauna and Flora*, ANT/5, 8 October, Extract from SCAR Bulletin No 41, May 1972, submitted by the UK

—— (1972d) *Purposes and designation of Special Areas*, ANT/9, 8 October, SCAR Paper presented by UK

—— (1972e) *Antarctic resources – effects of mineral exploration*, ANT/17, 18 October, Draft recommendation submitted by the UK

—— (1972f) *Mineral resources, effects of prospecting for minerals*, ANT/37, 31 October, Note of the Argentine Delegation

—— (1972g) *Antarctic resources: effects of mineral exploration*, ANT/44, 2 November, Note by Chile

—— (1975a) *Report of the Eighth Consultative Meeting, Oslo, 9–20 June 1975* (Oslo: Ministry of Foreign Affairs, 1976)

—— (1975b) *Review of Specially Protected Areas: SCAR Response to Antarctic Treaty recommendation VII-2*, ANT/1, April, submitted by Norway

—— (1977) Final Report of the Ninth Antarctic Treaty Consultative Meeting, 19 September to 7 October 1977, in W. M. Bush, *Antarctica and International Law* (London: Oceana Publications)

—— (1979a) *Final Report of the Tenth Consultative Meeting, Washington DC, September 17–October 5, 1979* (Washington DC: Department of State)

—— (1979b) *Hydrocarbon contamination of the Antarctic marine environment by scientific exploratory drillings*, ANT/X/23, 20 September, submitted by Chile

—— (1981) *Final Report of the Eleventh ATCM*, ANT/XI/34/Rev.3, 7 July

—— (1983a) *Final Report of the Twelfth Consultative Meeting, Canberra, 13–27 September 1983* (Canberra: Department of Foreign Affairs/Australian Government Publishing Service, 1984)

—— (1983b) *Man's impact on the Antarctic environment*, ANT/XII/1, draft recommendation submitted by the UK

—— (1983c) *Agreed Measures for the Protection of the Antarctic environment: Interim guidelines: Draft UK Recommendation (Item 6)*, ANT/XII/I/REV 1, 12 September, submitted by Australia

—— (1985a) *Final Report of the Thirteenth Antarctic Treaty Consultative Meeting, Brussels, 7–18 October 1985* (Brussels: Ministry of Foreign Affairs, External Trade and Co-operation in Development, 1986)

—— (1985b) *Item 9: Environmental Assessment Procedures*, ANT/XIII/INF 7, 9 October, Discussion paper submitted by the UK

—— (1987a) *Final Report of the Fourteenth Antarctic Treaty Consultative Meeting, Rio de Janiero, 5–16 October 1987* (Rio de Janiero: Ministry of External Relations, Federative Republic of Brazil)

—— (1987b) *Man's impact on the Antarctic Environment: Additional Protective Arrangements–Antarctic Protected Areas – examples of application*, ANT/XIV/WP/16, Working paper submitted by the Australian delegation

—— (1987c) *Man's impact on the Antarctic Environment: Additional Protection Arrangements*, ANT/XIV/WP/17, Working paper submitted by the Australian delegation

—— (1989a) *Statement* (unclassified telex message, 26 May 1989), UK Delegation, Preparatory Meeting, May

—— (1989b) *A working paper concerning a comprehensive scheme of protection for the Antarctic Environment*, PREP/WP/14, 11 May, submitted by the Delegation of Australia

—— (1989c) *A joint Australian/French proposal in the form of a paper including a draft recommendation for ATCM XV: comprehensive measures for the protection of the Antarctic environment and its dependent and associated ecosystems*, XV ATCM/WP/2

—— (1989d) *Franco-Australian draft working paper on possible components for a comprehensive convention for the preservation and protection of Antarctica*, XV ATCM/WP/3

—— (1989e) *Working Paper on the Protection of the Antarctic environment*, XV ATCM/WP/4, submitted by the New Zealand delegation

—— (1989f) *Comprehensive measures for the protection of the Antarctic environment and its dependent and associated ecosystems*, XV ATCM/WP/7, Working Paper submitted by Chile, annexed to the Final Report, pp. 227–36

—— (1989g) *Comprehensive measures for the protection of the Antarctic environment and its dependent and associated ecosystems*, XV ATCM/WP/8, Working Paper submitted by the US, annexed to the Final Report, pp. 237–42

—— (1989h) *Comprehensive measures for the protection of the Antarctic environment and its dependent and associated ecosystems: Proposed outline of a draft recommendation*, XV ATCM/WP/14, submitted by the Swedish delegation, annexed to Final Report, pp. 243–45

—— (1989i) *Comprehensive Measures for the protection of the Antarctic environment and dependent and associated ecosystems: Draft Principles*, put forward by the Chairman of Working Group 1, annexed to the Final Report, pp. 265–67

—— (1989j) *Final Report of the Fifteenth Antarctic Treaty Consultative Meeting, Paris, 9–20 October 1989*, Republic of France

—— (1991a) *Recommendations adopted at the XVIth Antarctic Treaty Consultative Meeting*, XVI ATCM/WP 57, 18 October, submitted by the Secretariat

—— (1991b) *Draft Final Report*, Federal Republic of Germany

—— (1992) *Draft Final Report of the XVIIth Antarctic Treaty Meeting*, XVII ATCM/WP 34, Submitted by the Secretariat, 19 November

ATCM/Special Consultative Meeting IV (SCM IV) (1983) Antarctic Mineral Resources Regime, Chairman's Informal Personal Report (Beeby I), January

—— (1984) Annex to Chairman's Informal Personal Report, Antarctic Mineral Resources Regime: Draft Articles, Revision, 29 March (Beeby II)

—— (1985) Statement by Ambassador Bo Johnson Theutenberg, Delegation of Sweden (Rio De Janiero, 26 February–8 March)

—— (1987a) Annex to Chairman's Informal Personal Report, Antarctic Mineral Resources: Draft Convention (MR 17 Revision III, April), (Beeby IV)

—— (1987b) Chairman's Informal Personal Report, Antarctic Mineral Resources: Draft Convention (MR/17 Revision IV, November), (Beeby V)

—— (1988a) *Meeting on Antarctic Mineral Resources, Wellington, 18–29 January 1988*, Press Release by Chairman

—— (1988b) *Adoption of Antarctic Minerals Convention*, Press Statement, 2 June

—— (1988c) *The Final Report and the Final Act of the Fourth Special Antarctic Treaty Consultative Meeting on Antarctic Mineral Resources, together with the Convention on the Regulation of Antarctic Mineral Resource Activities*, AMR/SCM/88/79, 2 June

ATCM/SCM XI (1990) *Interim Report of the Eleventh Antarctic Treaty Special Consultative Meeting, Viña del Mar, 19 November–6 December 1990*, Republic of Chile

—— (1991a) *Draft protocol to the Antarctic Treaty on Environmental Protection*, XI ATSCM/2/30, 29 April

—— (1991b) *Protocol on Environmental Protection to the Antarctic Treaty*, XI ATSCM/2, 21 June

—— (1991c) *Final Act of the Eleventh Antarctic Treaty Special Consultative Meeting*, 4 October

—— (1991d) *Final Report of the Eleventh Antarctic Treaty Special Consultative Meeting*, 3–4 October

Antarctic Treaty System (1959) *Conference on Antarctica: October 15 to December 1, 1959: Conference Documents, the Antarctic Treaty and related papers*, Washington DC

—— (1976a) *Establishment of Environment Protection Arrangements for Antarctica*, RPS-11, Document submitted by Australia to the Paris Special Preparatory Meeting, 28 June to 10 July

—— (1976 b) *Report of the Working Group on legal and political aspects of Minerals Exploration and Exploitation in Antarctica*, Paris Special Preparatory Meeting, 28 June to 10 July

—— (1989) *Handbook of the Antarctic Treaty System*, 6th edition (Cambridge: Polar Publications)

Antarctica Project (The) (1984) 'Status of Antarctic Minerals negotiations' *Antarctica Briefing*, No 5, 5 October (Washington DC: Antarctica Project and Greenpeace International)

—— (1987) 'Status of Antarctic Minerals Negotiations', *Antarctica Briefing*, No 13, 30 June (Washington DC: Antarctica Project and Greenpeace International)

—— (1989) Letter from representatives of 9 environmental organisations to President Bush, 23 June

Ary T. S. (n.d.) 'Comments on the Antarctic Minerals Convention', US Bureau of Mines, unpublished paper

Auburn, F. M. (1982) *Antarctic Law and Politics* (London: C. Hurst & Co)

Austin, Greg (1991) 'Bush comes around on Antarctic mining ban', *Sydney Morning Herald*, 5 July

Australian Mining Industry Council (1990) 'Exploration and mining in the Antarctic', *Mining Issues Paper*, Canberra, July

Axelrod, Robert (1984) *The evolution of cooperation* (New York: Basic Books)

Axelrod, Robert and Robert O. Keohane (1986) 'Achieving Cooperation under Anarchy: Strategies and Institutions', in Kenneth A. Oye (ed.), *Cooperation Under Anarchy* (Princeton: Princeton University Press)

Baker, J. M. (1983) *Impact of Oil Pollution on Living Resources*, IUCN Commission on Ecology Papers No 4 (Gland: IUCN)

Barceló, John J. (1986) 'The International Legal Regime for Antarctica', *Cornell International Law Journal*, 19 (2), Summer, pp. 155–61

Barnes, James N. (1982a) *Let's Save Antarctica* (Victoria, Australia: Greenhouse Publications)

—— (1982b) 'The emerging Convention on the Conservation of Antarctic Marine Living Resources: an attempt to meet the new realities of resource exploitation in the Southern Ocean', in Jonathan I. Charney (ed.), *The new nationalism and the use of common spaces* (New Jersey: Allanheld Osmun Publishers)

—— (1984) 'Non-governmental organisations: increasing the global perspective', *Marine Policy*, 8 (2), April, pp. 171–81

—— (1990a) 'Protection of the environment: are present regimes enough?', *International Challenges*, 10 (1), pp. 52–5

—— (1990b) 'What's wrong with CRAMRA?', *IUCN Bulletin*, 21 (2), June, pp. 19–20

—— (1990c) 'The Antarctic Treaty system in crisis: some observations and suggestions', paper delivered at Conference on *The Antarctic and the environment: future prospects?*, University of Louvain, Brussels, 9–10 October

—— (1990d) Statement on behalf of 14 environmental organisations to the US Congress House of Representatives Subcommittee on Oversight and Investigation, 14 March

Basson, Marinelle and John R. Beddington (1991) 'CCAMLR: the practical implications of an ecosystem approach', in Arnfinn Jorgensen-Dahl and Willy Ostreng (eds), *The Antarctic Treaty system in World Politics* (Basingstoke: Macmillan)

Beck, Peter J. (1983) 'Britain's Antarctic dimension', *International Affairs*, 59 (3), Summer, pp. 429–44

—— (1984) 'Antarctica: a case for the UN?', *The World Today*, 40 (4), April, pp. 165–72

—— (1986) *The International Politics of Antarctica* (London: Croom Helm)

—— (1988a) 'The Antarctic Minerals regime negotiations', *Polar Record*, 24 (148), January, pp. 59–61

—— (1988b) 'Another sterile annual ritual? the UN and Antarctica 1987', *Polar Record*, 24 (150), July, pp. 207–12

—— (1989a) 'Entering the age of the polar regions: the Arctic and Antarctic are no longer poles apart', *Ambio*, XVIII (1), pp. 92–4

—— (1989b) 'Convention on the Regulation of Antarctic Mineral Resource Activities: a major addition to the ATS', *Polar Record*, 25 (152), January, pp. 19–32

—— (1989c) 'A new polar factor in international relations', *The World Today*, 45 (4), April, pp. 61–8

—— (1989d) 'Antarctica at the UN 1988: seeking a bridge of understanding', *Polar Record*, 25 (155), pp. 329–34

—— (1990) 'Antarctica enters the 1990s: an overview', *Applied Geography*, 10, pp. 247–63

—— (1991a) 'The Antarctic resource conventions implemented: consequences for the sovereignty issue', in Arnfinn Jorgensen-Dahl and Willy Ostreng (eds), *The Antarctic Treaty system in World Politics* (Basingstoke: Macmillan)

—— (1991b) 'Antarctcia, Viña del Mar and the 1990 UN debate', *Polar Record*, 27 (162), pp. 211–16

Beddington, John (1987) 'The scientific requirements for the conservation of exploited living marine resources in the Antarctica', *Environment International*, 13 (1), pp. 115–119

—— and W. N. Bonner (1986) *Conservation in the Antarctic*, Report of the Joint IUCN/SCAR Working Group on long-term Conservation in the Antarctic (Cambridge: SCAR/IUCN)

Beeby, C. (1972) *The Antarctic Treaty* (Wellington: New Zealand Institute of International Affairs)

—— (1982) 'Towards an Antarctic minerals resources regime', *New Zealand International Review*, 7 (3), May/June, pp. 4–6

—— (1983a) 'A negotiators view', in New Zealand Institute of International Affairs, *The Antarctic: preferred futures, constraints and choices*, Pamphlet No 44 (Wellington: NZIIA)

—— (1983b) 'An overview of the problems which should be addressed in the preparation of a regime governing the mineral resources of Antarctica', in

Francisco Orrego Vicuña (ed.), *Antarctic resources policy: scientific, legal and political issues* (Cambridge: Cambridge University Press)

—— (1985) 'The Antarctic Treaty system and a mineral regime', *New Zealand Foreign Affairs Review*, 35 (1), January/March, pp. 18–26

—— (1987) 'The Minerals Negotiations: constraints and outcomes', paper delivered to the *Policies on Ice* Conference, 26–27 September, Wellington, New Zealand

—— (1990) 'The Antarctic Treaty system: goals, performance and impact', paper presented at the Nansen Institute Conference, *The Antarctic Treaty system in world politics*, Norway, May

Beitz, Charles R. (1979) *Political Theory and International Relations* (Princeton: Princeton University Press)

Bell, Karla (1982) 'Antarctica: getting into hot water', *Chain Reaction Supplement*, 29, Spring, pp. 22–8

Benninghoff, William S. (1987) 'The Antarctic Ecosystem', *Environment International*, 13 (1), pp. 19–25

—— and W. N. Bonner (1985) *Man's Impact on the Antarctic Environment: a procedure for evaluating impacts from scientific and logistic activities* (Cambridge: Scientific Committee on Antarctic Research)

Bergin, Anthony (1991) 'The politics of Antarctic minerals: the greening of white Australia', *Australian Journal of Political Science*, 26 (2), July, pp. 216–39

Bernhardt, J. Peter A. (1975) 'Sovereignty in Antarctica', *California Western International Law Journal*, 5 (2), Spring, pp. 297–349

Bertram, G. C. L. (1957) 'Antarctic Prospect', *International Affairs*, 33 (2), April pp. 143–53

—— (1958) *Antarctica: today and tomorrow* (Cambridge: Cambridge University Press)

Bilder, Richard B. (1982) 'The Present Legal and Political situation in Antarctica', in Jonathan I. Charney (ed.), *The new nationalism and the use of common spaces* (New Jersey: Allanheld Osmun Publishers)

Blay, S. K. N. and B. M. Tsanenyi (1989) *The Convention on the Regulation of Antarctic Mineral Resource Activities (CRAMRA): can a claimant veto it?*, Antarctic and Southern Ocean Law and Policy Occasional Papers No 1 (Tasmania: Faculty of Law, University of Tasmania)

—— (1990) 'Australia and the Convention for the Regulation of Antarctic Mineral Resource Activities (CRAMRA)' [sic], *Polar Record*, 26 (158), July, pp. 195–202

Blay, S. K. N., R. W. Piotrowicz and B. M. Tsamenyi (1989) *Antarctica after 1991: the legal and policy options*, Antarctic and Southern Ocean Law and Policy Occasional Papers No 2 (Tasmania: Faculty of Law, University of Tasmania)

Boardman, Robert (1990) *Global regimes and nation-states: environmental issues in Australian politics* (Ottawa: Carleton University Press)

Boczek, Boleslaw A. (1983–4) 'The protection of the Antarctic eco-system: a study in international environment law', *Ocean Development and International Law*, 13 (3), pp. 347–425

—— (1986) 'Specially Protected Areas as an instrument for the conservation of the Antarctic nature', in Rüdiger Wolfrum (ed.), *Antarctic Challenge II:*

conflicting interests, cooperation, environment protection, economic development (Berlin: Duncker and Humblot)

—— (1988) 'The legal status of visitors including tourists and non-governmental expeditions in Antarctica' in Rüdiger Wolfrum (ed.), *Antarctic Challenge III: conflicting interests, cooperation, environment protection, economic development* (Berlin: Duncker and Humblot)

Bohlen, Curtis (1990) (Assistant Secretary of State for Oceans and International Environmental and Scientific Affairs) Testimony to House of Representatives Committee on Science, Space and Technology, US House of Representatives, 12 July

Bonner, W. N. (1987) 'Antarctic science and conservation – the historical background', *Environment International*, 13 (1), pp. 19–25

—— (1989) 'Environmental assessment in the Antarctic', *Ambio*, XVIII (1), pp. 83–9

—— and D. W. H. Walton (eds) (1985) *Key Environments: Antarctica* (Oxford: Pergamon/IUCN)

Brewster, B. (1982) *Antarctic Wilderness at Risk* (Melbourne: Sun Books)

Broady, Paul (1991a) 'Antarctic Park', *New Scientist*, 130 (1765), 20 April, p. 2

—— (1991b) 'Antarctica more than ice and penguins' in Jagdish Patel and Susan Mayer (eds), *Antarctica: the scientists' case for a World Park* (London: Greenpeace UK)

Brown, Alan (1990a) 'The design of CRAMRA: how appropriate for the protection of the environment?' paper presented to the Conference on Antarctica, at the Fridtjof Nansen Institute, Oslo, 21 May

—— (1990b) 'The design of CRAMRA: how appropriate for the protection of the environment?' *International Challenges*, 10 (1), pp. 24–6

Brown, Lester R. (1988) 'Redefining National Security', in Charles W. Kegley Jr and Eugene R. Wittkopf (eds), *The Global Agenda: Issues and Perspectives*, 2nd Edition (New York: Random House)

Brown, Neville (1989) 'Climate, ecology and international security', *Survival*, XXXI (6), November/December, pp. 519–32

Brown, Seyom (1988) *New Forces, Old Forces and the Future of World Politics* (Illinois: Scott Foresman & Co)

Browne, Marjorie Ann (1987) 'Activity levels in Antarctica – statistical information', Memo to House Interior and Insular Affairs Committee, Subcommittee on National Parks and Public Lands, from Congressional Research Service, The Library of Congress, 22 December

Bull, Hedley (1977) *The anarchical society: a study of order in world politics* (Basingstoke: Macmillan)

Burgess, J. R. (1990) 'Comprehensive environmental protection of the Antarctic: new approaches for new times' in Grahame Cook (ed.), *The future of Antarctica: exploitation versus preservation* (Manchester: Manchester University Press)

Burke, Tom (1982) 'Friends of the Earth and the conservation of resources' in Peter Willetts (ed.), *Pressure groups in the global system* (London: Frances Pinter)

Burmester, Henry C. [n.d] 'Liability for Damage from Antarctic Mineral Resource Activities' [unpublished paper, photocopy]

Bush, W. M. (1982) *Antarctica and International Law: a collection of interstate and national documents*, Vols 1–3 (London: Oceana Publications Inc.)

—— (1990) 'The Antarctic Treaty System: a framework for evolution and the concept of a system' in R. A. Herr, H. R. Hall and M. G. Haward (eds), *Antarctica's future: continuity or change?* (Hobart: Australian Institute of International Affairs/Tasmanian Government Printing Office)

Campbell, I. B. and G. G. C. Claridge (1987) *Antarctica: Soils, Weathering Processes and Environment*, Development in Soil Science 16 (Amsterdam: Elsevier)

Caldwell, Lynton (1988) 'Beyond environmental diplomacy: the changing institutional structure of international cooperation' in John E. Carroll (ed.), *International Environmental Diplomacy* (Cambridge: Cambridge University Press)

Carroll, John E. (ed.) (1988) *International Environmental Diplomacy* (Cambridge: Cambridge University Press)

Center for Oceans Law and Policy (1987) *The Polar Regions: proceedings from the Eleventh Annual Seminar*, Oceans Policy Studies Series (Virginia: University of Virginia School of Law)

Cerone, Rudy (1978) 'Survival of the Antarctic Treaty: economic self-interest vs enlightened international co-operation', *Boston College International and Comparative Law Journal*, 2 (1), pp. 115–129

Charney, Jonathan I. (ed.) (1982) *The New Nationalism and Use of Common Spaces: Issues in Marine Pollution and the Exploitation of Antarctic* (New Jersey: Allanheld Osmun Publishers)

Child, Jack (1985) 'South American geopolitical thinking and Antarctica' *International Studies Notes*, 11 (3), Spring, pp. 23–8

—— (1988) *Antarctica and South American Politics: Frozen Lebensraum* (New York: Praeger)

Clark, Margaret (1988), 'Co-operation on Ice: the potential of collaboration in the southern cone', in Philip Kelly and Jack Child (eds), *Geopolitics of the Southern Cone and Antarctica* (Boulder: Lynne Rienner Publishers)

Clark, Pilita (1989) 'Richardson wavers on Antarctic Treaty' [sic], *Sydney Morning Herald*, 16 May, p. 4

Clifton, Jane (1989) 'Lange hits out at Australia's Antarctic plan', *Dominion*, 30 May, p. 2

Colson, David (1980) 'The Antarctic Treaty system: the mineral issue', *Law and Policy in International Business*, 12, pp. 841–902

—— (1986) 'The US position on Antarctica', *Cornell International Law Journal*, 19 (2), Summer, pp. 291–301

Conforti, Benedetto (1986) 'Territorial claims in Antarctica: a modern way to deal with an old problem', *Cornell International Law Journal*, 19 (2), Summer, pp. 249–58

Constable, Andrew (1990) 'Management protocol urgently needed – a non-government view', *ANARE News*, 64, December, p. 7

CCAMLR (1991) *Schedule of Conservation Measures in Force 1991/92*, Commission for the Conservation of Antarctic Marine Living Resources, Hobart

Cook, Grahame (ed.) (1990a) *The future of Antarctica: exploitation versus preservation* (Manchester: Manchester University Press)

Corn, M. Lynne et al. (1988) *Environmental effects of recent activities in Antarctica*, CRS Report for Congress, 88–439 SPR (Washington DC: Congressional Research Service, The Library of Congress)

Cousteau Society (1990a) *The Cousteau Society position on Antarctica*, November (Virginia, Norfolk: The Cousteau Society)

—— (1990b) 'Antarctica: the convergence of politics and science', *The Cousteau Watch*, 4 November

Cowhey, Peter F. and Edward Long (1983) 'Testing theories of regime change: hegemonic decline or surplus capacity', *International Organisation*, 37 (2), Spring, pp. 157–88

Crabbe, Michael (1985) 'Antarctica: oil potential for the next century', *Petroleum Economist*, 52 (1), October, pp. 365–67

Cross, Michael (1991a) 'Greenpeace to leave Antarctica without a trace', *New Scientist*, 130 (1765), 20 April, p. 11

—— (1991b) 'Antarctica: exploration or exploitation?', *New Scientist*, 130 (1774), 22 June, pp. 25–28

Croxall, J. P. (1987) 'The Status and Conservation of Antarctic Seals and Seabirds: a review', *Environment International*, 13 (1), pp. 55–70

Currie, Duncan (1989) *Antarctic mineral exploitation and international law*, unpublished paper prepared for Greenpeace International

Darby, Andrew (1989) 'US experts head south to battle huge oil spill', *Sydney Morning Herald*, 3 February

Davis, Bruce (1987) 'The UK, Australia and Antarctica: the background and some emergent issues', *Environment and Planning Law Journal*, 4 (1), March, pp. 26–32

Department of Foreign Affairs and Trade (1990a) *Australia and the Antarctic Environment*, Fact Sheet No 50, April (Canberra: DFAT, Overseas Information Branch)

Department of Foreign Affairs and Trade (1990b) *A comprehensive regime for the protection of the Antarctic environment and its dependent and associated ecosystems: a summary of Australian/French views*, 15 May (Canberra: Department of Foreign Affairs and Trade)

Department of State (1964) *US Policy and International Cooperation in Antarctica: a historical summary with emphasis on recent developments*, May (Washington DC: Historical Studies Division, Bureau of Public Affairs, Department of State)

—— (1974) *Environmental impact statement on the Convention for the Conservation of Antarctic Seals*, Washington DC, 8 August

Dey, Anita (1991) 'India in Antarctica: perspectives, programmes and achievements', *Polar Record*, 27 (161), April, pp. 87–92

Dingwall, Paul (1990) 'Antarctic tourism' *IUCN Bulletin*, 21 (2), June, pp. 15–16

Donnelly, Jack (1986) 'International human rights: a regime analysis', *International Organisation*, 40 (3), Summer, pp. 599–642

Doran, Brendon and John Handmer (1990) *New approaches to Antarctic conservation*, Report of a Seminar held on 26 April 1990 (Canberra: Australian National University, Centre for Resource and Environmental Studies)

Drewry, D. J. (1988) 'The challenge of Antarctic science', *Oceanus*, 31 (2), Summer, pp. 5–10

Ecofund Australia (1986) *Antarctic conservation convention*, Second Draft, August

Elder, Danny and Katya Michos-Ederer (1990) 'IUCN and Antarctica: what next?' *IUCN Bulletin*, 21 (2), June, pp. 24–5

Elliot, David H. (1977) *A framework for assessing environmental impacts of possible Antarctic mineral development: Part 1* (prepared for the US Department of State) Institute of Polar Studies, Ohio State University

—— (1988) 'Antarctica: is there any oil and natural gas', *Oceanus*, 31 (2), Summer, pp. 32–8

Elliott, Lorraine M. (1991) *Continuity and change in cooperative international regimes: the politics of the recent environment debate in Antarctica*, Working Paper 1991/3 (Canberra: Department of International Relations, ANU)

—— (1993) *Protecting the Antarctic environment: Australia and the Minerals Convention* (Canberra: Australian Foreign Policy Publications Program, Australian National University)

El-Sayed, Sayed Z. (1988) 'Living resources: the BIOMASS Program' *Oceanus*, 31 (2), Summer, pp. 75–9

Elzinga, Aant and Ingemar Bohlin (1989) 'The politics of science in polar regions', *Ambio*, XVIII (1), pp. 71–6

Erize, Francisco J. (1987) 'The Impact of Tourism on the Antarctic environment', *Environment International*, 13 (1), pp. 133–6

Evans, Senator Gareth (Minister for Foreign Affairs and Trade) (1990) *Protecting Antarctica: an example of courage*, address at the official welcome for the Trans-Antarctic 1990 International Expedition, Sydney, 19 March 1990, News Release M49

—— (1991) 'Antarctica: an international environment win for Australia', Speech to celebrate the opening for signature of the Environment Protocol to the Antarctic Treaty, 4 October 1991, *Ministerial Document Service*, No. 65/91–92, 9 October, pp. 2071–76

Evans, Gareth and Bruce Grant (1991) *Australia's foreign policy in the world of the 1990s* (Melbourne: Melbourne University Press)

Fairclough, A. J. (1991) 'Global environmental and natural resource problems – their economic, political and security implications', *The Washington Quarterly*, 14 (1), Winter, pp. 81–97

Falk, Richard (1983) *The end of world order: essays on normative international relations* (New York: Holmes and Meier)

—— (1990) 'Antarctic Treaty system: are there viable alternatives', *International Challenges*, 10 (1), pp. 91–3

Fanshawe, John (1990) 'Antarctica: mining struggle just tip of iceberg', *Canberra Times*, 27 June

Farman, Joe (1991) 'Antarctica: a bone of contention' in Jagdish Patel and Susan Mayer (eds), *Antarctica: the scientists' case for a World Park* (London: Greenpeace UK)

Fifield, Richard (1987) *International Research in the Antarctic* (Oxford: Oxford University Press)

Finlayson, Jock A. and Mark W. Zacher (1981) 'The GATT and the regulation of trade barriers: regime dynamics and functions', *International Organisation*, 35 (4), Autumn, pp. 561–602

Fletcher, Susan R. (1989) *Antarctica: environmental protection issues*, Summary of a CRS Research Workshop, 10 April 1989, CRS Report for Congress

89–272 ENR (Washington, DC: Congressional Research Service, The Library of Congress)

Foreign and Commonwealth Office (FCO) (1988) *Convention on the Regulation of Antarctic Mineral Resource Activities: a commentary on environmental aspects*, prepared by the Polar Regions Section, South America Department for the UK IUCN Committee, 15 October

—— (1989a) *The Antarctic minerals convention: a memorandum*, Polar Regions Section, South America Department, 13 June

—— (1989b), *UK Policy on the protection of the Antarctic environment*, Polar Regions Section, South America Department, 15 June

Fouéré, Erwan (1988) 'Emerging trends and international environmental agreements' in John E. Carroll (ed.), *International Environmental Diplomacy* (Cambridge: Cambridge University Press)

Francioni, Francesco (1986) 'Legal Aspects of mineral exploitation in Antarctica', *Cornell International Law Journal*, 19 (2), Summer, pp. 163–83

Frank, Ronald F. (1983–84) 'The Convention on the Conservation of Antarctic Marine Living Resources', *Ocean Development and International Law*, 13 (3), pp. 291–345

Frankel, Joseph (1988) *International Relations in a Changing World*, 4th edition (Oxford: Oxford University Press)

Gardam, Judith G. (1985) 'Management regimes for Antarctic marine living resources – an Australian perspective', *Melbourne University Law Review*, 15 (2), December, pp. 279–312

Goldsworthy, Lyn (1984) 'World Park on Ice', *Chain Reaction*, 37, April/May, pp. 36; 40

—— (1990a) 'Make the Antarctic a World Park!', *IUCN Bulletin*, 21 (2), June, pp. 22–3

—— (1990b) 'Antarctica: winning a cold war', *Greenpeace Australia News*, 1 (1), July/August/September, p. 12

—— (1990c) 'Antarctica: closer to World Park' *Greenpeace Australia News*, 1 (2), October/November/December, p. 4

—— (1991) 'Antarctica: keeping on the pressure', *Greenpeace Australia News*, 2 (2), Autumn, p. 7

Gould, Laurence M. (1971) 'The History of the Scientific Committee on Antarctic Research (SCAR)', in Louis O. Quam (ed.), *Research in the Antarctic*, Publication No 93 (Washington, DC: American Association for the Advancement of Science)

Gourevitch, Peter (1978) 'The second image reversed: the international sources of domestic politics', *International Organisation*, 43 (4), Autumn, pp. 881–912

Greenpeace (n.d.(a)) *The Antarctic Declaration*, Washington DC

—— (n.d.(b)) *Greenpeace Antarctica: the World Park option for Antarctica*

—— (n.d.(c)) *The World Park Option for Antarctica*, Greenpeace Briefing Paper (London: Greenpeace UK)

—— (n.d.(d)) *Second thoughts on CRAMRA* (Lewes: Stichting Greenpeace Council)

—— (n.d.(e)) *Why Greenpeace continues to work towards a World Park*, Background paper, Greenpeace Australia

—— (1986) *The World Park Option for Antarctica: Background for a fourth UN debate*, 17 November (Lewes: Greenpeace International, Stichting Greenpeace Council)

—— (1987a) *Chronology of the Airstrip issue*, ATCM 13/NGO/3, Greenpeace International

—— (1987b) *The French airstrip: what the ATCPs should do*, ATCM 13/NGO/5, Greenpeace International

—— (1988a) *Political solutions to the Antarctic seals problem*, CCAS/NGO/2, Stichting Greenpeace Council

—— (1988b) *Ratification strategy*, Report of a campaigners' meeting, 22 July

—— (1989a) *No mining in Antarctica: a minerals convention is not needed* (Amsterdam: Greenpeace International)

—— (1989b) *Report on the visit to Dumont D'Urville, Greenpeace Antarctic Expedition 1988/89*, Greenpeace International, 9 March

—— (1989c) *The Antarctic Minerals Bill: background to the current Antarctica debate*, Greenpeace UK (campaign briefing paper)

—— (1990) *Report of fishing vessels monitored during Greenpeace 1989/90 Antarctic expedition*, NGO Information Paper for CCAMLR IX, CCAMLR-IX/NGO/3, 31 October

Greig, D. W. (1978) 'Territorial sovereignty and the status of Antarctica', *Australian Outlook*, 32 (2), August, pp. 117–29

Grieco, Joseph M. (1988) 'Anarchy and the limits of cooperation: a realist critique of the newest liberal institutionalism', *International Organisation*, 42 (3), Summer, pp. 485–507

Groom, A. J. R. (1988) 'The advent of international organisation' in Paul Taylor and A. J. R. Groom (eds), *International Institutions at work* (London: Pinter Publishers)

Grunberg, Isabelle (1990) 'Exploring the "myth" of hegemonic stability', *International Organisation*, 44 (4), Autumn, pp. 431–77

Gulland, J. A. (1988) 'The management regime for living resources' in Christopher C. Joyner and Sudhir K. Chopra (eds), *The Antarctic legal regime* (Dordrecht: Martinus Nijhoff)

Gurtov, Mel (1988) *Global Politics in the Human Interest* (Boulder: Lynne Rienner Publishers)

Guyer, Roberto O. (1983) 'Antarctica's role in international relations', in Francisco Orrego Vicuña (ed.), *Antarctic resources policy: scientific, legal and political issues* (Cambridge: Cambridge University Press)

Haas, Ernst B. (1980) 'Why collaborate? issue-linkage and international regimes', *World Politics*, XXXII (3), April, pp. 357–405

—— (1983) 'Regime decay: conflict management and international organisations, 1945–1981', *International Organisation*, 37 (2), Spring, pp. 189–256

—— (1990a) 'Reason and change in international life: justifying a hypothesis', *Journal of International Affairs*, 44 (1), Spring/Summer, pp. 209–40

—— (1990b) *When knowledge is power: three models of change in international organisations* (Berkeley: University of California Press)

Haas, Peter M. (1989) 'Do regimes matter: epistemic communities and Mediterranean pollution control', *International Organisation*, 43 (3), Summer, pp. 377–403

—— (1990) 'Obtaining international environmental protection through epistemic consensus', *Millennium*, 19 (3), Winter, pp. 347–63

—— (ed.) (1992) 'Knowledge, power and international policy coordination', *International Organisation*, 46 (1), Winter

Hagelin, Lena and Paul Bogart (1989) 'A call for a World Park Antarctica', *Ambio*, XVIII (1), pp. 95–6

Haggard, Stephan and Beth A. Simmons (1987) 'Theories of international regimes', *International Organisation*, 41 (3), Summer, pp. 491–517

Halliday, Fred (1991) 'International relations: is there a new agenda?', *Millennium*, 20 (1), Spring, pp. 57–72

Hambro, Edvard (1974) 'Some notes on the future of the Antarctic Treaty collaboration', *American Journal of International Law*, 68 (2), April, pp. 217–26

Hamer, Mick (1990) 'Britain shifts position on Antarctica', *New Scientist*, 128 (1744), 24 November, p. 5

Hampson, Fen Osler (1989–90) 'Climate change: building coalitions of the like-minded', *International Journal*, XLV (1), Winter, pp. 36–74

Hamzah, B. A. (ed.) (1987) *Antarctica in International Affairs* (Kuala Lumpur: Institute of Strategic and International Studies)

Handmer, John (ed.) (1989) *Antarctica: Policies and policy development*, Resource and Environmental Studies No 1 (Canberra, Australian National University: Centre for Resource and Environmental Studies)

Hanessian, John (1960) 'The Antarctic Treaty 1959', *International and Comparative Law Quarterly*, 9, July, pp. 437–80

Harris, Stuart (ed.) (1984) *Australia's Antarctic policy options*, CRES Monograph 11 (Canberra, Australian National University: Centre for Resource and Environmental Studies)

—— (1988) 'Australian foreign policy and its links with defence policy', *Australian Foreign Affairs Record*, 59 (5), May, pp. 181–5

—— (1990) *The influence of the United Nations on the Antarctic system: a source of erosion or cohesion?*, Working Paper 1990/10 (Canberra: Australian National University, Department of International Relations)

Hart, Paul Dudley (1988) 'The growth of Antarctic tourism', *Oceanus*, 31 (2), Summer, pp. 93–100

—— (1989a) Speech at a dinner hosted by the Minister for Industry and Regional Development, M. Roger Fouroux and the Associate Minister for Foreign Affairs, Mme Avice, Paris, 19 June 1989, in *Backgrounder*, No 662, 21 June 1989

—— (1989b) Speech to a dinner hosted by the Lord Mayor of London, Mansion House, London 21 June 1989, in *Backgrounder*, No 662, 21 June 1989

—— (1990a) 'Australia's policy in Antarctica' in R. A. Herr, H. R. Hall and M. G. Haward (eds) *Antarctica's future: continuity or change?* (Hobart: Australian Institute of International Affairs/Tasmanian Government Printing Office)

—— (1990b) 'Antarctic policy', Statement by the Prime Minister, 23 November

—— , Senator Gareth Evans (Minister for Foreign Affairs and Trade) and Senator Graham Richardson (Minister for the Arts, Sports, Environment, Tourism and Territories), 'Protection of the Antarctic Environment', Joint Statement, 22 May

—— and Michel Rocard (1989) Joint Statement on International Environmental Issues: agreed by the Prime Ministers, Canberra, 18 August

Hay, John E., Alan D. Hemmings and Norman G. Thom (eds) (1990) *Antarctica 150: scientific perpsectives – policy futures* (Auckland: University of Auckland)

Hayashi, Moritaka (1986) 'The Antarctica question in the United Nations', *Cornell International Law Journal*, 19 (2), Summer, pp. 275–90

Hayton, Robert D. (1956) 'The "American" Antarctic', *The American Journal of International Law*, 50 (1956), pp. 583–610

—— (1960a) 'The Antarctic settlement of 1959', *American Journal of International Law*, 54 (2), April, pp. 348–71

—— (1960b) 'The Nations and Antarctica', *Österr. Zeitschrift für öffentliches Recht*, X (3–4), pp. 368–412

Headland, Robert K. (1989) *Chronological list of Antarctic expeditions and related historical events* (Cambridge: Cambridge University Press)

Heap, J. A. (1983) 'Co-operation in the Antarctic: a quarter of a century's experience', in Francisco Orrego Vicuña (ed.), *Antarctic resources policy: scientific, legal and political issues* (Cambridge: Cambridge University Press)

—— (1987) 'The Antarctic Treaty system: environmental protection, conservation and the question of competing use', *Environment International*, 13 (1), pp. 15–18

—— (1990) 'A stronger Antarctic Treaty?', *IUCN Bulletin*, 21 (2), June, pp. 20–1

—— (1991) 'Antarctic politics and Antarctic science – are they at loggerheads', *Antarctic science*, 3 (2), June, p. 123

Hemmings, Alan (1989) 'Why the minerals convention must go', *The Dominion*, 3 October, p. 12

—— (1990) 'Is Antarctica demilitarised?' in R. A. Herr, H. R. Hall and M. G. Haward (eds), *Antarctica's future: continuity or change?* (Hobart: Australian Institute of International Affairs/Tasmanian Government Printing Office)

Henrikson, Alan K (1986) 'The last place on earth, the first place in heaven', in Imagining Antarctica, *Antarktis – Vorstellung und Wirklichkeit* (Massachusetts: Polarities Incorporated)

Herr, R. A., R. Hall and B. W. Davis (eds) (1982) *Issues in Australia's Marine and Antarctic Policies*, Public Policy Monograph (University of Tasmania: Department of Political Science)

Herr, R. A., H. R. Hall and M. G. Haward (eds) (1990) *Antarctica's future: continuity or change?* (Hobart: Australian Institute of International Affairs/ Tasmanian Government Printing Office)

Heyward, Peter (1990) 'An Australian perspective on CCAMLR IX', *ANARE News*, 64, December, pp. 4–5

Higgott, Richard A. and Andrew Fenton Cooper (1990) 'Middle power leadership and coalition building: Australia, the Cairns Group and the Uruguay Round of trade negotiations', *International Organisation*, 44 (44), Autumn, pp. 589–632

Holdgate M. W. and Jon Tinker (1979) *Oil and other minerals in Antarctica: the environmental implications of possible mineral exploration and exploitation in Antarctica*, The Bellagio Report (Cambridge: Scientific Committee on Antarctic Research)

Holsti, K. J. (1972) 'The concept of power in the study of international relations', in Robert L. Pfaltzgraff Jnr (ed.), *Politics in the International System*, 2nd edition (Philadelphia: J. P. Lippincott Co)

—— (1985) *The dividing discipline: hegemony and diversity in international theory* (Winchester, Mass.: Allen & Unwin)

—— (1988), 'Interdependence, Integration or Fragmentation: Scenarios for the Future', in Charles W. Kegley Jr and Eugene R. Wittkopf (eds), *The Global Agenda: Issues and Perspectives*, 2nd Edition (New York: Random House)

Honnold, E. (1978) 'Thaw in International Law? Rights in Antarctica under the Law of Common spaces', *The Yale Law Journal*, 87 (4), March, pp. 804–59

Houweling, Suzanne (1989) 'Cabinet to say no to Antarctic mining', *The Australian*, 22 May, pp. 1–2

Hope, Deborah (1989) 'Antarctica: US thinks we'll cave in', *Sydney Morning Herald*, 13 September

House, Richard (1988) 'Carve up at the Pole', *South*, 94, August, pp. 111–12

Howard, Matthew (1989) 'The Convention on the Conservation of Antarctic Marine Living Resources: a five year review', *International and Comparative Law Quarterly*, 38, Part 1, January, pp. 104–49

Hunt, Paul (1991) 'Impasse freezes Antarctic protection hopes', *The Australian*, 20 June, p. 6

Hunter Miller, David (1927) 'National Rights in the Antarctic', *Foreign Affairs*, 5 (3), pp. 508–10

Hurrell, Andrew and Benedict Kingsbury (eds) (1992) *The international politics of the environment* (Oxford: Clarendon Press)

Hurwich, Evelyn and James N. Barnes (1989) Letter to William K. Reilly, The Administrator, US Environmental Protection Agency, 30 June

Iggulden, David (1990) 'The role and effect of NGOs in Australia' in José Yañez and María Eugenia Zañartu (eds), *ler Taller Latinoamericano de Organizaciones No Gubernamentales Sobre Conservación Antártica* (Santiago, Chile: CODEFF)

Imber, Mark (1988) 'International institutions and the common heritage of mankind: sea, space and polar regions' in Paul Taylor and A. J. R. Groom (eds), *International Institutions at work* (London: Pinter)

International Institute for Environment and Development (IIED) (1986) *Comments for the updated study of the United Nations Secretary General on the Question of Antarctica*, Washington, 5 May 1986

International Union for the Conservation of Nature and Natural Resources (IUCN) (1981b) *Resolutions*, 15th session of the General Assembly of IUCN, Christchurch, New Zealand, 11–23 October

—— (1984a) *Resolutions*, 16th session of the General Assembly of IUCN, Madrid, Spain, 5–14 November

—— (1984b) *Conservation and Development of Antarctic ecosystems*, IUCN, Gland, Switzerland

—— (1987) *Position paper on Proposed Antarctic Minerals regime*, December

—— (1989) *Antarctica: resolutions from IUCN General Assemblies* (typescript) September (Gland, Switzerland: IUCN)

—— (1990) *Resolutions of the 18th General Assembly, Perth, Australia, 28 November to 5 December 1990* (typescript)

—— (1991) *A Strategy for Antarctic Conservation* (Gland: IUCN)

—— /SCAR (1986) *Conservation in the Antarctic*, Report of the Joint IUCN/ SCAR Working Group on long-term conservation in the Antarctic (Gland, Switzerland: IUCN)

Jervis, Robert (1988) 'Realism, game theory, and cooperation', *World Politics*, XL (3), April, pp. 317–49

Johnson, Eric (1976) 'Quick before it melts: toward a resolution of the jurisdictional morass in Antarctica', *Cornell International Law Journal*, 10 (1), December, pp. 173–98

Jones, Barry (1985) 'Antarctica under threat', *Australian Foreign Affairs Record*, 56 (9), September, pp. 839–42

Jorgensen-Dahl, Arnfinn and Willy Ostreng (eds) (1991) *The Antarctic Treaty system in World Politics* (Basingstoke: Macmillan)

Jouventin, Pierre and Vincent Bretagnolle (1990) 'French biologists reply to Greenpeace report', *Polar Record*, 26 (156), January, p. 54

Joyce, Christopher (1991) 'Asbestos found in Antarctic waste dump', *New Scientist*, 130 (1764), 13 April, p. 8

Joyner, Christopher (1986) 'Protection of the Antarctic environment: rethinking the problems and prospects', *Cornell International Law Journal*, 19 (2), Summer, pp. 259–73

—— (1987) 'The Antarctic Minerals negotiating process', *American Journal of International Law*, 81 (4), October, pp. 888–905

—— (1988) 'The evolving minerals regime for Antarctica' in Christopher Joyner and Sudhir K. Chopra (eds) *The Antarctic legal regime* (Dordrecht: Martinus Nijhoff Publishers)

—— (1990) 'Antarctica and the Indian Ocean states: the interplay of law, interests and geopolitics', *Ocean Development and International Law*, 21 (1), pp. 41–70

—— and Sudhir K. Chopra (eds) (1988) *The Antarctic legal regime* (Dordrecht: Martinus Nijhoff Publishers)

Kaiser, Karl (1971) 'Transnational relations as a threat to the democratic process', in Robert O. Keohane and Joseph S. Nye Jr (eds), *Transnational relations and world politics* (Cambridge, Mass: Harvard University Press)

Kamieniecki, Sheldon (1991) 'Political mobilisation, agenda building and international environmental policy', *Journal of International Affairs*, 44 (2), Winter, pp. 339–58

Karns, Margaret P. and Karen A. Mingst (1991) 'Domestic sources of international cooperation: a conceptual framework', prepared for the Annual Meeting of the International Studies Association, 20–23 March 1991, Vancouver

Keage, P. L. (1986) *Antarctic protected areas: future options*, Environmental Studies Occasional Paper No 19 (Hobart: Centre for Environmental Studies, University of Tasmania)

——, P. R. Hay and J. A. Russell (1988) *Proposals to improve Antarctic management plans*, Discussion paper, SCAR Subcommittee on Conservation, 24–26 August

Keating, Colin (1988) (NZ Ministry of External Relations and Trade) Letter to Professor Euan Young, Zoology Department, University of Auckland, 7 December

Keating, P. J. (1988) (The Treasurer, Australia), Letter to Senator the Hon Gareth Evans, Minister for Foreign Affairs and Trade (re signature of CRAMRA), 21 September

—— (1989) Letter to the Prime Minister, Bob Hawke, 27 April

Kegley Jr, Charles W. and Eugene R. Wittkopf (eds) (1988) *The Global Agenda: Issues and Perspectives*, 2nd edition (New York: Random House)

Kelly, Philip and Jack Child (eds) (1988) *Geopolitics of the Southern Cone and Antarctica* (Boulder: Lynne Rienner Publishers)

Keohane, Robert O. (1984) *After hegemony: cooperation and discord in the world political economy* (Princeton: Princeton University Press)

—— (1986a) 'Reciprocity in international relations', *International Organisation*, 40 (1), Winter, pp. 1–27

—— (ed.) (1986b) *Neorealism and its critics* (New York: Columbia University Press)

—— (1989) *International Institutions and State Power: Essays in International Relations Theory* (Boulder: Westview Press)

—— (1990) 'Multilateralism: an agenda for research', *International Journal*, XLV (4), Autumn, pp. 731–64

—— and Joseph S. Nye (eds) (1971) *Transnational relations and world politics* (Cambridge, Mass: Harvard University Press)

——, —— (1977) *Power and Interdependence: World Politics in Transition* (Boston: Little Brown & Co)

——, —— (1987) 'Power and Interdependence revisited', *International Organisation*, 41 (4), Autumn, pp. 725–53

Kerry, K. R. and D. J. Tranter (1978) 'Biology of the Southern Ocean', *Habitat*, 5 (5), January, pp. 22–7

Kimball, Lee (1983) *Report on Antarctic events: 1983* (Washington DC: International Institute for Environment and Development)

—— (1985a) 'We have met the enemy and they are us: diplomats meet scientists at the South Pole: The Antarctic Treaty system meets the United Nations', paper delivered to the Antarctican Society, Washington DC, 19 March

—— (1985b) *Report on Antarctica* (Washington DC: IIED)

—— (1985c) *Report on Antarctica* (Washington DC: IIED)

—— (1986) 'Policy impact on polar environmental issues: summary' prepared for the National Science Board on the National Science Foundation Role in Polar Regions, 18 December 1986, IIED, Washington DC

—— (1987a) *Future directions for the management of Antarctic science* (Washington DC: IIED)

—— (1987b) *Report on Antarctica* (Washington DC: IIED)

—— (1987c) *Report on Antarctica* (Washington DC: IIED)

—— (1988a) *Report on Antarctica: the Antarctic minerals convention negotiations* (Washington DC: IIED)

—— (1988b) *Special Report on the Antarctic minerals convention* (Washington DC: IIED/WRI)

—— (1988c) 'The Antarctic Treaty system', *Oceanus*, 31 (2), Summer, pp. 14–19

—— (1988d) 'The role of non-governmental organisations in Antarctic affairs' in Christopher Joyner and Sudhir K. Chopra (eds), *The Antarctic legal regime* (Dordrecht: Martinus Nijhoff Publishers)

—— (1990) *Southern exposure: deciding Antarctica's future* (Washington DC: World Resources Institute)

—— (1991a) 'CRAMRA and other environmental regimes in the ATS: how well does it fit?' in Arnfinn Jorgensen-Dahl and Willy Ostreng (eds), *The Antarctic Treaty system in World Politics* (Basingstoke: Macmillan)

—— (1991b) *Report on Antarctica* (Washington DC: World Resources Institute)

Kindleberger, Charles P. (1986) 'Hierarchy versus inertial cooperation', *International Organisation*, 40 (4), Autumn, pp. 841–47

Kish, John (1973) *The Law of International Spaces* (Leiden: A. W. Sijthoff)

Kitney, Geoff (1989) 'US stand on Antarctica may spur anti-mining opinion', *Financial Review*, 26 June, p. 5

Krasner, Stephen D. (ed.) (1983a) *International Regimes* (Ithaca: Cornell University Press)

—— (1983b) 'Structural causes and regime consequences: regimes as intervening variables' in Stephen D. Krasner (ed.), *International Regimes* (Ithaca: Cornell University Press)

—— (1985) *Structural conflict: the Third World against global liberalism* (Berkeley: University of California Press)

Kratochwil, Friedrich (1982) 'On the notion of "interest" in international relations', *International Organisation*, 36 (1), Winter, pp. 1–30

Kremenyuk, Victor A. (1991) 'The emerging system of international negotiation' in Victor A. Kremenyuk (ed.), *International negotiation: analysis, approaches, issues* (San Francisco: Jossey-Bass Publishers)

Kriwoken, Lorne and Peter Keage (1989) 'Antarctic environmental politics: protected areas', in John Handmer (ed.), *Antarctica: policies and policy development*, CRES Resource and Environmental Studies No 1 (Canberra: Centre for Resource and Environmental Studies, Australian National University)

Lagoni, Rainer (1984) 'Convention on the conservation of marine living resources: a model for the use of a common good', in Rüdiger Wolfrum (ed.), *Antarctic challenge: conflicting interests, cooperation, environmental protection, ecomonic development* (Berlin: Duncker and Humblot)

Laughlin, Thomas L. (1989), 'The environmental aspects of the new Antarctic Mineral Convention', paper given at the International Studies Association Conference, 1 April 1989, London

Laws, Richard (1983) 'Antarctica: a convergence of life', *New Scientist*, 99 (1373), 1 September, pp. 608–16

—— (1985) 'International Stewardship of the Antarctic: Problems, Successes and Future Options', *Marine Pollution Bulletin*, 16 (2), February, pp. 49–55

—— (1987) 'Science, the Treaty and the Future', in D. W. H. Walton (ed.), *Antarctic Science* (Cambridge: Cambridge University Press)

—— (1990) 'Science as an Antarctic resource', in Grahame Cook (ed.), *The future of Antarctica: exploitation versus preservation* (Manchester: Manchester University Press)

—— (1991a) 'Unacceptable threat to Antarctic science', *New Scientist*, 129 (1762), 30 March, p. 4

—— (1991b) 'Antarctic politics and science are coming into conflict', *Antarctic science*, 3 (3), September, p. 231

Lee, Seo-Hang (1988) 'Third World approaches to Antarctica', in *Third World Affairs 1988* (Third World Foundation for Social and Economic Studies)

Lee, Wei-Chin (1990) 'China and Antarctica: so far and yet so near', *Asian Survey*, June, pp. 576–86

Lemonick, Michael A. (1990) 'Antarctica', *Time*, 135 (3), 15 January, pp. 30–6

Lentz, Sally Ann (1990) *Effective legal mechanisms for the protection of the Antarctic environment*, unpublished paper submitted to Greenpeace International, Amsterdam, October

Lewis, Richard S. and Philip M. Smith (eds) (1973) *Frozen future: a prophetic report from Antarctica* (New York: Quadrangle Books)

Lipps, Jere H. (1978) 'Man's Impact along the Antarctic Peninsula', in Bruce C. Parker (ed.), *Environmental Impact in Antarctica* (Blacksburg: Virginia Polytechnic Institute and State University)

Lorius, Claude (1990) 'Science in the Antarctic', *IUCN Bulletin*, 21 (2), June, pp. 13–14

Luard, Evan (1977) *International Agencies: the emerging framework of interdependence* (London: Macmillan)

Lucas, P. H. C. and P. R. Dingwall (1987) 'Protected Areas and Environmental Conservation in Antarctica and the Southern Ocean', in J. G. Nelson, R. Needham and L. Horton (eds), *Proceedings*, Arctic Heritage Symposium, Banff, Alaska, 24 August 1985 (Ottawa: Association of Canadian Universities for Northern Studies)

Luke, Anthony (1991a) 'Madrid meeting bans Antarctic mining for 50 years', *New Scientist*, 130 (1767), 4 May, p. 6

—— (1991b) 'America backtracks on Antarctic mining ban', *New Scientist*, 130 (1775), 29 June, p. 7

Lyons, D. J. (1990) *Antarctic hypotheticals: a summary report* (Hobart: Institute of Southern Ocean and Antarctic Studies, University of Tasmania)

MacKenzie, Debora (1990) 'Talks fail to secure Antarctica's future', *New Scientist*, 128 (1747), 15 December, p. 4

MacNeill, Jim (1989–90) 'The greening of international relations', *International Journal*, XLV (1), Winter, pp. 1–35

McNickle, R. K. (1949) 'Antarctic Claims', *Editorial Research Reports*, II (19), 12 November

Manheim Jr, Bruce S. (1988) *On Thin Ice: the failure of the National Science Foundation to protect Antarctica* (Washington DC: Environmental Defense Fund)

Mann, Dean E. (1991) 'Environmental learning in a decentralised world', *Journal of International Affairs*, 44 (2), Winter, pp. 301–37

Mansukoski, Riitta (1987) 'The Antarctic activities of Finland', *Navigator International*, 87, pp. 15–16

Marchal, Anne (1989) 'Convention for the Conservation of Antarctic Seals: 1988 review of operations', *Polar Record*, 25 (153), April, p. 143

Marcoux, J. Michel (1971) 'Natural resource jurisdiction on the Antarctic continental margin', *Virginia Journal of International Law*, 11 (3), pp. 374–405

May, John (1989) *The Greenpeace Book of Antarctica: a new view of the seventh continent* (Australia: Child and Associates)

Merriam, Clark Lee (1990) 'Cleaning up Antarctica', *Calypso Log*, 17 (2), April, pp. 19–20

Millar, T. B. (ed.) (1986a) *Australia, Britain and Antarctica* (London: Institute of Commonwealth Studies, University of London)

Mitchell, Barbara (1977) 'Resources in Antarctica: potential for conflict', *Marine Policy*, 1 (2), April, pp. 91–101

—— (1980) 'The Politics of Antarctica', *Environment*, 22 (1), January/February, pp. 12–41

—— (1981) 'Cracks in the Ice', *The Wilson Quarterly*, V (4), Autumn, pp. 69–87

—— (1982) *The Management of Antarctic Mineral Resources* (London: IIED)

—— (1983) *Frozen Stakes: the future of Antarctic minerals* (London: IIED)

Morris, Michael (1990) *Great power relations in Argentina, Chile and Antarctica* (Basingstoke: Macmillan)

Mosley, Geoff (1986a) 'Antarctica: peace out of the cold', *Peace Studies*, 2, April/May, pp. 8–11; 48

—— (1986b) *Antarctica: our last great wilderness* (Hawthorn, Melbourne: Australian Conservation Foundation)

Müller, Scharnhorst (1984) 'The impact of UNCLOS III on the Antarctic regime', in Rüdiger Wolfrum (ed.), *Antarctic challenge: conflicting interests, cooperation, environmental protection, ecomonic development* (Berlin: Duncker and Humblot)

Muller-Schwarze, D. and Paul Belanger (1978) 'Man's Impact on Antarctic Birds', in Bruce C. Parker (ed.), *Environmental Impact in Antarctica* (Blacksburg: Virginia Polytechnic Institute and State University)

Mullins, Janet (1990) (Assistant Secretary, Legislative Affairs, Department of State), Letter to Morris Udall, Chair of House Committee on Interior and Insular Affairs, 3 January

Murphy, Kevin (1990) 'Australians in Antarctica: what price a presence', *The Bulletin*, 112 (5726), 10 July, pp. 42–8

Mussack, Irmi E. L. (1988) *An approach to the management of tourism in the Antarctic*, Thesis for MSc, Centre for Resource Management, Canterbury University (Christchurch, New Zealand)

Myers, Charles E. et al. (1980) 'Environmental assessment of Antarctic research', *Environmental Science and Technology*, 14 (6), June, pp. 668–72

Myhre, J. D. (1986) *The Antarctic Treaty system: politics, law and diplomacy* (Boulder: Westview Press)

Nadelmann, Ethan A. (1990) 'Global prohibition regimes: the evolution of norms in international society', *International Organisation*, 44 (4), Autumn, pp. 479–526

Nansen Foundation (1973) *Antarctic Resources, Report from the informal meeting of experts, 30 May–9 June 1973* (Polhogda, Norway: Fridtjof Nansen Foundation)

National Science Foundation (NSF) (1987) *The Role of the National Science Foundation in Polar Regions: a Report to the National Science Board*, Committee on the NSF Role in Polar Regions, 19 June (Washington DC: National Science Board, National Science Foundation)

—— (1989a) *Transcript of public hearing on Antarctic tourism and enforcement procedures, 15 July 1988*, NSF, Washington DC

—— (1989b) *A National Science Foundation Strategy for Compliance with Environmental Laws in Antarctica*, A report to the Director from the Office of the General Counsel, NSF, September 1989 [draft]

Natural Environment Research Council (1989) *Antarctica 2000: NERC Strategy for Antarctic Research* (UK: NERC)

Neales, Sue (1988) 'Treaty members carve up the Antarctic', *Financial Review*, 28 November, p. 13

Neilson, James (1977) 'Chile and Argentina collide in the Antarctic', *Observer Foreign News Service*, No 35733, 16 February

Ness, Gayl D. and Steven R. Brechin (1988) 'Bridging the gap: international organisations as organisations', *International Organisation*, 42 (2), Spring, pp. 245–73

Newman, Paul E. (1987) 'The Antarctic Minerals Convention: developments from the October 1986 Tokyo meeting of the Antarctic Treaty Consultative Parties', *Denver Journal of International Law and Policy*, 15 (2/3), Winter/Spring, pp. 421–33

Newsom, David D. (1988/89) 'The new diplomatic agenda: are governments ready?', *International Affairs*, 65 (1), Winter, pp. 29–41

New Zealand Institute of International Affairs (1984) *The Antarctic: preferred futures, constraints and choices*, Pamphlet No 44 (Wellington: NZIIA)

Nicholson, Ian E. (1986) 'Antarctic Tourism – the need for a legal regime?', *Maritime Studies*, 29, May–June, pp. 1–7

Nicol, Stephen (1990) 'No agreement yet on krill fishery', *ANARE News*, 64, December, p. 6

Nossiter, Bernard D. (1977) 'Antarctica's rich resources threaten 13-nation treaty', *International Herald Tribune*, 21 September

Nye, Joseph S. (Jnr) (1987) 'Nuclear learning and US-Soviet security regimes', *International Organisation*, 41 (3), Summer, pp. 371–402

—— (1988) 'Neorealism and neoliberalism', *World Politics*, XL (2), January, pp. 235–51

Office of Technology Assessment (1989) *Polar Prospects: A Minerals Treaty for Antarctica*, OTA-O-428 (Washington DC: US Congress/US Government Printing Office)

Ophuls, William (1988) 'The International State of Nature and the politics of scarcity' in Charles W. Kegley Jr and Eugene R. Wittkopf (eds), *The Global Agenda: Issues and Perspectives*, 2nd edition (New York: Random House)

Orrego Vicuña, Francisco (ed.) (1983) *Antarctic Resources Policy: Scientific, Legal and Political Issues* (Cambridge: Cambridge University Press)

—— (1988) *Antarctic mineral resources: the emerging legal framework* (Cambridge: Cambridge University Press)

—— (1991) 'The effectiveness of the decision-making machinery of CCAMLR: an assessment' in Arnfinn Jorgensen-Dahl and Willy Ostreng (eds), *The Antarctic Treaty system in World Politics* (Basingstoke: Macmillan)

Oxman, Bernard (1978) 'The Antarctic Regime: an Introduction', *University of Miami Law Review*, 33 (2), December, pp. 285–97

—— (1983) 'Summary of the Law of the Sea Convention' in Bernard Oxman, David D. Caron and Charles L. O. Buderi (eds), *Law of the Sea: US policy dilemma* (San Francisco: ICS Press)

—— (1986) 'Antarctica and the new Law of the Sea', *Cornell International Law Journal*, 19 (2), Summer, pp. 211–47

Oye, Kenneth A. (ed.) (1986) *Cooperation under Anarchy* (Princeton: Princeton University Press)

Pain, Stephanie (1990) 'New Zealand's U-turn threatens Antarctic mining treaty', *New Scientist*, 125 (1707), 10 March, p. 7

Pallemaerts, Marc (1989) E-mail communication, 'Belgium joins Australia and France in opposing CRAMRA', 5 July

Pallone, Frank (1978) 'Resource exploitation: the threat to the legal regime of Antarctica', *International Lawyer*, 12 (3), Summer, pp. 547–61

Parker, Bruce C. (ed.) (1978) *Environmental Impact in Antarctica* (Blacksburg: Virginia Polytechnic Institute and State University)

Parliament of Australia (1989) *Commonwealth Parliamentary Debates*, 'Antarctica' Senate, Vol S.133, pp. 1647–69

Parliament of New Zealand (1990) *Hansard*, Questions for oral answer: Antarctic Minerals Convention, 27 February, pp. 147–8

—— (1990) *Hansard*, Debate: urgent public matter – Ratification of the Convention on the Regulation of Antarctic Mineral Resource Activities, 27 February, pp. 156–77

Parliament of the UK (1989a) House of Lords, *Official Record* (Antarctic Minerals Bill, Second Reading) 20 April, pp. 926–46

—— (1989) House of Commons, *Official Record* (Antarctic Minerals Bill, Second Reading) 4 July, pp. 212–51

Palmer, Geoffrey (Prime Minister of New Zealand) (1990) 'Antarctica', Press Statement, 26 February

Parsons, Sir Anthony (1987) *Antarctica: the next decade* (Cambridge: Cambridge University Press)

Pascoe, Susan (1990) 'The French and Australian joint initiative about the Antarctic environment' in José Yañez and María Eugenia Zañartu (eds), *1er Taller Latinoamericano de Organizaciones No Gubernamentales Sobre Conservación Antártica* (Santiago, Chile: CODEFF)

Patel, Jagdish (1991) 'Mining Antarctica', *New Scientist*, 129 (1755), 9 February, p. 2

—— and Susan Mayer (eds) (1991) *Antarctica: the scientists' case for a World Park* (London: Greenpeace UK)

Peake, Ross (1989) 'ALP turns on Antarctic mining', *The Age*, 12 May, p. 6

Pearce, David (1990) 'Economics and the global environmental challenge', *Millenium*, 19 (3), Winter, pp. 365–87

Peterson, M. J. (1980) 'Antarctica: the last great land rush on earth', *International Organisation*, 34 (3), Summer, pp. 377–403

—— (1986) 'Antarctic implications of the Law of the Sea', *Ocean Development and International Law*, 16 (2), pp. 137–81

—— (1988) *Managing the Frozen South: the creation and evolution of the Antarctic Treaty system* (Berkeley: University of California Press)

Philips, Cassandra (1989) 'Call for Antarctic Protection', *WWF News*, 60, July/ August, p. 8

Pinto, M. C. W. (1978) 'The International Community and Antarctica', *University of Miami Law Review*, 33 (2), December, pp. 475–87

Pirages, Dennis (1978) *The new context for international relations: global ecopolitics* (North Scituate, Mass.: Duxbury Press)

—— (1988) 'The Ecological Perspective on International Politics' in Charles W. Kegley Jr and Eugene R. Wittkopf (eds), *The Global Agenda: Issues and Perspectives*, 2nd edition (New York: Random House)

Polar Research Board (1986a) *Antarctic Treaty system: an assessment* (Proceedings of a workshop held at Beardmore South Field Camp, Antarctica, 7–13 January 1985) (Washington DC: National Academy Press)

—— (1986b) *US Research in Antarctica in 2000 AD and beyond: a preliminary assessment* (Washington DC: National Academy Press)

de Poorter, Maj and Sabine Schmidt (1990) 'Greenpeace environment and scientific programme in Antarctica' in John E. Hay, Alan D. Hemmings and Norman G. Thom (eds), *Antarctica 150: scientific perpsectives – policy futures* (Auckland: University of Auckland)

Porritt, Jonathon (1990) 'Britain – the dirty man of Antarctica', *Daily Telegraph*, 3 November

Porter, Gareth and Janet Welsh Brown (1991) *Global environmental politics* (Boulder: Westview Press)

Potter, Neal (1969) *Natural Resource Potentials of the Antarctic*, The American Geographical Society, Occasional Publication No 4 (Burlington, Vermont: Resources for the Future)

Powell, D. L. (1990) 'Antarctic Marine Living Resources and CCAMLR' in R. A. Herr, H. R. Hall and M. G. Haward (eds) *Antarctica's future: continuity or change?* (Hobart: Australian Institute of International Affairs/ Tasmanian Government Printing Office)

Prins, Gwyn (1990) 'Politics and the environment', *International Affairs*, 66 (4), October, pp. 711–30

Puchala, Donald J. and Raymond F. Hopkins (1983) 'International regimes: lessons from inductive analysis' in Stephen D. Krasner (ed.), *International Regimes* (Ithaca: Cornell University Press)

Puissochet, Jean-Pierre (1991) 'CCAMLR – a critical assessment' in Arnfinn Jorgensen-Dahl and Willy Ostreng (eds), *The Antarctic Treaty system in World Politics* (Basingstoke: Macmillan)

Purver, Ron (1984) 'Security and arms control at the poles', *International Journal*, 39 (4), Autumn, pp. 888–910

Putnam, Robert D. (1988) 'Diplomacy and domestic politics: the logic of two-level games', *International Organisation*, 42 (3), Summer, pp. 427–60

Quigg, Philip W. (1983) *A Pole Apart: the emerging issue of Antarctica* (New York: McGraw Hill)

Quilty, Patrick C. (1986–87) 'Formulating the Future', *Australian Natural History*, 22 (3), Summer, pp. 116–18

Redgwell, Catherine (1990) 'Antarctica', *International and Comparative Law Quarterly*, 39, April, pp. 474–81

Reich, Rosamunde J. (1980) 'The development of Antarctic tourism', *Polar Record*, 20 (126), pp. 203–14

Reuters (1989) 'Antarctic Treaty', Bltn.17, 13/REUTER/AAP/AFP/JM/bt, 21 April 1989

Reynolds, Anne and Kaye Dyson (1991) 'A signing of sanity', *Greenpeace Australia News*, 2 (4), Spring, pp. 7–9

Rigg, Kelly (1990a) 'The Antarctic future: a Greenpeace Perspective', *IUCN Bulletin*, 21 (2), June, pp. 21–2

—— (1990b) 'Environmentalists' perspective on the protection of Antarctica' in Grahame Cook (ed.), *The future of Antarctica: exploitation versus preservation* (Manchester: Manchester University Press)

Roberts, B. B. (1966) 'Wildlife conservation in the Antarctic', *Oryx*, Journal of the Fauna Preservation Society, XIII (4), April

—— (1977) 'Conservation in the Antarctic', *Phil. Trans. Royal Society London*, 279, pp. 97–104

Rosecrance, Richard (1981) 'International theory revisited', *International Organisation*, 35 (4), Autumn, pp. 691–713

Rosenau, James N. (1980) *The study of global interdependence: essays in the transnationalisation of world affairs* (London: Frances Pinter)

—— (1986) 'Before cooperation: hegemons, regimes, and habit-driven actors in world politics', *International Organisation*, 40 (4), Autumn, pp. 849–89

—— (1989) 'Global changes and theoretical challenges: toward a Postinternational politics for the 1990s' in Ernst-Otto Czempiel and James N. Rosenau (eds), *Global changes and theoretical challenges* (Massachusetts: Lexington Books)

Ross Dependency Research Committee (1985) *Future Directions in NZ Antarctic Research* (Wellington: Department of Scientific and Industrial Research)

Rothwell, Donald R. (1990a) *A World Park for Antarctica? foundations developments and the future*, Antarctic and Southern Ocean Law Policy Occasional Paper No 3 (Hobart: Institute of Antarctic and Southern Ocean Studies, University of Tasmania)

—— (1990b) 'The Antarctic Treaty system: resource development, environmental protection or disintegration?', *Arctic*, 43 (3), September, pp. 284–91

Rowland, J. R. (1988) 'The Treaty regime and the politics of the consultative parties' in Christopher C. Joyner and Sudhir K. Chopra (eds), *The Antarctic legal regime* (Dordrecht: Martinus Nijhoff Publishers)

Rowlands, Ian (1991) 'The security challenges of global environmental change', *The Washington Quarterly*, 14 (1), Winter, pp. 99–114

Rudolph, Emanuel D. (1970) 'Conserving the Antarctic Terrestrial Ecosystem', *Biological Conservation*, 3 (1), October, pp. 52–4

Ruggie, John Gerard (1983) 'Continuity and transformation in the world polity: toward a neorealist synthesis', *World Politics*, XXXV (2), January, pp. 261–85

Ryzhkov, His Excellency Mr Nikolai I. (1990) [Chairman of the Council of Ministers of the Union of Soviet Socialist Republics] Speech on the occasion of the parliamentary dinner given by the Prime Minister of Australia, 14 February 1990

Sands, Phillipe J. (1989) 'The environment, community and international law', *Harvard International Law Journal*, 30 (2), Spring, pp. 393–420

Schatz, Gerald S. (1978) 'The polar regions and human welfare: regimes for environmental protection', in Edmund A. Schofield (ed.), *Earthcare: global protection of natural areas*, Proceedings of the 14th Biennial Wilderness Conference (Boulder: Westview Press)

Schofield, Edmund (1976) 'Antarctic up for Grabs', *Sierra Club Bulletin*, 61 (6), pp. 17–20

Scientific Committee on Antarctic Research (SCAR)/EAMREA (1979) *Possible environmental effects of mineral exploration and exploitation in Antarctica* (an adaptation of a Report by the Group of Specialists on the Environmental Impact Assessment of Mineral Resource Exploration and Exploitation in the Antarctic – EAMREA) (Cambridge: SCAR)

—— (1985) *Conservation areas in the Antarctic*, A review prepared by the subcommittee of the Conservation Working Group on Biology (Cambridge: SCAR/ICSU)

——/AEIMEE (1986) *Reports of the SCAR Group of Specialists on Antarctic Environmental Implications of possible mineral exploration and exploitation (AEIMEE)* (Cambridge: SCAR)

—— (1987) *The Protected Area system in the Antarctic, Report of the SCAR ad hoc Group on Additional Protective Measures*, Response of SCAR to Antarctic Treaty Consultative Meeting Recommendation XIII–5, Cambridge, UK, June

—— (1989) *Waste disposal in the Antarctic*, Report of the SCAR Panel of Experts on Waste Disposal (Tasmania: Australian Antarctic Division on behalf of SCAR/ICSU)

Scott, David Clark (1989) 'Australia advocates no mining in Antarctica', *Christian Science Monitor* (World Edition Weekly), 1–7 June

Scully, R. Tucker (1986) 'The evolution of the Antarctic Treaty system – the institutional perspective' in Polar Research Board, *Antarctic Treaty system: an assessment* (Washington DC: National Academy Press)

—— (1988a) 'The Antarctic mineral resource negotiations' *Oceanus*, 31 (2), Summer, pp. 20–1

—— (1988b) 'The institutional development of the Antarctic Treaty system: the question of a secretariat' in Rüdiger Wolfrum (ed.), *Antarctic Challenge III: conflicting interests, cooperation, environment protection, economic development* (Berlin: Duncker and Humblot)

—— (1989) Letter to the Sierra Club, summary of position taken in responding to the Government of Australia, 11 July

—— and Lee Kimball (1989) 'Antarctica: is there life after minerals: the minerals treaty and beyond', *Marine Policy*, 13 (2), April, pp. 87–98

Shackleton, (Lord) Edward (1989) 'Cold realism must prevail', *The Australian*, 23 November

Shapley, Deborah (1985) *The Seventh Continent: Antarctica in a Resource Age* (Washington DC: Resources for the Future)

—— (1988) 'Antarctica: why success?' in Alexander L. George, Philip J. Farley and Alexander Dallin (eds), *US-Soviet Security cooperation: achievement, failures, lessons* (New York: Oxford University Press)

Sherman, Kenneth and Alan F. Ryan (1988) 'Antarctic Marine Living Resources', *Oceanus*, 31 (2), Summer, pp. 59–63

Shusterich, Kurt M. (1984) 'The Antarctic Treaty System: history, substance and speculation', *International Journal*, 39 (4), Autumn, pp. 800–27

Simma, Bruno (1986) 'The Antarctic Treaty as a Treaty providing for an objective regime', *Cornell International Law Journal*, 19 (2), Summer, pp. 189–209

Skegestad, Gunnar (1975) 'The frozen frontier: models for international cooperation', *Cooperation and Conflict*, 10, pp. 167–87

Smith, Roger K. (1987) 'Explaining the non-proliferation regime: anomalies for contemporary international relations theory', *International Organisation*, 41 (2), Spring, pp. 253–81

Snidal, Duncan (1990) 'IGOs, regimes and cooperation: challenges for international relations theory', in Margaret P. Karns and Karen A. Mingst (eds), *The US and multilateral institutions* (Boston: Unwin Hyam)

Sollie, Finn (1977) *The political problems of Antarctica*, Press Briefing Seminar, 25 July (London: Earthscan)

—— (1984) 'Polar politics: old games in new territories or new patterns in political development', *International Journal*, 39 (4), Autumn, pp. 695–720

Soroos, Marvin S. (1986) *Beyond sovereignty: the challenge of global policy* (Columbia: University of South Carolina Press)

Spain (1990) Senate, Parliamentary Question, 27 February [Translation prepared by Australian Department of Foreign Affairs and Trade]

Spivak, Jonathan (1974) 'Frozen Assets', *Wall Street Journal*, 21 Feburary

Splettstoesser, John F. (1985) 'Antarctic geology and mineral resources', *Geology Today*, March–April, pp. 41–5

Stein, Arthur A. (1983) 'Coordination and Collaboration: regimes in an anarchic world', in Stephen D. Krasner (ed.), *International Regimes* (Ithaca: Cornell University Press)

—— (1990) *Why nations cooperate: circumstance and choice in international relations* (Ithaca: Cornell University Press)

Stephen, Sir Ninian (1990) 'The environmental protection of the Antarctic', Milthorpe Memorial Lecture, 24 October 1990, *The Monthly Record*, 61 (10), October, pp. 721–29

Stewart, Duncan (1963) 'Antarctic Mineralogy', *Antarctic Geology, SCAR Proceedings 1963* (Cambridge: SCAR)

Stonehouse, Bernard (1990) 'A traveller's code for Antarctic visitors' *Polar Record*, 26 (156), January, pp. 56–8

Strange, Susan (1983) 'Cave! Hic dragones: a critique of regime analysis', in Stephen D. Krasner (ed.), *International Regimes* (Ithaca: Cornell University Press)

—— (1987) 'The persistent myth of lost hegemony', *International Organisation*, 41 (4), Autumn, pp. 551–74

Sun, Marjorie (1988) 'Antarctica Pact could open way for mining', *Science*, 240 (4859), 17 June, p. 1612

Susskind, Lawrence and Connie Ozawa (1992) 'Negotiating more effective international environmental agreements', in Andrew Hurrell and Benedict Kingsbury (eds), *The international politics of the environment* (Oxford: Clarendon Press)

Suter, Keith D. (1980) *World Law and Law Wilderness*, 2nd edition (Sydney: Friends of the Earth)

—— (1991) *Antarctic: private property or public heritage?* (Leichhardt: Pluto Press)

Talboys, Rt Hon B. E. (1978) 'New Zealand and Antarctica', speech to the New Zealand Antarctic Society, 26 April

Taylor, Phillip (1984) *Non-state actors in international politics: from transregional to substate organisations* (Boulder: Westview Press)

Thacher, Peter S. (1991) 'Multilateral cooperation and global change', *Journal of International Affairs*, 44 (2), Winter, pp. 433–55

Thomson, Michael and Charles Swithinbank (1985) 'The prospects for Antarctic minerals', *New Scientist*, 107 (1467), 1 August, pp. 31–5

Thornton, Brenda Sue (1990) 'A new model for Antarctic conservation', *Polar Record*, 26 (159), October, pp. 329–30

Tierney, Trevor (1978) 'Antarctica – the last wild continent', *Habitat*, 5 (5), January, pp. 6–13

Tinker, Jon (1979) 'Antarctica: towards a new internationalism', *New Scientist*, 83 (1172), 13 September, pp. 799–801

Toma, Peter A. (1956) 'Soviet attitude towards the acquisition of territorial sovereignty in the Antarctic', *American Journal of International Law*, 50, pp. 611–26

Triggs, Gillian D. (1984) 'A minerals regime for Antarctica: compromise or conflict?', *Australian Mining and Petroleum Yearbook*,: pp. 525–54

—— (1985) 'The Antarctic Treaty regime: a workable compromise or a "purgatory of ambiguity"?', *Case Western Reserve Journal of International Law*, 17 (2), Spring, pp. 195–228

—— (1986) *International Law and Australian sovereignty in Antarctica* (Melbourne: Legal Books)

—— (ed.) (1987) *The Antarctic Treaty regime: law environment and resources* (Cambridge: Cambridge University Press)

Udall, Morris (et al.) (1989) [Chair of the House Committee on Interior and Insular Affairs, with Chairs of Sub-committees], Letter to President Bush, 9 November

US Congress (1990a) A Bill 'Antarctic Protection and Conservation Act of 1990' (HR 3977), 101st Congress, 2d session, 7 February 1990

—— (1990b) House of Representatives, *Congressional Record*, 4 April, pp. E989–91 [Statement by Hon Bruce F. Vento, Antarctic World Park and Protection Bill]

—— (1990c) A Bill 'Antarctica World Park and Protection Bill', (HR 4514), 101st Congress, 2d Session, April 4

—— (1990d) Senate, *Congressional Record*, 3 May, pp. S5658–9 [Statement by Senator Gore on Antarctic Environmental Protection Act of 1990, S2571]

US Marine Mammal Commission (1987) *Annual Report to Congress*, Washington DC

Vaughan, Chris and Ian Anderson (1989) 'Senators champion polar park', *New Scientist*, 124 (1685), 7 October, p. 8

Walker, R. B. J. (1988) *One World, Many Worlds: Struggles for a Just World Peace* (Boulder: Lynne Rienner Publishers)

Wallace, Catherine (1988) 'Property rights where sovereignty is uncertain: the Antarctic Minerals Convention as a pre-condition for development', unpublished paper

—— (1991) *Protocol to the Antarctic Treaty on Environmental Protection: a critique*, unpublished paper,

Waller, Deborah Cook (1989) 'Death of a treaty: the decline and fall of the Antarctic Minerals Convention', *Vanderbilt Journal of Transnational Law*, 22 (3), pp. 631–68

Walton, D. W. H. (ed.) (1987a) *Antarctic Science* (Cambridge: Cambridge University Press)

—— (1987b), 'Science and Antarctic politics', in D. W. H. Walton (ed.), *Antarctic Science* (Cambridge: Cambridge University Press)

Waltz, Kenneth N. (1979) *Theory of International Politics* (New York: Random House)

—— (1990) 'Realist thought and neorealist theory', *Journal of International Affairs*, 44 (1), Spring/Summer, pp. 21–37

Ward, Veronica (1991) 'Regime norms as "implicit" third parties: explaining the Anglo-Argentine relationship', *Review of International Studies*, 17 (2), April, pp. 167–92

Westermeyer, William (1982) 'Resource allocation in Antarctica: a review', *Marine Policy*, 6 (9), October, pp. 303–25

—— (1984) *The politics of mineral resource development in Antarctica: alternative regimes for the future* (Colorado: Westview Press)

Wijkman, Per Magnus (1982) 'Managing the global commons', *International Organisation*, 36 (3), Summer, pp. 511–36

Wilkinson, Peter (1988) 'Expedition 88', *Greenpeace*, 13 (3), May–June

Willetts, Peter (1982) 'Pressure groups as transnational actors' in Peter Willetts (ed.), *Pressure groups in the global system* (London: Frances Pinter)

Wilson, Roger (1983) 'Antarctica: the last continent faces exploitation', *The Ecologist*, 13 (2/3), pp. 74–83

—— (1986) 'An environmental critique of the Antarctic Treaty system', in T. B. Millar (ed.), *Australia, Britain and Antarctica* (London: Institute of Commonwealth Studies, University of London)

de Wit, Maarten J. (1985) *Minerals and Mining in Antarctica* (Oxford: Clarendon Press)

Wolfrum, Rüdiger (ed.) (1984) *Antarctic challenge: conflicting interests, cooperation, environmental protection, ecomonic development* (Berlin: Duncker and Humblot)

—— (ed.) (1986), *Antarctic Challenge II: conflicting interests, co-operation, environmental protection, economic development* (Berlin: Duncker and Humblot)

—— (ed.) (1988) *Antarctic Challenge III: conflicting interests, cooperation, environmental protection, economic development* (Berlin: Duncker and Humblot)

—— (1991) 'The unfinished task: CRAMRA and the question of liability' in Arnfinn Jorgensen-Dahl and Willy Ostreng (eds), *The Antarctic Treaty system in World Politics* (Basingstoke: Macmillan)

Wolk, S. (1958) 'The basis of Soviet claims in the Antarctic', *Bulletin*, Institute for the Study of the USSR, V (4), April, pp. 43–8

Woolcott, Richard (1985) 'The interaction between the Antarctic Treaty system and the UN system', *Australian Foreign Affairs Record*, 56 (1), January, pp. 17–25

Wright, N. A. and P. L. Williams (1974) *Mineral resources in Antarctica*, US Geological Survey Circular No 705 (Virginia: USGS)

Wyndham, R. H. (1980) 'The Antarctic' in Coral Bell (ed.), *Agenda for the Eighties: Contexts of Australian choices in foreign and defence policy* (Canberra: Australian National University Press)

Yañez, José and María Eugenia Zañartu (eds) (1990) *1er Taller Latinoamericano de Organizaciones No Gubernamentales Sobre Conservación Antártica* (Santiago, Chile: CODEFF)

Yap, Nonita T. (1989–90) 'NGOs and sustainable development', *International Journal*, XLV (1), Winter, pp. 75–105

Young, Oran R. (1978) 'Anarchy and social choice: reflections on the international polity', *World Politics*, XXX (2), January, pp. 241–63

—— (1980) 'International regimes: problems of concept formation', *World Politics*, XXXII (3), April, pp. 331–56

—— (1982) *Resource Regimes: Natural Resources and Social Institutions* (Berkeley: University of California Press)

—— (1983) 'Regime dynamics: the rise and fall of international regimes' in Stephen D. Krasner (ed.), *International regimes* (Ithaca: Cornell University Press)

—— (1989a) *International Cooperation: Building Regimes for Natural Resources and the Environment* (Ithaca: Cornell University Press)

—— (1989b) 'The politics of international regime formation: managing natural resources and the environment', *International Organisation*, 43 (3), Summer, pp. 349–75

—— (1990a) 'Global environmental change and international governance', *Millennium*, 19 (3), Winter, pp. 337–46

—— (1990b) 'Political leadership and regime formation: the emergence of international institutions in international society', Paper prepared for the Annual Conference of the International Studies Association, Washington DC, 10–14 April 1990

—— (1991) 'Political leadership and regime formation: on the development of institutions in international society', *International Organisation*, 45 (3), Summer, pp. 281–308

Zacher, Mark W. (1987) 'Trade gaps, analytical gaps: regime analysis and international commodity trade regulation', *International Organisation*, 41 (2), Spring, pp. 173–202

—— (1990) 'Toward a theory of international regimes', *Journal of International Affairs*, 44 (1), Spring/Summer, pp. 139–57

Zegers, Fernando (1978) 'The Antarctic system and the utilisation of resources', *University of Miami Law Review*, 33 (12), December, pp. 427–53

Zeller, E. J. and E. E. Angino (1978) 'The disposal of radioactive wastes in the Antarctic ice sheet: an environmental impact assessment', in B. C. Parker (ed.), *Environmental Impact in Antarctica* (Blacksburg, Virginia: VPI Press)

Ziolkowski, Heidi (1990) 'Tourism in Antarctica', *Calypso Log*, 17 (2), April, pp. 18; 20

Zorn, Stephen A. (1984) 'Antarctic minerals: a common heritage approach', *Resources Policy*, 10 (1), March, pp. 2–18

Zumberge, J. H. (1979) 'Mineral resources and geopolitics in Antarctica', *American Scientist*, 67, January/February, p. 68

—— (1982) 'Potential mineral resource availability and possible environmental problems in Antarctica' in Jonathan I. Charney (ed.), *The new nationalism and the use of common spaces* (New Jersey: Allanheld Osmun Publishers)

—— (1987) 'The Scientific Committee on Antarctic Research, the Antarctic Treaty and conservation in Antarctica', *Environment International*, 13 (1), pp. 3–7

Newspapers and Newsletters Cited in the Text

Antarctic News (ASOC: New Zealand Organising Committee)

No 2, November 1984
No 3, January 1986
No 4, 21 October 1986
No 5, May 1987
No 6, September 1987
No 7, January 1988
No 8, October 1988
No 9, November 1989

Campaigners Antarctic Notes

No 5, 30 January 1984
No 27, 1 April 1985
No 68, 3 March 1989

The Times

21 October 1985; 1 February 1986

The Guardian

23 July 1983; 18 February 1984; 20 February 1989

International Herald Tribune

25 November 1980; 31 December 1981; 17 August 1988

Financial Times

21 January 1983; 12 May 1983; 26 August 1983; 13 March 1985

Christian Science Monitor

6–8 June 1970

Dawn

22 November 1983

Observer News Service

No 43008, 8 June 1982

Index